Modern Instrumentation:
A Computer Approach

Modern Instrumentation: A Computer Approach

Gordon Silverman
Manhattan College, New York, NY, USA

Howard Silver
Fairleigh Dickinson University, Teaneck, NJ, USA

Institute of Physics Publishing
Bristol and Philadelphia

© IOP Publishing Ltd 1995

All rights reserved. No part of this publication may be reproduced, stored in a retrieval system or transmitted in any form or by any means, electronic, mechanical, photocopying, recording or otherwise, without the prior permission of the publisher. Multiple copying is permitted in accordance with the terms of licences issued by the Copyright Licensing Agency under the terms of its agreement with the Committee of Vice-Chancellors and Principals.

British Library Cataloguing-in-Publication Data

A catalogue record for this book is available from the British Library

ISBN 0 7503 0298 4

Library of Congress Cataloging-in-Publication Data are available

Consultant Editor:
J G Webster, University of Wisconsin, Madison, WI, USA

IOP Publishing Ltd and the author have attempted to trace the copyright holders of all the material reproduced in this publication and apologize to copyright holders if permission to publish in this form has not been obtained.

Published by Institute of Physics Publishing, wholly owned by The Institute of Physics, London
Institute of Physics Publishing, Techno House, Redcliffe Way, Bristol BS1 6NX, UK
US Editorial Office: Institute of Physics Publishing, The Public Ledger Building, Suite 1035, Independence Square, Philadelphia, PA 19106, USA

Typeset in TeX using the IOP Bookmaker Macros
Printed in the UK by J W Arrowsmith Ltd, Bristol

Contents

Preface		ix
Copyright, Product Names and Trade Names		xi
PART 1		
Fundamentals of Data Acquisition and Instrument Control		1
1	**Introduction**	**3**
	1.1 The role of computer-based instrumentation in measurement and process control	3
	1.2 Examples of computer-based instrumentation	6
	1.3 Architectures of computer-based instrument systems	14
	References	17
2	**Information and its Representation**	**18**
	2.1 Measuring information content in laboratory environments	18
	2.2 What to consider when selecting alpha-numeric codes	29
	2.3 Representation of abstract information	36
	2.4 What limits information?	38
	2.5 System capacity and sampling rate	44
	2.6 Summary	46
	References	47
	Additional reading	47
3	**Hardware Architecture of PC-based Instrument Systems**	**49**
	3.1 Functional components of an instrument platform	49
	3.2 Multiple-processor systems	86
	3.3 Architecture of instruments for automated environments	88
	3.4 The complete 'computer-on-a-chip' and portable instrumentation	97
	3.5 Choosing a PC platform	99
	References	104

CONTENTS

4 Software for Instrument Systems — 106
- 4.1 Computer operating systems — 107
- 4.2 The nature of programming — 120
- 4.3 Applications programming languages and packages — 138
- References — 144

5 Data-processing Considerations — 145
- 5.1 Computer-based instrument capacities — 145
- 5.2 Organizing data (data structures) — 148
- 5.3 Time or frequency basis of modelling — 155
- 5.4 Software architectures for input/output — 162
- References — 171

6 Data Acquisition and Instrument Control Resources — 172
- 6.1 Transducers — 173
- 6.2 Signal conditioning — 196
- 6.3 Telemetry — 200
- 6.4 Data conversion — 202
- 6.5 Instrument control: digital I/O — 212
- References — 215

PART 2
Applied Instrumentation Automation — 217

7 Design Aids—SPICE — 220
- 7.1 Introduction — 220
- 7.2 Elementary circuit examples — 221
- 7.3 Operational amplifier subcircuits and applications — 228
- 7.4 Device modelling — 243
- 7.5 Fourier analysis — 251
- 7.6 Use of the PROBE utility — 254
- References — 260

8 Design Aids—MathCAD, MATLAB — 261
- 8.1 Introduction — 261
- 8.2 Elementary operations — 261
- 8.3 Graphing with MathCAD — 263
- 8.4 Equation solving — 266
- 8.5 Fourier series and discrete Fourier transforms — 269
- 8.6 Signal detection in noise — 275
- 8.7 Data analysis techniques — 277
- 8.8 File transfer between MathCAD and a spreadsheet — 286
- 8.9 An introduction to MATLAB — 288
- References — 297

9 Design Aids—DSPlay — 298

9.1 Introduction — 298
9.2 DSPlay features — 298
9.3 Spectral analysis of a simple signal — 300
9.4 Adding and multiplying signals — 305
9.5 Convolution and filtering — 312
9.6 Signal detection in noise — 315
References — 321

10 Spreadsheets: Lotus 1-2-3 — 322

10.1 Introduction — 322
10.2 Lotus 1-2-3 features — 323
10.3 Tabulating and graphing a formula — 325
10.4 Data analysis techniques — 330
10.5 Signal detection in noise — 335
References — 343

11 A Graphical User Interface Development Tool: LabWindows — 344

11.1 Introduction — 344
11.2 Data analysis and data presentation examples — 345
11.3 Data-acquisition examples — 358
References — 372

12 The Windows Operating Environment — 374

12.1 Introduction — 374
12.2 Windows application programs — 374
12.3 Data transfer between Windows applications — 379
References — 391

13 Fully Integrated Applications — 392

13.1 Introduction — 392
13.2 Data acquisition using Lotus and Lotus Measure — 392
13.3 Measurement of optical fibre bandwidths using LabWindows — 398

14 New Tools for Laboratory Environments — 407

14.1 Introduction — 407
14.2 Artificial intelligence and expert systems — 409
14.3 Neural nets — 431
14.4 Future developments — 443
References — 444

Index — 447

Preface

The history of instrument development parallels and complements the growth and improvements of components and technology. Instrument design has progressed from (early) vacuum tube architecture, to discrete transistor circuitry, to integrated circuit implementations, to the most recent computer-based developments in which programming plays a key role. In the latter case, design emphasis is on the needs of instrument interfacing as well as high-level language (HLL) applications. In response to these trends a number of excellent texts and monographs have been published whose emphasis has been on various aspects of instrument design, particularly in the broad area of biomedical engineering. These books fall into several categories including: transducer (or measurement) technology, computer interfacing and, more recently, computer-based signal processing. This text addresses the emerging instrumentation needs and technology within the biomedical community and the broad scientific and engineering communities. This endeavour brings together new, 'more efficient' methods for experimental control and data acquisition. It combines underlying theoretical material such as what constitutes and limits information, the basics of computer-based hardware architecture, advanced technologies for laboratory automation, software tools for the analysis of data, and includes at least one technology that has the potential to revolutionize laboratory and industrial instrumentation, namely that of neural nets. At present, instrument design is accomplished at high levels of abstraction, making use of graphical interfaces for such purposes (e.g. instrument design in a Windows environment). Increasingly, instrument design will be supported by methods of artificial intelligence, and object-oriented programming based not on pointing devices but on verbal commands.

The text is divided into two main sections. Part 1 is entitled 'Fundamentals of Data Acquisition and Instrument Control', and discusses the fundamentals of data acquisition and analysis, computer-based instrument systems, and instrument control. Part 2 emphasizes applied instrumentation using various computer tools for instrument design and development such as circuit analysis, mathematical equation solvers, spreadsheet and graphical user interface software packages; its title is 'Applied Instrumentation Automation.'

In general this text is intended to facilitate communication between individuals whose primary expertise may not be in instrumentation—the broad scientific community—and specialists who may be asked to implement specific designs. From the engineers' point of view it will enable them to appreciate the automation tasks needing to be carried out in the laboratory. This will be accomplished, in part, through a judicious choice of examples (e.g. mass spectroscopy). This text can be adopted for use in either a senior-level technical elective course (in either instrumentation or biomedical engineering) or a similar first-year graduate course, and has been used in manuscript form to support such a course at Fairleigh Dickinson University. In addition, parts have been used as educational background material for scientists at The Rockefeller University.

There are a number of people who must be recognized for their contributions. In particular, the authors are grateful to graduate students Edward Carter, David Ivaldi, Stephen K Komorowski, and Miltiade Serbos. They are responsible for carrying out a number of experiments, especially those described in chapter 13. Professor Peter Schaeffer of Fairleigh Dickinson University's Department of Physics not only supervised these students but also supplied many innovative suggestions. Professor Stephen Dobrow of Fairleigh Dickinson University's Department of Electrical Engineering is using the manuscript to develop a graduate course in computer-based instrumentation, and has provided useful feedback to the authors. We are most grateful to Evan Brody, a computer systems analyst, who shared his expertise on Windows and provided a driver program used to demonstrate the important concept of dynamic data exchange in chapter 12. Dr Bernard Harris, Associate Professor of Electrical Engineering at Manhattan College, reviewed the work in its entirety and suggested many helpful improvements. In addition, the authors had the unqualified support and understanding of their wives, Roslyn and Ronnie, whose patience was essential to the ultimate completion of the work.

Gordon Silverman and
Howard Silver
New York, NY
1 July 1994

Copyright, Product Names and Trade Names

Material related to DSPlay is reproduced by permission of Intelligent Instrumentation.

Material related to Lotus® 1-2-3® is used with permission of Lotus Development Corporation. Lotus® and 1-2-3® are registered trademarks of Lotus Development Corporation.

Material related to MathCAD® is used with permission of Mathsoft, Inc., 101 Main Street, Cambridge, MA 02142, USA (telephone: 1-800-MATHCAD). MathCAD® is a registered trademark of Mathsoft, Inc.

Material related to MATLAB® is used with permission of The MATHWORKS, Inc.

Material related to PSpice® is used with permission of MicroSim Corporation.

Material related to LabWindows® is used with permission of National Instruments Corporation.

Part 1

Fundamentals of Data Acquisition and Instrument Control

Describing fundamental elements of computer-based instrument systems is the focus of chapters 1 to 6. These include:

- The computer's role in automated instrumentation.
- The nature of information and measurement.
- Hardware and software architectures and components.

Developing an appreciation of the role of the computer in measurement and instrumentation is achieved by describing several important applications including mass spectroscopic and physiological data acquisition as well as the use of a robot in automated analysis environments. This is done in chapter 1.

Fundamental aspects of experimental environments are discussed. Specific topics describe: the way in which to compute the information content of experimental data; and factors that constrain or limit information such as noise, resolution, and measurement error. This comprises the goal of chapter 2.

The basic functional elements of a computer are introduced in chapter 3; these building blocks are then combined into more complex architectural arrangements suitable specifically to measurement and control environments. The roles of the computer's central processing unit (CPU), memory, its input/output system, and its communication bus are described with particular emphasis on systems employing more than one processor ('multiprocessor' environments) for improved information processing capability. Standard interprocessor communication protocols (e.g. the GPIB standard) are shown to support computer-based instrumentation intended for automated environments. (Additional practical applications are described in part 2.)

The need for system software and high-level language programming in computer-based instrument systems is introduced in chapter 4. The computer's organized collection of control programs for providing an environment in which to execute programs—its operating system (OS)—is described with particular emphasis on DOS. This OS is targeted at the 'single-user' (general-purpose) personal computer which is normally the key hardware element of an automated instrument system.

Follow-up to this basic material is found in chapter 5 where data-processing considerations are noted. This chapter includes material that a user needs in order to make important system design decisions. Three methods of controlling the experimental instruments are described:

- Polling or programmed control in which the computer queries each instrument in sequence and accepts data if available.
- Interrupt-driven architectures in which the instrument signals the computer that data are available; until such times the computer continues with other processing tasks.
- Direct memory access (DMA) organizations where the computer temporarily transfers control of the system to another element or instrument for the purpose to transferring data.

In addition, this chapter describes two methods for analysing (linear) systems, either on the basis of time responses or on the basis of the frequency of the signals that the source generates.

Part 1 concludes with a description of the components needed to produce information for the computer. Included in chapter 6 are operating principles of a variety of instrument transducers as well as a functional description of the way in which information is converted from analogue (continuously variable) form into digital (discrete) form.

1

Introduction

Observing and measuring the environment has always been part of human behaviour and development—first in order to gain understanding and then to be able to control events. The tools that have been employed for these purposes have grown more complex with time. With the continuing maturing of computer technology, the instruments of measurement and control have created the potential for vastly increased human productivity. (This has both positive and negative implications for society; the quality of our lives will be greatly improved, but what will become of those who are displaced or who do not learn the new technologies?) Perhaps within the next few years we will be able to provide verbal instructions to our instruments and tools in natural language and these will be translated into the many detailed instructions needed to carry out complex operations.

This chapter provides an overview of computer-based instruments, describes some examples of their use, and discusses some generic instrument architectures (functional organization). Various terms that are introduced informally will be defined more precisely in subsequent chapters.

1.1 THE ROLE OF COMPUTER-BASED INSTRUMENTATION IN MEASUREMENT AND PROCESS CONTROL

Computers are instruments that process information and thus aid the user in recording, classifying, and summarizing information [1]. These capabilities encompass a number of specific instrument tasks:

- Data handling: acquisition, compression, reduction, interpretation and record keeping.
- Control of instruments, system resources (e.g. printers), and processes.
- Procedure and experiment development.
- Operator interface: control of the course of the experiment or process.
- Documentation and report production.

The computer-based instrument environment is summarized in figure 1.1. Uniquely definable events are generated by the experiment or process under

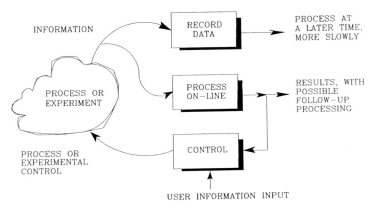

Figure 1.1 Information processing and instrument systems.

control and detected by the instrumentation system†. A number of alternatives regarding the disposition and processing of this information are possible:

- Simple recording of the data for archiving or processing at a later time. (Processing of data without the need to obtain results in any prescribed time limit is referred to as an *off-line* operation.)
- *Real-time* or *on-line* processing of the information that allows the user to modify the experiment. Such real-time or on-line operations require the instrument to produce results within a prescribed or guaranteed time limit. The data may also be recorded for subsequent (off-line) processing and for documentation of results.
- On-line modification of the experiment is possible by using on-line results together with user input to alter one or more of the parameters of the experiment.

1.1.1 Data handling

As noted above, data handling may include one or more distinct operations; each having a well defined purpose:

- Acquisition of data, which includes the detection and classification of events. Event classification requires the coding or representation of that event.
- Data compression, which is used to compact the data, improving the efficiency with which data are recorded or processed. This is important for reducing the storage requirements and/or increasing the rate at which the instrument system can process jobs or tasks.
- Reduction of data, which eliminates spurious or irrelevant information.

† Throughout the text the term *experiment* shall refer to either an experiment or a process under control.

- Interpretation of data, which involves deciding whether the data satisfy the constraints of a theory or conform to the intended or anticipated outcome of the experiment. Emerging data interpretation techniques, such as those involving neural nets or fuzzy logic, are gradually augmenting human interpretation of data.
- Documentation of the experiment, which may use computer-based instrumention to maintain a record (or *journal*) of the entire experiment, including user responses, automatically. These records can form the basis of a report which is generated after the experiment has been completed.

1.1.2 Process control and program development

To control experimental instruments and system resources (e.g. printers, plotters, data storage devices) the computer generates commands that specify what is to be done as well as when particular operations are to be carried out; these commands form the *program*.

The user creates such programs by providing the computer with a series of statements. These statements and the usage rules (syntax) form a 'language.' Numerous languages are available (Pascal, Fortran, Ada, Basic, C, C++, Modula, ...) and choosing the most appropriate one to use depends on several factors including how fast the computer can execute the statements as well the compatibility of the language with the experiment to be completed (see chapter 4). For ease of modification it is desirable that the language is chosen close to 'natural language'. This may help to make the program more readable and simplify modification when needed. As yet, computers by themselves do not recognize English-like statements. The computer subsequently translates the program into those (electrical) signals that perform the required operations (e.g. energizing an instrument relay at a given time).

The process of successfully creating such programs (as well as the associated instrument system) is referred to as the *development phase* and the computer system that is used is called a *development station*. At times, and in some circumstances, it is not possible to use the computer for other purposes such as experimental control while a program is under development.

1.1.3 Computer interface

Computer interfaces make up the connections between the computer and other parts of the instrument system as well as the connections between the computer and the user. The connection may or may not be physical. For example, images displayed to the user to convey information are referred to as *graphical interfaces*. (Many modern instrument systems employ graphical images or *icons*, rather than text, for conveying operating system information to the user.)

1.1.4 Documentation

The final task of an instrument system is to provide support for report production. The report usually includes a concise description of the experiment and its results, together with the user's conclusions. Repeated experiments produce different data and results but their protocols do not change; in such circumstances the computer can include a template into which experimental values may be entered in a convenient manner without the necessity for the user to make extensive descriptive textual entries. Spreadsheets (see chapter 10) are particularly suited for such purposes. Such application packages permit data to be entered directly from experimental records to appropriate places in the report. They often include graphing capabilities, so results can be conveniently presented in graphical form directly from the acquired experimental data.

1.2 EXAMPLES OF COMPUTER-BASED INSTRUMENTATION

The following three examples show how the computer has become an integral part of the instrumentation. The first describes a case of data acquisition and analysis. The second indicates circumstances in which the computer controls various parameters of the experiment as well as acquiring data. The third is a case in which on-line modification of the experiment occurs.

Example 1.1.   On-line acquisition, analysis, and presentation of neurophysiological data

Electrophysiological data are produced from electrochemical reactions within an animal's nervous system, generally within a neuron, in the tissue immediately surrounding the neurons, or at the boundary surface (e.g. skin) of the organism. Such information, in the form of electrical signals (current or voltage), often reflects the state of the organism as a result of behavioural changes or experimental manipulation of the preparation such as chemical, electrical, or mechanical stimulation. The data are essential in understanding how the organism functions and testing theories about the organization of living things.

In this application, the data are inherently transitory in nature and therefore the instrumentation must be able to 'capture' the information when it is present. A typical system for such applications is shown in figure 1.2 [2]. (While discussion of this system will be confined to physiological environments, with equal validity it can be considered as a system for recording transient information of many kinds.)

The experiment produces electrochemical or other signals to be acquired and processed. These are converted into electrical form using a sensor. (Sensors are discussed in chapter 6.) Often, signals produced by the sensor are weak and may contain unwanted contaminants. The 'external preprocessor' consisting of the amplifier and filter elements are included to intensify the signal (AMPLIFIER)

EXAMPLES OF COMPUTER-BASED INSTRUMENTATION

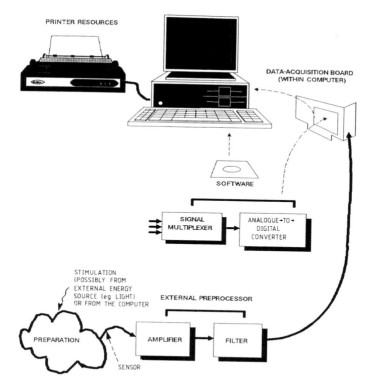

Figure 1.2 Representative physiological data-acquisition, analysis and presentation system.

and reduce unwanted contaminants (FILTER). Signals from the preprocessor are transmitted to the 'data-acquisition board' located within the computer. (In some configurations the data-acquisition hardware may be found external to the computer; in such cases additional signals from the computer are needed to control the functions carried out by the data-acquisition board.) The data-acquisition board includes: a signal multiplexer (for sharing or combining information from multiple sites on the preparation or from several preparations); an analogue-to-digital converter (which transforms information from one form (*analogue*) into a form that can be readily processed by the computer (*digital*)).

The computer contains numerous instructions and commands for controlling the experiment, collecting the data, storing results, analysing what has been collected, and documenting the experiment and the results. Two representative capabilities of such systems are shown in figures 1.3 and 1.4.

Figure 1.3 depicts the display that the user sees and uses prior actually to acquiring data. This display shows the various parameters associated with the experiment. (Displays of this type, which use icons to represent various

8   INTRODUCTION

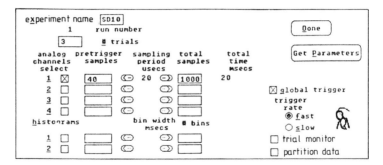

Figure 1.3   A typical menu for data acquisition. (Reproduced by permission of Elsevier Science Publishers.)

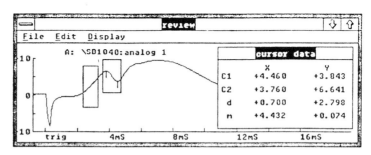

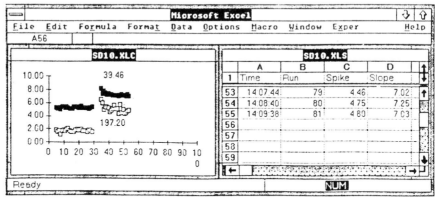

Figure 1.4   An example of the presentation of experimental results. (Reproduced by permission of Elsevier Science Publishers.)

switches, are referred to as *graphical user interfaces* or GUIs.) The software that was developed by the user and that is being executed by the computer simulates the front panel of the dedicated electronic instrument. (A dedicated instrument has a fixed, unmodifiable, purpose.) However, it is a 'virtual' front panel because the controls only appear as images on the computer's monitor.

Using this display the user can specify such things as: how many channels of information are to be combined; how often information is to be acquired or sampled; and the total number of samples. A user interacts with this display in one of two ways:

- The user selects ('points to') a parameter to be adjusted using a manipulanda (e.g. a *mouse* or *joystick*) and then enters its value from the keyboard. (Information entered from the keyboard is normally validated or confirmed once the user depresses the 'Enter' key ($\hookleftarrow$).)
- The user selects the parameter to be changed using specialized keys (*function keys*) or a sequence of keystrokes. He or she then enters its value from the keyboard.

Figure 1.4 shows how the experimental results can be displayed. (In this case the results of an experimental run are displayed—the panel marked 'review'— as well as data that have been automatically analysed.) Commercially available software programs, such as a 'spreadsheet' (e.g. Lotus® 1-2-3® or Microsoft Excel), is useful for carrying out extended analysis of the results as well as for generating graphs. The icon in the upper left-hand corner of the figure is a reminder to the user that the acquisition panel (figure 1.3) is still available (or active) if needed. The user can manipulate a set of markers (cursors) on the display—possibly by using a mouse—in order to specify which data are to be analysed; these are shown as 'boxes' within the panel marked 'review'.

Each panel and its associated display is referred to as a 'window'. These windows are controlled by yet another GUI named 'Windows' (chapter 12) which manages the computer's resources (printer, display, mouse, ... etc). The Windows program can execute programs concurrently and share the computer's display. A most important part of the Windows application package is the Dynamic Data Exchange (DDE) protocol or clipboard. The computer may be readily reconfigured for entirely new experimental purposes thus making it a 'flexible dedicated instrument'. The panels shown in this example were developed with the help of another software package (Windows System Development Kit).

Example 1.2. Mass spectrometry

Mass spectrometers are instruments used to measure the ratio of mass to charge of ions that are generated from a sample material. Results are in the form of a ratio, m/z, where m is the mass of the ions and z is the charge of the ions that are produced from the sample material [3, 4]. Such measurements allow a user to determine, with great sensitivity, the structures of complex molecules. Analysers of this kind can be found in scientific laboratories, industrial environments (e.g. stack emissions), and chemical process control plants.

The basic operating principles are depicted in figure 1.5. A test sample is introduced into an ionizer which produces the sample ions. These ions then pass

10   INTRODUCTION

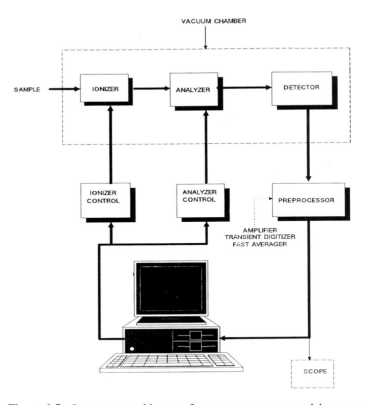

Figure 1.5 Instrument architecture for a mass spectrometry laboratory.

Table 1.1 Mass spectrometry analysis methods.

Type	Method
Quadrupole	A radiofrequency field causes ions to oscillate; only those with a given mass/charge compatible with the field reach the detector. Changing the field filters a new group of ions.
Magnetic-sector	Ions pass through a transverse magnetic field; the field forces them to follow a curved trajectory; detectors located at appropriate points filter only those ions with a given mass/charge ratio.
Time-of-flight	A pulsed beam of ions strikes the sample; secondary ions are accelerated and reach a velocity based on mass/charge; the time it takes to reach the detector is a function of mass.

to the analyser which 'sorts' the ions according to the mass. Specific sorting methods include those shown in table 1.1.

In each of the techniques cited in table 1.1, sorting of ions according to their mass is accomplished by changing the way in which the ions interact

EXAMPLES OF COMPUTER-BASED INSTRUMENTATION

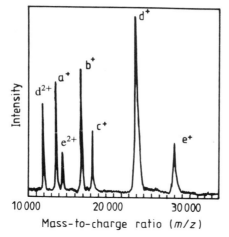

Figure 1.6 The mass spectrum obtained from a mixture of five proteins. Approximately 0.5 pmol of each protein were present in the mixture added to the probe tip. The proteins were bovine pancreatic ribonuclease A (a), equine skeletal muscle myoglobin (b), bovine milk β-lactoglobulin (c), bovine pancreatic trypsinogen (d), and bovine erythrocyte carbonic anhydrase II (e). The peaks labelled 'x^1' and 'x^2' refer to the singly and doubly protonated protein-ion species, respectively. (Reproduced by permission of the authors.)

with the incident energy. This is done by either changing a magnetic field or a voltage and/or frequency of a signal applied to one of the control elements of the analyser. These changes are accomplished in an orderly manner. The control element is 'swept' through a sequence of values while the output from the detector is measured, indicating the abundance of ions at a given mass-to-charge ratio. The result is a graph similar to the one shown in figure 1.6.

The computer plays a key role in the mass spectrometer. Software within the computer sequences both the ionizer and the analyser while it collects data from the detector. The program may have GUIs similar in purpose to those described in example 1.1. (In some time-of-flight systems where a laser is used to ionize the sample and create the ions, time-of-flight measurements occur too rapidly for the computer to control the experiment as well as keep track of the detector output. In such instances the computer performs data acquisition or controls external hardware that scans the sample.)

Example 1.3: Automated chemical analysis using a robotic arm

Robotic (articulated) arms emulate the motions of a human arm/hand combination. A variety of arrangements for accomplishing this are possible and figure 1.7 shows one simple example. The five independent motions are summarized as follows:

1. The entire arm can rotate on the base plate.

12 INTRODUCTION

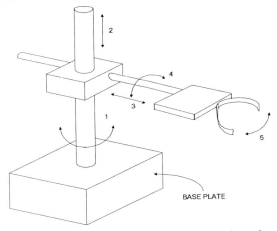

Figure 1.7 A representative articulated arm with five independent motions.

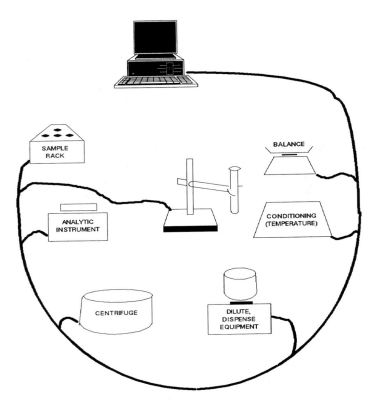

Figure 1.8 A computer-controlled automated substance analysis system using a robotic arm.

2. The arm can be raised or lowered.
3. The hand (grippers) can be extended or withdrawn.
4. The hand can pronate (rotate).
5. The fingers of the hand can open and close (yaw).

Consider this robotic arm integrated into the system shown in figure 1.8 [5, 6]. All of the instrumentation necessary for the automated analysis is located within the operational range of the robotic arm. In addition, the instruments are also under computer control. Instruments in such cases may include: a centrifuge (e.g. for analysis of blood samples); an analytic instrument (e.g. a spectrophotometer or chromatograph); a rack to hold the samples; a balance; a conditioning unit (possibly a stirrer or temperature oven); an instrument for dispensing, extracting and/or diluting chemicals.

A program within the computer can be used to define precisely the steps taken by the robotic arm to carry out a routine test. This program must also take into consideration the tasks to be carried out by each instrument. When the computer does not obtain ongoing, continuous, status information from an instrument, the resultant arrangement is referred to as *open-loop* control of the particular instrument. Then the program must include appropriate time delays.

The computer can also receive signals from the various instruments that advise the program of their status; this is referred to as *closed-loop* control which is generally more desirable than the aforementioned open-loop control.)

A complete description of the motion of the robotic arm requires many detailed steps, even for simple tasks. This can include literally thousands of operations or instructions (e.g. pronate the hand 10° followed by closing the fingers 30°, etc). For the user to complete such a programming task prior to successful operation of the program would take a prohibitively long time; it requires the user to predict accurately the precise coordinates of all axes of motion for each desired position. Instead, robotic arms usually come with a series of programs that greatly simplify the task. These programs are often called training programs. A training program tracks the position of each member of the robotic arm. Users position the arm and its members manually, using either a manipulanda or reserved keystrokes in combination with numerical information. When the user is satisfied with the successive motion that the robotic arm is to make, the data—as tracked by the computer—are stored in the computer's memory (an open file). In this way a complex sequence of motions can be generated. These motions coupled with commands to the individual instruments produce an automated procedure for chemical analysis or processing.

Widely recognized industry protocols have been developed to standardize commands to individual instruments. Examples of these standard protocols include RS-232, and IEEE 488 (chapter 3). These protocols can be integrated directly into computer programming languages (chapter 4) thereby simplifying the programming task particularly if high-level languages are used. (Such protocols are informally referred to as *handshaking*.)

14   INTRODUCTION

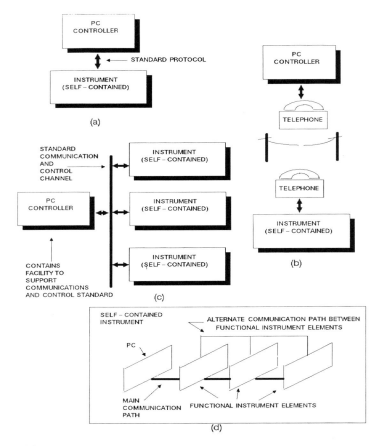

Figure 1.9   Single-PC instrument architectures. (a) Dedicated system. (b) Remotely controlled architecture. (c) Multiple instrument. (d) Tightly coupled.

Partial results obtained during the course of the analysis might prompt the user to alter the program of tests. If the program is designed as a series of individual tests organized in a modular fashion then the user has the ability to alter the sequence of tests as the experiment evolves.

1.3   ARCHITECTURES OF COMPUTER-BASED INSTRUMENT SYSTEMS

The functional organization or arrangement of the defining elements of an instrument is called its *architecture* [7]. The instrument and/or computer architecture identifies instrumental operations to be performed as well as the relationships, including order or timing, between these operations. Such descriptions may be applied to either the physical components (*hardware*) of the system or to the program of operations contained therein (*software* architecture). System architecture can be effectively represented in diagrams such as those shown in figure 1.9. When appropriate, additional functional refinements (architectures) such as those for the computer will be described in subsequent chapters.

A single-purpose (dedicated) computer-controlled instrument is shown in figure 1.9(a). This simple form of instrumentation is convenient because it is consistent with existing (building) wiring, particularly the telephone system. Standard communication protocols such as RS-232 allow manufacturers to develop instruments to accepted standards. This architecture is often limited to a single computer† and a single instrument, and the distance between the PC and the instrument is also limited. (By adding more communication lines, additional instruments can be controlled from the same PC.) Remote control of the instrument is possible by addition of resources called modems (chapter 3) between the PC and the telephone line as well as between the line and the instrument; this architecture is shown in figure 1.9(b). Real-time operations using either of these configurations is difficult. This follows from the fact that time is needed for communication between the PC and the instrument, placing significant limits on the ability of the system to obtain complete results in a prescribed time interval. Complete processing of data are slowed in part because the data must pass through a functional element—the controller—which is included to perform the operations required by the standard protocol.

With the development of (and need for) instruments with greater capacity new architectures have emerged. One configuration is shown in figure 1.9(c) and includes a single PC together with multiple instruments. The PC itself includes a resource called an IEEE 488 controller which functionally supports the IEEE 488 communication (and instrument control) performance standard. (See chapter 3 for a description of the IEEE 488 protocol.) While this architecture is flexible and new instruments can be easily added, the speed of operation can deteriorate to the point where the capabilities of the PC are exceeded. The speed of operation is reduced because of competition for system (PC) services from the increased number of instruments; for example, a data-acquisition instrument and an oscilloscope display may both make intense demands on the PC secondary-storage system.

The architecture shown in figure 1.9(d) is a 'tightly coupled' instrument system. In such cases the instruments are integral to the PC itself. Communication between the PC and the instrument is rapid. Real-time (on-line) operation of the instrument is facilitated by a direct communication path between the instrument and other critical parts of the PC such as its memory. No controller is necessary and consequently the time delays associated with such

---

† Throughout this book the term *computer* will be associated with the personal computer (or PC) which is the type of computer most likely to be found in instrument systems.

16 INTRODUCTION

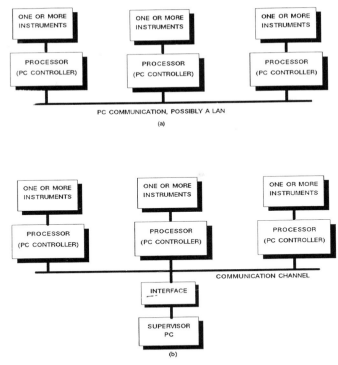

Figure 1.10 Multiprocessor instrument architectures. (a) Shared communication channel, (b) Master/slave organization.

functional elements do not exist. By varying the functional combinations, the system can be readily reconfigured for new applications.

Each of the architectures shown in figure 1.9 includes a single PC. Additional operating speeds are possible if more than one PC is included in the instrumental configuration. Such architectures are called multiprocessor (-based) instrument systems—two examples are depicted in figure 1.10. Each PC or processor is capable of carrying out program instructions in its own right. (A measure of a system's capability is the number of instructions it can carry out in one second. Because PCs work at very high speeds, the unit associated with this measure is 'millions of instructions per second' or MIPS. Multiprocessor systems may perform at rates of 10–100 MIPS. Recent developments, including greatly reduced (smaller-sized) circuit dimensions, and 'massive parallelism', anticipate systems capable of 1000 MIPS ('teramips'). The speed of operation of such systems is greatly increased; however, the activities or functions performed by the individual PCs must be coordinated; one PC may be designated as a supervisor (master) while the others carry out measurements and control under instruction of the master; these PCs may be designated as slaves. A number of advantages are possible:

- Flexibility: the system can readily be reconfigured to run a new experiment or execute a mature experiment while development of new programs can occur at the same time.
- Improved reliability: if one of the PCs fails during an experiment, another can take over its responsibilities without the need to interchange the components physically.
- Compatibility with multiuser environments: the processors can often execute independent tasks. This is distinct from applications in which a single PC is shared among a number of users (time sharing).

In figure 1.10(a) the processors share a single communication channel. Communication between cooperating processors is rapid when the processors are interconnected via a bus dedicated to that purpose (*tight coupling*). In another form the communication channel is known as a *local-a*rea *n*etwork or LAN. Communication between processors is slower because the system adheres to the LAN standard of communication and this introduces time delays in order to achieve uniformity between different installations.

figure 1.10(b) explicitly identifies the supervisor. The master contains an interface that supports the communication between cooperating processors. This communication channel might be the IEEE 488 standard mentioned above.

In addition to the increased cost of such systems, it should be noted that as more processors are added the information-handling capabilities of the communication channel becomes a limiting factor or bottleneck. If a processor makes a request for system resources, it may be forced to wait until the communication channel can accept its message. This limits the overall operating speed of the instrument. (The overall operating ability of the instrument is measured by its throughput—how much information it can completely process in a given time increment. Communication protocol requirements *may* reduce throughput.)

REFERENCES

[1] Silverman G 1988 Computers in the biological laboratory *Encyclopedia of Medical Devices and Instrumentation* ed J G Webster (New York: Wiley-Interscience)

[2] Stromquist B R, Pavlides C and Zelano J A 1990 On-line acquisition, analysis and presentation of neurophysiological data based on a personal microcomputer system *J. Neurosci. Methods* **35** 215–22

[3] Beavis B C and Chait B T 1990 Rapid, sensitive analysis of protein mixtures by mass spectrometry *Proc. Natl. Acad. Sci., USA* **87** 6873–7

[4] Beavis B C and Chait B T 1989 Factors affecting the ultraviolet laser desorption of proteins *Mass Spectrometry* **3** 7

[5] Cerda V and Ramis G 1990 *An Introduction to Laboratory Automation* (New York: Wiley-Interscience)

[6] Grandsard P, Magargle R and Markelov M 1991 Robotics and sample preparation automation in contemporary labs *Sci. Comput. Automation* June

[7] Van de Goor A J 1989 *Computer Architecture and Design* (New York: Addison-Wesley)

2

Information and its Representation

The fundamental outcomes of experiments are events, and quantifying events is essential to the design of any instrument system. Questions, such as 'How fast must the computer operate?' or 'How much storage is required?' are answered by how events are quantized or measured. This chapter describes the fundamental concepts of information, representation of such information, limits to information, and system capacity. All of these are important parameters for choosing or designing an instrument system.

2.1 MEASURING INFORMATION CONTENT IN LABORATORY ENVIRONMENTS

An experiment generates a set of outcomes which can be designated $S_1, \ldots, S_n$. These comprise the domain of the experiment. Subsets of this domain are called events. Experiments can only generate a finite number of distinct events. While an infinite number is theoretically possible, many events are indistinguishable from one another and this limits the ultimate number of uniquely recognizable events.

The *information content* of the outcomes is determined by two factors: the number of distinct events and the probabilities associated with these events. For purposes of discussion, the probability of an event will be regarded as its relative frequency of occurrence in the experiment. This is not a formal definition of the probability of an event, but it is a reasonable heuristic 'working definition.'

Example 2.1. One way of looking at information content of a data source

Consider an experiment with eight outcomes all of which are equally likely to occur. The outcomes will be designated by the decimal digits 1 to 8. One way of looking at the information content of such an experiment is to ask, 'How many questions with yes or no answers must I ask in order to determine uniquely which experimental outcome has occurred?'. An efficient algorithm (method of solution) is to ask a sequence of questions like those shown in figure 2.1. Note that a maximum of three questions with 'yes' or 'no' responses

INFORMATION CONTENT IN LABORATORY ENVIRONMENTS

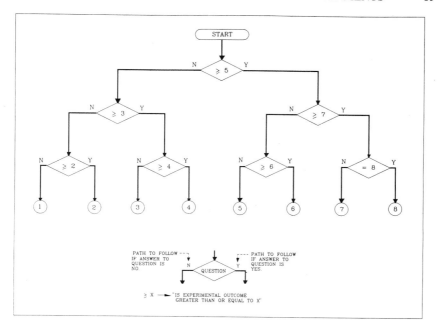

Figure 2.1 A flow diagram for an algorithm to identify an experimental outcome.

uniquely identifies any of the eight outcomes. Each answer reduces the number of possibilities by one half. This experiment is considered to have three 'units' of information—only three questions are required to arrive at a correct answer.

The basic *unit of information* is the bit; the experiment considered above has three bits of information. In general, for experiments with equally likely outcomes the amount of information is given by

$$2^H = n \qquad (2.1)$$

where H is the amount of information in bits and n is the number of distinct outcomes. (Each time another question in the sequence is required the number of potential outcomes increases by a power of 2.) This equation is normally written as

$$H = \log_2 n \qquad (2.2)$$

and is derived from equation (2.1) by taking the logarithm of both sides of the equation; this logarithm is taken to the base 2. Note that

$$\log_b x^y = y \log_b x$$

and

$$\log_b b = 1$$

where b is the *base*.

Considerations outlined above were restricted to the case where all outcomes were equally likely or equally probable. What happens when this is not the case? If, for example, one of the experimental events occurred with a probability of 0.99—probabilities range in value from 0 to 1—it would be prudent to 'guess' that this outcome had occurred rather than to use the algorithm depicted in figure 2.1. Such a guess would be correct 99% of the time. If the guess was incorrect, a logical second choice would be the next most probable outcome. This implies that high-probability events do not convey as much information as events with somewhat lower probabilities as the former occur with greater frequency. Since we expect the high-probability event to occur very frequently, its outcome is 'not unusual' while the occurrence of the low-probability event might cause the experimenter to take special note. The sun always rises in the East and thus conveys little information; consider the response upon seeing the sun rising in the West.

The implications of differing probabilities can be reflected as a change in the definition of information. Equation (2.2) includes n 'equally probable' outcomes. The probability of any one outcome is therefore $1/n$; call this quantity p_i. The information associated with the single experimental outcome i is

$$h_i = \log_2 \frac{1}{p_i} = \log_2 1 - \log_2 p_i = 0 - \log_2 p_i = -\log_2 p_i.$$

While actual experiments may have different information content, what is of interest is the *average* information content. Finding an average may be interpreted in a way that is revealed by the following example.

Example 2.2. Looking at the average information content

Consider a series of experimental outcomes represented by the following symbols, with repeated letters indicating the same (numerical) outcome:

$$a, b, b, c, d, a, d, e.$$

The arithmetic average is obtained by summing the results and dividing by the number of outcomes:

$$\text{average} = \frac{a+b+b+c+d+a+d+e}{8} = \frac{2a+2b+c+2d+e}{8}$$

$$= \frac{2}{8}a + \frac{2}{8}b + \frac{1}{8}c + \frac{2}{8}d + \frac{1}{8}e.$$

Note that the fractional quantity represents the (relative frequency) probability of occurrence of a, b, c, d, and e. Thus, to find the average value of a series of numbers, sum the outcomes multiplied by their probabilities as shown in the equation below:

$$\text{average} = \sum_i p_i i.$$

INFORMATION CONTENT IN LABORATORY ENVIRONMENTS

Table 2.1 Experimental outcomes and assocoated probabilities.

Outcome	Probability	$-\log_2 p_i$	$-p_i \log_2 p_i$
1	0.40	1.322	0.519
2	0.05	4.322	0.216
3	0.05	4.322	0.216
4	0.05	4.322	0.216
5	0.10	3.322	0.332
6	0.05	4.322	0.216
7	0.15	2.737	0.411
8	0.15	2.737	0.411

Applying this to the calculation of information content produces the following result:

$$H = -\sum_i p_i \log_2 p_i. \quad (2.3)$$

Example 2.3. Direct evaluation of the information content of an experiment when the probabilities of the outcomes are known

An experiment produces eight distinct outcomes with some known associated probabilities. This information is summarized in table 2.1 together with the information content. The information content of this experiment is calculated by summing the values in the last column. The result is 2.537 bits of information. Compare this case to an experiment with eight equally likely outcomes that has three bits of information. It has less information content. Indeed, it can be shown that the information content of an experiment is at its maximum when all outcomes are equally likely.

2.1.1 Information formats

In experimental environments there may be many variables that affect the information content. However, in many situations the results depend on two variables. One is defined as the independent variable and the second, which is functionally related to the independent variable, is called the dependent variable. The experimenter normally manipulates the independent variable and measures the resulting dependent variable.

Examples of independent variables—by no means exhaustive—include time, voltage, current, light intensity, temperature, distances or angles (from some origin), frequency of the stimulating energy source, and chemical concentration. Examples of dependent variables may also be taken from this list.

Two alternative formats are possible for the independent variables of an experiment: their values may either be continuous or discrete. 'Continuous' means that the outcomes occur from a continuum of values while 'discontinuous'

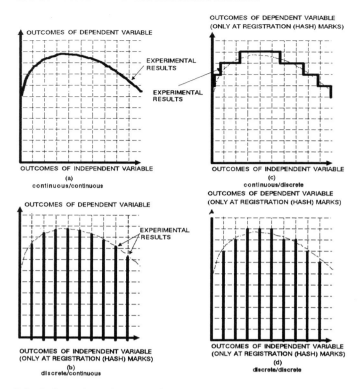

Figure 2.2 Information formats for experiments with one independent and one dependent variable (independent/dependent). (a) Continuous/continuous. (b) Discrete/continuous. (c) Continuous/discrete. (d) Discrete/discrete.

means that the outcomes are restricted to a finite set of outcomes. In the discussion that follows, the variables will be limited to a single independent and a single dependent variable. However, the basic concepts remain valid for those cases with more than one independent and dependent variable.

Continuous or discrete outcomes are also possible for the dependent variables in an experiment. Consequently there are four possible combined formats for the dependent and independent variables. These are summarized in figure 2.2. In figure 2.2(a) both variables can assume a continuum of informational values. In figure 2.2(b) the independent variable can have discrete outcomes while the dependent variable provides continuous information. Figures 2.2(c) and 2.2(d) show the cases for the other possibilities. (Dotted and/or dashed lines are included in the figure in order to provide a reference on the basis of which the case can be compared with its alternative. For discrete cases, the permissible outcomes are shown at equal intervals although it is possible for these outcomes to have unequal separations.)

Continuous formats are useful when describing models of the system under investigation but they are only convenient fictions; any physical outcome must ultimately take discrete values. Contaminants (e.g. noise) usually limit the continuum of values to a finite number. Experiments may generate a continuum of values but subsequent processing reduces these to discrete outcomes, particularly when a computer forms part of the architecture. Almost all instances of information from instrument systems are ultimately discrete in nature. (A strip chart recording is one example in which the outcome is continuous.) In figure 2.2, the solid lines represent actual possible outcomes that might be observed. Unless noted, for the remainder of this chapter the discussion will be restricted to the format shown in figure 2.2(d) (discrete/discrete).

2.1.2 Representing information in natural number systems

One of the simplest schemes for assigning numbers to an experimental outcome is to use the natural decimal number system. In addition to identifying particular events such a system has the advantage of providing relative measures or ordering of events—some events may be 'greater' than others. These relative values are convenient for implementing arithmetic, and calculating differences (or sums) of experimental outcomes.

The natural decimal numbering system may considered to be a 'translation' between experimental events and symbols by which these outcomes are identified. The decimal number system is a generally recognized scheme for representing information because it allows experimenters to report, compare, and interpret instrumental results.

The decimal number system (for representing information) includes:
- Ten distinct symbols or decimal digits: 0, 1, 2, 3, 4, 5, 6, 7, 8, 9.
- A system of weights that is based on the position of the symbol within the number. The standard format for this symbolism is

$$\cdots + D_2(10^2) + D_1(10^1) + D_0(10^0) + D_{-1}(10^{-1}) + D_{-2}(10^{-2}) + \cdots.$$

For convenience, the powers of ten are omitted as well as the '+' signs. The number appears as

$$\cdots D_2 D_1 D_0 . D_{-1} D_{-2} \cdots$$

with each 'D' replaced by one of the allowed symbols (decimal digits) and the 'dot' (or decimal point) separating the positive powers of ten from the negative powers of ten.

As a consequence of the discrete nature of experiments, only a finite number of digits are required for each experimental outcome. This defines or limits the range of outcomes. For example, an experiment with the following format:

$$D_2 D_1 D_0 . D_{-1}$$

would provide for 10 000 outcomes from 000.0 to 999.9. (The 10 000 outcomes might also represent experimental values from -500.0 to $+499.9$.)

24 INFORMATION AND ITS REPRESENTATION

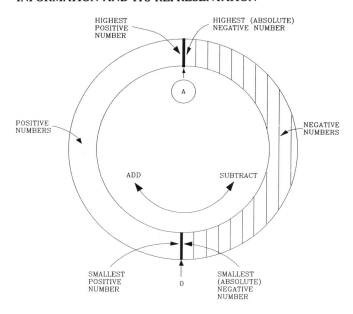

Figure 2.3 Number system representations.

If the range of outcomes provided by the instrument system is limited to the values from 000.0 to 999.9, how might it respond to an input whose value is the equivalent of 1000.0 (or −000.1)? Normally, computer-based instruments have a 'circular' organization or number format (see figure 2.3). As a consequence of this arrangement, ordinary addition and subtraction can yield an erroneous answer. Addition or subtraction will 'advance' the resultants in the directions shown in the figure. Any addition or subtraction operation that crosses the boundary (labelled) 'A' will generate an incorrect answer.

2.1.3 Number systems and arithmetic operations

While the decimal number system lends itself to convenient human interpretation of results, it is a cumbersome information coding scheme for instrumentation systems. Internal operations within computers require a numbering scheme that includes only two symbols in order to process experimental data. Designation of data and operations is more compatible with a binary number system—designated binary because it uses only two symbols, 0 and 1. The numbering scheme parallels the decimal system in the sense that it includes a combination of symbols and positional weights. The standard format for the binary numbering scheme is

$$\cdots D_2(2^2) + D_1(2^1) + D_0(2^0) + D_{-1}(2^{-1}) + D_{-2}(2^{-2}) \cdots$$

and its normal representation is

INFORMATION CONTENT IN LABORATORY ENVIRONMENTS

$$\ldots D_2 D_1 D_0 . D_{-1} D_{-2} \ldots$$

Each symbol (D) used in the binary numbering system is referred to as a bit which comes from *b*inary dig*it* and can have a value of 0 or 1. (In a sense this is an unfortunate terminology because it can be confused with a unit of information; the two terms have different meanings.) The least significant bit is the symbol at the right of the standard form while the most significant bit is the symbol to the left.

Example 2.4. Representing negative integers in the binary number system

When representing negative numbers in the binary number system one does not normally place a negative sign in front of a positive number. (Within the computer it is possible to organize information that includes a binary symbol (1) to represent a negative sign but this requires an additional level of coding.) If the concept of counting is used to progress from one binary number to the next, just as it is in the decimal number system, then the four-bit number 1111 would represent -1, as adding 1 to this produces (1)0000 which is the next logical number in the sequence. (The '(1)' indicates the carry, which is ignored.)

The rules of binary addition ($A + B$) are shown in table 2.2.

Table 2.2 Summary of the rules for binary addition.

A	B	Sum	Carry
0	0	0	0
0	1	1	0
1	0	1	0
1	1	0	1

The binary numbers between -8 and $+7$ are shown below:

0111	1111
0110	1110
0101	1101
0100	1100
0011	1011
0010	1010
0001	1001
0000	1000

26 INFORMATION AND ITS REPRESENTATION

(these binary numbers are listed in tabular form; within the computer the number 1000 is considered 'adjacent' to 0111; refer to figure 2.3).

Example 2.5. An algorithm for finding negative binary numbers

To negate a binary number, follow these rules:

1. Start with the least significant bit.
2. Copy that bit.
3. If the bit is 0, pass to the next most significant bit and repeat step 2, or else copy the bit (1) and go to step 4.
4. Complement the remaining bits in the number. To complement a bit, change a 1-bit to a 0 and a 0-bit to a 1.

Carrying this out for the binary number 01100 will produce 10100. (An equivalent result is obtained by complementing all bits in the number and adding 1 using the rules of binary addition.)

Example 2.6. Subtracting binary numbers

Subtraction in the binary number system can be performed using one of two methods. In the first method a table of rules similar to the one for binary addition is used; this is given as table 2.3.

Table 2.3 Rules for binary subtraction $(A - B)$.

A	B	Difference	Borrow
1	0	1	0
1	1	0	0
0	0	0	0
0	1	1	1

The second method involves the following steps:

1. Obtain the ('twos') complement of the B; to carry this out use the algorithm cited above for negating a binary number.
2. *Add* the result to A.

An example of this is shown below (decimal equivalents are shown in parentheses).

A	0100	(4)
B	0110	(6)
Twos complement of B	1010	
Add this to A	1110	(-2)

On occasion, the experimenter will be required to convert results from one number system to another. Most often this conversion is needed between decimal and binary numbers (or between decimal and hexadecimal numbers). Conversion from one number base to another is discussed in the next example.

Example 2.7. Converting numbers from one base to another

An algorithm for performing this conversion is based on the analysis shown below. The number N_1 (starting from base 1) is to be converted to a number N_2 (in base 2). N_1 and N_2 must be equivalent ($N_1 = N_2$). (The discussion below is restricted to whole numbers; for fractional quantities the division operation would be replaced by a multiplication operation.)

$$N_1 = N_2$$
$$a_{1n}b_1^n + a_{1(n-1)}b_1^{n-1} + \cdots + a_{10} = a_{2m}b_2^m + a_{2(m-1)}b_2^{m-1} + \cdots + a_{20}.$$

Divide both sides by b_2:

$$\frac{a_{1n}b_1^n}{b_2} + \frac{a_{1(n-1)}b_1^{n-1}}{b_2} + \cdots + \frac{a_{10}}{b_2} = a_{2m}b_2^{m-1} + a_{2(m-1)}b_2^{m-2} + \cdots + \frac{a_{20}}{b_2}.$$

The results of this operation (on both sides) consist of a whole number and a remainder. Whatever the remainder on the left-hand side (the original number), it must be equal to the remainder on the right-hand side. The coefficient a_{20} is one of the unknown coefficients that we seek and is also the remainder on the right-hand side. The algorithm for number conversion proceeds as follows:

1. Divide the original number by the base (radix) of the number to be determined; the calculations are performed using arithmetic operations (rules) of base b_1.
2. The result consists of a whole number and a remainder. The remainder is the lowest-order coefficient not yet determined. The conversion is complete if the whole number (result) is 0; otherwise proceed to step 3.
3. Reserve the remainder—it is one of the required coefficients; go back to step 1 with the whole number—the result of step 2—as the new 'original' number.

Table 2.4 Conversion of the decimal number 25 into binary.

Step	Current number	Whole result	Remainder
1	25	12	1
2	12	6	0
3	6	3	0
4	3	1	1
5	1	0	1

Table 2.4 shows a conversion of the decimal number 25 into its binary equivalent. The binary equivalent of the decimal number 25 is 11001 (write the last column in reverse order from bottom to top). This can be verified by noting the following:

$$1(2^4) + 1(2^3) + 0(2^2) + 0(2^1) + 1(2^0) = 16 + 8 + 0 + 0 + 1 = 25.$$

The equation just cited also demonstrates a method for converting binary numbers into equivalent decimal numbers. Simply multiply each bit by an appropriate power of 2 and sum the results. It is somewhat easier—more familiar—than using the algorithm just described where the conversion would require arithmetic operations in the binary system.

Example 2.8. The hexadecimal number system

The binary numbers encountered in modern instrument systems are often unwieldy and difficult to interpret without effort. One compromise that has become widely accepted is to use the hexadecimal number system particularly when some results are to be displayed to the user. (The octal number system is also found in such systems but will not be described; understanding octal numbers is readily inferred from the discussion of the hexadecimal system.)

As with all number systems, the hexadecimal number system includes an appropriate number of symbols and weights. The hexadecimal number system includes the sixteen symbols 0, 1, ..., 9, A, B, C, D, E, and F. The standard form of hexadecimal integers is

$$S_n(16^n) + S_{n-1}(16^{n-1}) + \cdots + S_0$$

which is normally written as

$$S_n S_{n-1} \cdots S_0.$$

The hexadecimal equivalent of the decimal number 29 is 1D. (In some computer languages hexadecimal numbers include a prefix to 'signal to the computer' that the symbols that follow form a hexadecimal number. For example, the hexadecimal number 1D might be designated as &H1D with the 'H' designating hexadecimal.) An important property of a hexadecimal number is the ease with which it can be converted into a binary number (and vice versa). The list below shows the sixteen hexadecimal numbers and their equivalent binary representations:

Hexadecimal	Binary	Hexadecimal	Binary
0	0000	8	1000
1	0001	9	1001
2	0010	A	1010
3	0011	B	1011
4	0100	C	1100
5	0101	D	1101
6	0110	E	1110
7	0111	F	1111

To convert a binary number into its equivalent hexadecimal value proceed as follows:

1. Starting with the least significant bit, arrange the binary number into groups of four digits.
2. Replace each group with its corresponding hexadecimal number.

As an example, consider the following:

binary number :	1101100101111011			
after step 1 :	1101	1001	0111	1011
after step 2 :	D	9	7	B

(Conversely, if the original number was hexadecimal (or 'hex'), replacing each hex digit with its binary equivalent would complete the conversion.)

When converting decimal numbers to binary equivalents it is sometimes easier to first convert the decimal number to its hexadecimal equivalent and subsequently from hex to binary.

The conversions described above are intended to provide some insights into the representations of information as well as the conversions that are, at times, required of the experimenter. As with the instrumentation tasks themselves, these conversions have been highly automated; calculators are readily available that perform machine conversions; the computer itself can be programmed to perform these conversions. (Some computer languages include attributes for number conversion; a function can be invoked to perform the conversion.) As a practical matter, however, there are occasions when the experiment requires 'manual' conversions, either because the calculator is not handy or because the computer must be diverted from its immediate tasks, something that is not always possible.

2.2 WHAT TO CONSIDER WHEN SELECTING ALPHA-NUMERIC CODES

Conventional natural number systems illustrate one way in which information can be coded or represented. There are a variety of additional coding systems that employ arbitrary information representations. Several methods and techniques are notable.

Example 2.9. Other codes

Number systems include weighted representations with the weights formed by using increasing (and decreasing) powers of the radix or base of the number system; other systems are possible. Table 2.4 shows a variety of binary codes with their corresponding decimal equivalents. It includes only a few of the possible codes for representing the decimal digits from 0 to 9. When the decimal

Table 2.5 Some binary codes and their decimal equivalents. To find the decimal equivalent of each code add the products of each bit and its corresponding weight. The decimal number is found in the parentheses.

Pure binary	8421 weights	2421 weights	5211 weights	842'1' weights[a]	Excess-3 weights
0000	0000 (0)	0000 (0)	0000 (0)	0000 (0)	
0001	0001 (1)	0001 (1)	0001 (1)		
0010	0010 (2)	0010 (2)			0011 (0)
0011	0011 (3)	0011 (3)	0011 (2)		0100 (1)
0100	0100 (4)	0100 (4)		0100 (4)	
0101	0101 (5)		0101 (3)	0101 (3)	0101 (2)
0110	0110 (6)			0110 (2)	0110 (3)
0111	0111 (7)		0111 (4)	0111 (1)	0111 (4)
1000	1000 (8)		1000 (5)	1000 (8)	1000 (5)
1001	1001 (9)		1001 (6)	1001 (7)	1001 (6)
1010				1010 (6)	1010 (7)
1011		1011 (5)	1011 (7)	1011 (5)	1011 (8)
1100		1100 (6)			1100 (9)
1101		1101 (7)	1101 (8)		
1110		1110 (8)			
1111		1111 (9)	1111 (9)	1111 (9)	

[a] Subtract 2 or 1 if 1-bit appears in the corresponding code position to obtain decimal equivalent.

number has more than one digit, each decimal digit is coded using table 2.5. (Note that all codes in the table (except the 8421 weight) have a 1 in the most significant digit for the decimal numbers 5 or greater. This can be an advantage in certain situations.) There are some 76 million codes possible using four bits. Codes that include more than four bits have been used, and this greatly increases the number of coding possibilities.

Example 2.10. Unit-distance codes

Other information coding schemes may use a series of 'logical operations' to calculate the code; coding is performed 'on the fly', or as needed, rather than maintaining a ('look-up') table of equivalences. One example is shown below:

Decimal number	Gray code	Decimal number	Gray code
0	00000	9	01101
1	00001	10	01111
2	00011	11	01110
3	00010	12	01010
4	00110	13	01011
5	00111	14	01001
6	00101	15	01000

WHAT TO CONSIDER WHEN SELECTING ALPHA-NUMERIC CODES

Decimal number	Gray code	Decimal number	Gray code
7	00100	16	11000
8	01100		
⋮		⋮	

The pattern indicated for the code can be extended for numbers beyond 16. The code above (called the 'Gray' code) is one example of codes referred to as unit-distance codes. These codes take their name from the fact that only one bit position changes between successive codes. An advantage can be appreciated by referring to figure 2.4; sections of the Gray code (figure 2.4(a)) and the binary code (figure 2.4(b)) are included. When converting positional information from an experiment into a coded form, a series of 'brushes' or contacts travel on the pattern. (Lights and photodiodes, or magnetic strips may also be used.) The position of the brushes—the reading line in the figure—is directly converted into coded form. In the case of the pure binary pattern, any misalignment of

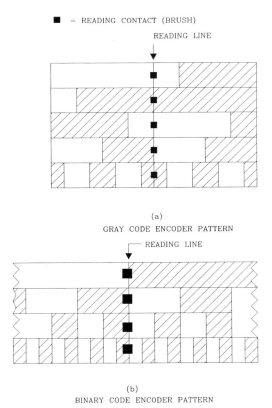

Figure 2.4 Encoder patterns for information representation. (a) Gray code encoder pattern. (b) Binary code encoder pattern.

32 INFORMATION AND ITS REPRESENTATION

Table 2.6 Gray code generation via logical operations.

First bit of operation	Second bit of operation	Result
0	0	0
0	1	1
1	0	1
1	1	0

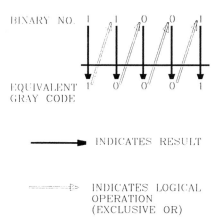

Figure 2.5 Conversion from binary to Gray code.

the brushes can cause serious errors in the resultant code. For example, at the position shown in the figure, misalignment of the brushes can result in any code from 0000 to 1000 (depending on which brushes are misaligned). For the Gray code, however, brush misalignment can be tolerated without significant error. For the case shown, misalignment of the most significant bit (brush) will produce a result that is one number removed from its proper value; this is normally within accepted standards for the experiment.

As noted, the Gray code can be generated starting with the binary form of the code, and processing this number using a series of logical operations. Consider, first, the logical operation defined in the table 2.6. (To aid in interpretation, examine the first row. If the first bit in the operation is 0 and the second bit in the operation is 0 then the result will be 0. The other rows are similarly interpreted.)

Now, refer to figure 2.5. The binary number 11001 is converted to a Gray code using the following rules:

1. Copy the most significant bit of the binary number into the answer (solid arrow on the left).

WHAT TO CONSIDER WHEN SELECTING ALPHA-NUMERIC CODES

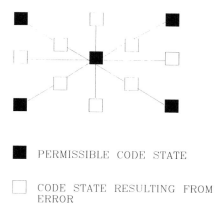

■ PERMISSIBLE CODE STATE

☐ CODE STATE RESULTING FROM ERROR

Figure 2.6 The state space for information codes including redundancy (error detection only).

2. Perform the logical operation shown in the table using the result from step 1 and the next most significant bit of the binary number (empty arrow on the left). The result of the logical operation (called 'exclusive OR') generates the next bit in the code.

3. Continue in this fashion until all bits of the binary number have been coded.

The binary number in the example is equivalent to the decimal number 25. Thus the Gray code for decimal 25 is 10001.

2.2.1 Redundancy and error checking/correction

Experimental information undergoes many operations between event detection and final analysis and/or interpretation. The coding process noted above is one such operation. Others include transmission, storage, playback, and analysis; this list is not exhaustive. Errors can arise in the course of these operations. However, the coding process may include elements that can detect and sometimes correct such errors. In general, this is accomplished with the introduction of additional bits in the code. These bits do not directly enhance the information content (of the event). They are 'redundant'; however, they can be used to detect and/or correct errors. Refer to figure 2.6.

The filled squares represent states of a code that can be generated by the experimental outcomes. The empty squares are additional states of the code, which do not result from the experiment but may arise from an error that is introduced during subsequent information processing (e.g. transmission of the information to the computer). If a computer detects one of these additional states, it can decide that an error has occurred.

If allowable states are separated by only one redundant state then the most reasonable conclusion is that an error has occurred, but nothing can be said regarding the correct state—which state was correct before the error? If each legitimate state is separated by two redundant states then one could conclude with some justification that the legitimate state closest to the redundant state is the correct code. (Clearly, if more than one error occurs the conclusion is incorrect, but statistically two errors are far less likely than a single error.) The redundant states require additional bits in the code and this is a 'cost' that must be absorbed by the instrument system. (It might be reflected in higher required system capacity, which is discussed below.)

Example 2.11. A case of an error-detection coding scheme

One widely recognized error-detecting scheme requires the addition of one additional bit to the code. There are two forms of this method referred to as either odd parity or even parity. The bit to be added depends on the number of bits in the word to be coded and is summarized in the table below:

Number of bits in source code	Added bit in odd-parity scheme	Added bit in even-parity scheme
Odd	0	1
Even	1	0

In other words the bit that is added will make the total number of bits odd (under odd parity) or even (under even parity).

2.2.2 Alpha-numeric coding (ASCII)

Some information consists of alpha-numeric characters. For example, some instruments require such data when they are under the operational control of a computer. The code that is often used for such purposes is the American Standard Code for Information Interchange (ASCII). It is a binary code and may be transmitted with parity bits as described above.

2.2.3 Coding efficiency and data compaction

As discussed above, the number of bits of information generated by an experiment depends in part on the probabilities of the underlying events. The codes that have been described up to this point have not taken the probabilistically variable information associated with individual events into account. All of these codes have the same number of bits independently of the frequency of the underlying event. Such coding schemes can be inefficient; they may waste valuable storage space and, when processing data, fewer numbers of bits can be handled in shorter time intervals. A simple example of such codes is the familiar Morse code; in this code the most frequently occurring letter in the English language (e) has the simplest code (a dot). The logic behind this choice takes into account the time that is required to send the letter by a telegrapher.

WHAT TO CONSIDER WHEN SELECTING ALPHA-NUMERIC CODES   35

Example 2.12. Coding for efficiency

A certain experiment can generate a series of outcomes consisting of either the presence or absence of a (constant-amplitude) signal in each of 100 time intervals. The probability of the signal occurring in each time interval is 0.9. (The probability of the absence of the signal is therefore 0.1.) Suppose that the experimenter is interested in recording what transpired during the entire epoch (100 time intervals), in other words the exact signal (or absence) sequence. The amount of information in each time interval—use equation (2.3)—is 0.47 bits. Thus, to represent the entire epoch (100 time intervals) will require an average of 47 bits.

How might such a record be coded? (Remember, the bits of a code are distinct from the information content of an experiment.) One possibility is to record one binary bit for each outcome—1 for the signal, if present, and 0 if there is no signal for each time interval. This, of course, would require a sequence of 100 bits, far in excess of the actual information contained in the experiment. A more efficient coding system is possible:

Break down the sequence of events into groups of seven. (Since the absence of a signal is a rare event, the code will deal with absences.) A group of seven may contain any number of absences from 0 to 7—a table of possible outcomes together with the probability of each is shown below.

Number of missing signals	Probability
0	0.478
1	0.372
2	0.124
3	0.023
4	0.0026
5	0.000 67
6	0.000 0063
7	0.000 0001

(notice that almost fifty per cent of the time (0.478) each group will consist of the presence of all seven signals.) To generate an efficient code use a simple symbol or code to stand for frequently occurring groups. Use a 0 to correspond to a group of seven in which no signals are present. The signal 1 means that the group of seven contains the presence of some signals. A 1 must be followed by additional symbols indicating the number and location of signals within the group. Encode the location of signals within the group as follows:

Code	Location
000	1
001	2
010	3
011	4
100	5
101	6
110	7

The code 111 would be reserved for those cases where there was more than one signal in the sequence of seven. This sequence (111) appears in the code as many times as a signal occurred in the seven being coded, followed by an appropriate number of location symbols. (Whatever the coding scheme, each code must be unique and not confused with any other message.) These three-bit location codes would indicate the serial positions of either the signals or their absence, whichever are fewer. The outcome for a group of seven is summarized below:

Number of absences in a group of 7	Number of binary digits used in the code
0	1
1	4
2	10
3	16
4	19
5	19
6	19
7	19

The average length of a message—the code describing seven outcomes—may be calculated by multiplying the number of required binary digits in each coded message by the probabilities specified in the previous table and then summing all products. This comes to 3.62 binary digits to encode seven outcomes. This corresponds to an average of 0.52 bits per outcome which is not greatly in excess of the 0.47 bits of information represented by the experiment and obtainable with a perfect code—one that would be 100% efficient. (It would require an average of 52 bits to encode the entire epoch using the scheme just cited as opposed to 47 bits if the coding scheme was totally efficient.) It is to be noted that 'efficient' codes do not have any room for redundancy and thus any processing errors seriously alter the record.

2.3 REPRESENTATION OF ABSTRACT INFORMATION

Individual experimental outcomes may not provide the most meaningful information. Rather, it is often the relationships between the events that enable us to draw conclusions. Some examples:

REPRESENTATION OF ABSTRACT INFORMATION

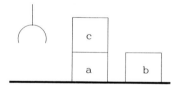

PREDICATE STATEMENTS DESCRIBING
THE WORLD SCENE

on_table(block_a)
on_table(block_b)
on(block_a,block_c)
clear(block_b)
clear(block_c)
empty(robot_gripper)

Figure 2.7 Abstract representation of information.

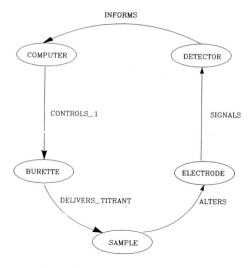

Figure 2.8 A conceptual graph representation of an automatic titration system.

- When trying to diagnose hypertension we are interested in the time course or tendencies of blood pressure and not any single reading.
- When designing a 'pick-and-place' robot for automating a process, we are not so much interested in the light and dark regions of an object's image as the relationships between such regions; these permit us to conclude whether or not the object has a given shape.

Information of this kind is more abstract and the representations considered to this point will not suffice. A number of schemes have gained recognition [1].

- *Logical representation*: this method uses expressions in formal logic. (Information in this form can be used as inputs to 'logical-inference engines' that implement the rules of inference.) An example of information representation using predicate analysis is shown in figure 2.7. The predicate statements in the figure summarize abstract information regarding the scene.
- *Representation of information as procedures*: these include a set of problem-solving instructions. Knowledge may be in the form of a set of rules such as:

$$\text{If } \cdots \text{ then } \cdots .$$

A modern instrumentation system might include 'expert advice' which is available to the experimenter. An example of a simple procedural rule is:

```
If (data_rate(experiment) > high_limit)
    then print('system capacity warning')
```

Other applications of such techniques to instrument systems include: data interpretation, projecting consequences of given experimental situations, determining malfunctions in complex situations, configuring system components to meet performance goals, devising actions to achieve a set of goals, comparing observed and expected behaviour of an experiment, and controlling complex environments.

- *Data representation by networks*: such schemes include a graph in which nodes are objects and directed arcs show relationships to other objects (nodes) in the data-base. Figure 2.8 shows one network description of an instrument environment.
- *Structured representation*: each object has a system of values associated with it including such information as numeric, symbolic, references to other objects, or procedures for performing a particular task. Examples of such structured representations include frames and scripts.

Instrument systems that incorporate such information representation schemes are relatively new developments in measurement and control technology. They can be expected to appear with greater frequency in the future. A special form of such systems (neural nets) is discussed in chapter 14.

2.4 WHAT LIMITS INFORMATION?

The experimenter must have some basis for specifying how many digits and/or distinct codes are needed to satisfy the needs of the experimental environment. In order to make such decisions there must be agreement regarding (data) parameter definitions as well as understanding the limitations imposed by contaminants (noise).

WHAT LIMITS INFORMATION? 39

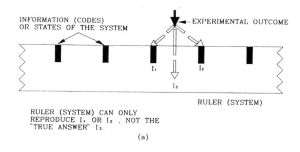

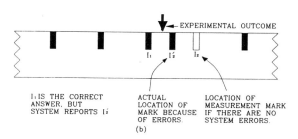

Figure 2.9 Resolution and accuracy in instrument systems. (a) Resolution concerns the ability of the system to repeat the required measurement. (b) Accuracy is a measure of the system's ability to repeat the correct measurement.

2.4.1 Resolution, accuracy and significant figures

Data resolution describes the system's ability to repeat a given measurement. Accuracy is concerned with the system's ability to reproduce the correct or true result. The number of significant figures comes from the system's ability to perform calculations.

Figure 2.9 demonstrates the difference between the concepts of resolution and accuracy. The measurement system is represented by a ruler and the event to be measured is indicated by an arrow. Figure 2.9(a) depicts system limitations imposed by resolving power. The registration marks are the codes (numbers) that the system can reproduce; in the case of figure 2.9(a) they correspond perfectly to results that can be generated by the experiment (no errors). An experimental outcome occurs and is equivalent to I_3. The system cannot reproduce I_3 and must report either I_2 or I_1 depending on which is closer to the true value. Resolution may be reported as a percentage of the range of measurement values; it may also be reported in terms of the number of distinct codes or states of the system. (The term '12-bit resolution' corresponds to 4096 states or a resolution of 1 part in 4096.)

Accuracy limitations come from errors that are introduced by the measurement system itself; figure 2.9(b) shows the effect. If the system was perfect it would locate the registration mark at the position marked 'I_2'. Because of accumulated tolerances within the system I_2 is really located at I'_2. For the experimental outcome indicated, the system reports I'_2 which is now 'close' to the position of the outcome presented to the system. Had the system not included errors, it would have reported I_1 as it should have, being naturally closer (by virtue of resolution) to the true value. System error is also indicated as a percentage of the range of the instrument.

A warning is to be noted at this point. Consider the following example: a manufacturer reports a system accuracy of 0.1% of the full scale (say the FS voltage is 10 V); this appears to be a 'high-quality' instrument. However, the experimenter wishes to measure quantities as low as 5×10^{-3} V (5 mV). The absolute value of system error is 0.001×10 or 10×10^{-3} V (10 mV). The absolute error is thus twice the value of the minimum signal to be measured! When considering accuracy it is important to note the absolute system errors and compare these to the requirements of the experiment.

Normally, the accuracy of an instrument system should be better than the system's resolution. A typical specification might have the following form:

Resolution: x parts per million or y per cent of FS where x and y are values supplied by the manufacturer of the instrument.
Accuracy: error less than plus or minus half the least significant digit (or bit).

When processing data—restricted to decimal numbers for this discussion—the significant digits determine the accuracy [2]. They are defined as follows:

- All non-zero digits are counted as significant.
- Any zeros that are not used as placeholders for the decimal point are significant.

For example, 0.213 has three significant digits because the leading 0 is a placeholder (by convention) and is therefore not significant. The number 0.050 10 has four significant figures because the first two are merely placeholders while the other zeros are not. When carrying out arithmetic operations the following rules should be observed:

1. When adding or subtracting, resolution is limited to the least precise number.
2. When multiplying or dividing, the result is only as accurate (number of significant figures) as the least accurate number.
3. Roots of a number are limited to the accuracy of the original number.
4. Exact numbers do not change the accuracy or precision of a calculation.

2.4.2 Noise limitations

During the course of processing experimental outcomes the signals may become contaminated with noise (unwanted and/or unexpected events). At times, two information sources can interfere with each other even though they are coherent

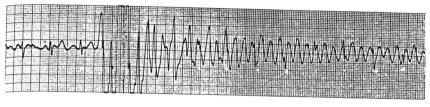

Figure 2.10 A movement artefact can mimic serious arrhythmias. This noise-contaminated tracing was misinterpreted as ventricular fibrillation in a real clinical setting.

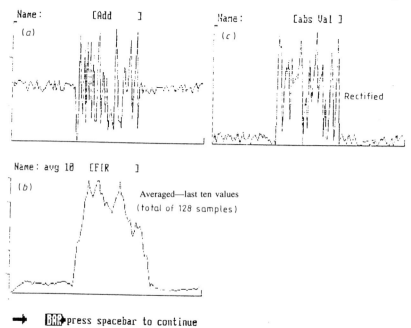

Figure 2.11 Simulated EMG activity during rapid elbow extension. (a) Time record. (b) Full-wave rectified. (c) Subsequent integration of the signal.

when considered separately (e.g. electromyographic information is sometimes corrupted with cardiac artefacts).

Noise, of course, can mask information or may even generate false interpretations. Figure 2.10 shows what can happen when noise—in this case coming from the experimental data—interferes with the information [3]. An observer (either human or machine) would incorrectly classify this result as a case of ventricular fibrillation.

In other cases information that appears to be contaminated with noise is, in fact, meaningful when processed in an appropriate manner. Consider figure 2.11 which represents a typical electromyographic (EMG) record—simulated using

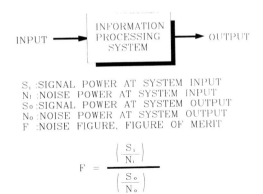

Figure 2.12 The figure of merit for a system contaminated by noise.

S_i : SIGNAL POWER AT SYSTEM INPUT
N_i : NOISE POWER AT SYSTEM INPUT
S_o : SIGNAL POWER AT SYSTEM OUTPUT
N_o : NOISE POWER AT SYSTEM OUTPUT
F : NOISE FIGURE, FIGURE OF MERIT

$$F = \frac{\left(\frac{S_i}{N_i}\right)}{\left(\frac{S_o}{N_o}\right)}$$

the DSPlay® software package (see chapter 9)—which might be generated from the agonist triceps during a rapid elbow extension. Figure 2.11(a) shows a time record of the signal. In this state it is difficult to interpret; however, the information becomes considerably more meaningful after appropriate processing which includes full-wave rectification (figure 2.11(b)) and then integration (figure 2.11(c)). The processing is summarized by

$$\text{result} = \int |(\text{potential})|\, dt.$$

On calculating the integral of (electrical) potential the result is a measure of work or effort being made. The effort summarizes the behaviour (and the information in this instance) far more clearly than the original record.

Noise consists of unwanted and random contamination. As such it may be difficult to quantify and must be considered to be a random phenomenon (process) characterized by statistical or other mathematical representations.

Noise can be considered to be made up of sinusoids of varying frequency and amplitude. (Analysis of information in the frequency domain is considered in more detail in chapter 5.) The effect of these sinusoids on information is included by calculating their associated power. (In this case power is obtained by calculating E^2 where E is the amplitude of the sinusoid under consideration. Total power is obtained by summing the mean (average) power of each relevant sinusoid [4]). If S represents the signal power and N represents the noise power then the effect of noise can be determined by calculating various ratios as shown in figure 2.12. The system itself can alter both the signal power and the noise power—the information processing system may introduce additional noise not present at its input. The noise figure, F, is a figure of merit used to characterize the effect of a system on information that it processes. The signal-to-noise ratio (S/N) is an important parameter in specifying system requirements including the one discussed in the next section.

2.4.3 Quantizing information and 'companding'

The discrete information levels previously discussed—information formats, section 2.1.1—can, if properly chosen, minimize or eliminate the effects of noise. To do so will require compromises based on the signal-to-noise ratio.

Consider an experiment that can generate M distinct states (messages). These messages have associated noise whose magnitude has a value of n (measured as its root mean square or RMS value). The message levels are chosen to be separated by αn, where α is a number chosen to minimize the probability of an error. It can be shown [4] that the number of quantizing (message) levels is limited by the signal-to-noise ratio as well as the factor α, which is the safety factor chosen to minimize system errors:

$$M = \sqrt{\left(1 + \frac{12}{\alpha^2} \frac{S}{N}\right)}.$$

Introducing large differences between quantizing levels reduces the probability of an error that may result from noise. However, such differences introduce another kind of difficulty, referred to as the quantizing error. (Figure 2.13 shows the quantizing levels.) All experimental outcomes between $k - (S/2)$ and $k + (S/2)$ are represented (coded) as k. Unless the experimental outcome is exactly k an error is introduced, namely the difference between the experimental outcome (in the range) and k. Assuming that the experimental outcomes are uniformly distributed (equally likely) in the range $(k - (S/2), k + (S/2))$, an average or mean quantizing error can be calculated. This value turns out to be $S^2/12$. If S is chosen to be large to minimize noise error then a large quantizing error is introduced.

The problem just cited can be partially addressed by using a system with non-uniform quantizing levels. The table below summarizes the constraints under the assumption of constant signal-to-noise ratio for both small and large signals:

	Large S	Small S
Large signal	Small noise error	Large noise error
Small signal	Large quantizing error	Small quantizing error

To achieve both small quantizing errors and small noise errors would suggest small quantizing levels for small signals and large quantizing levels for large signals. Such a system is referred to as data compression. One system [5] proposes a compression scheme as sketched in figure 2.14. Input signals are scaled as shown in the figure before being quantized. (Scale the smaller signals by a greater amount—multiply small signals by the inverse of the scale factor shown on the ordinate.) This scaling (distortion) must be reversed when the data are to be interpreted by multiplying the information by the scale factor shown on the ordinate of figure 2.14. The quantized levels must be expanded. The entire process is known as companding—taken from *com*pression and ex*pand*ing.

44 INFORMATION AND ITS REPRESENTATION

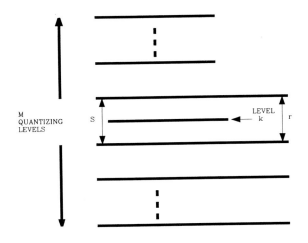

r IS THE RANGE OF EXPERIMENTAL OUTCOMES
CODED AS LEVEL k.

Figure 2.13 Quantizing error.

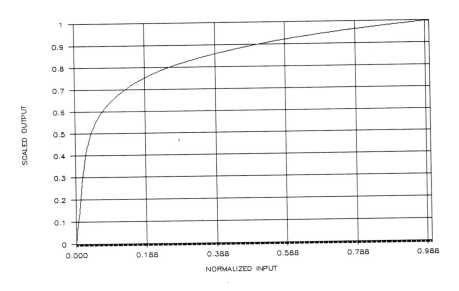

Figure 2.14 A companding curve.

2.5 SYSTEM CAPACITY AND SAMPLING RATE

Up to this point in the discussion of the nature and representation of information very little has been said regarding the speed at which information must be handled. Processing speed is taken into account by specifying the system capacity, which is defined as

$$\text{capacity} = \frac{H}{T}$$

where H is the amount of information to be processed and T is the time that the system needs (or has) to process this information.

Once again, the experimenter is faced with compromises that must be made with regard to system capacity and noise. Figure 2.15 includes two possibilities for representing a given amount of information. In the first instance (figure 2.15(a)) four possible quantizing levels are shown; the time required to process the entire message—one of the four levels—is given as T. In figure 2.15(b) the same information is coded using a sequence consisting of two elements with each element of the total message restricted to two allowable levels. If the total amplitude in both cases is A, then the tolerance (based on considerations of noise, accuracy, etc) for the system shown in figure 2.15(a) may be taken as $|A/8|$ while the tolerance for the system in figure 2.15(b) is $A/2$. However, the capacity of an instrument system that processes data as shown in figure 2.15(a) has a capacity of $2/T$ while the capacity of a similar system for the data in figure 2.15(b) is $2/2T$ or $1/T$—in other words it has only half the capacity. To make the systems truly equivalent would require the lower-capacity system to process data at twice its present rate. This normally results in a more costly system that will consume more power.

2.5.1 Sampling rate

System capacity limits the rate at which the dependent variable of an experiment can be quantized. There is a most important lower limit on this sampling rate. The sampling rate is constrained by the Nyquist sampling rate—sometimes referred to as the Shannon sampling rate—which requires that the dependent information is sampled at more than twice the highest frequency contained in the information source. If it is not sampled at this rate at least, the original information cannot be recovered. This is summarized as follows [6]:

$$C = 2B \log_2 M$$

which is the Nyquist formula. C is the system capacity in bits per second, B is the bandwidth of the system in hertz (see chapter 5), and M is the number of allowable levels per signal element in the message. (It is common in instrument systems to use two levels per signal element in which case the formula reduces to $C = 2B$.)

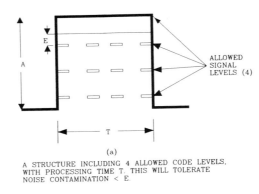

(a)

A STRUCTURE INCLUDING 4 ALLOWED CODE LEVELS, WITH PROCESSING TIME T. THIS WILL TOLERATE NOISE CONTAMINATION < E.

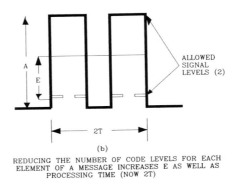

(b)

REDUCING THE NUMBER OF CODE LEVELS FOR EACH ELEMENT OF A MESSAGE INCREASES E AS WELL AS PROCESSING TIME (NOW 2T)

Figure 2.15 Information processing requires compromises between noise tolerance and the system capacity. (a) A signal that includes four code levels, and requires T seconds to process. This will tolerate noise contamination $< E$. (b) Reducing the number of code levels for each element of a message increases E as well as processing time (now $2T$).

2.6 SUMMARY

The amount of information is a fundamental measure of an experiment or process. For those instances in which the experiment is concerned with functional relationships between two variables the information is characterized by the number and relative frequency of events generated by the experiment. These outcomes are restricted to a finite number because noise and/or other contaminants and considerations (e.g. instrument bandwidth) limit an instrument's ability to discriminate between events. If certain absolute limits are observed (e.g. sampling rate) then the information present in the experiment can be recovered. Information is represented as numbers and in some cases can be summarized in higher (English-like) abstract form. The representation (coding) of information can be structured to minimize instrument processing errors. It can also be arranged to increase the efficiency with which data are processed. System resolution and accuracy are important parameters to consider when specifying instrument system requirements; they must be compatible with the underlying information. System capacity determines the speed at which information is processed. It must be consistent with the amount of information to be processed as well as the time it takes to process the data—some results may be required in a fixed time in order to control the process or experiment in timely fashion (real-time systems).

REFERENCES

[1] Luger G F and Stubblefield W A 1989 *Artificial Intelligence and the Design of Expert Systems* (New York, NY: Benjamin–Cummings)
[2] Kramer A D 1986 *Mathematics for Computers* (New York, NY: McGraw Hill)
[3] Mark R G and Moody G B 1988 Automated arrhythmia analysis *Encyclopedia of Medical Devices and Instrumentation* ed J G Webster (New York, NY: Wiley–Interscience)
[4] Taub H and Schilling D L 1971 *Principles of Communication Systems* (New York, NY: McGraw-Hill)
[5] Mazda F 1985 *Electronic Engineer's Reference Book* (Oxford: Butterworths)
[6] Halsall F 1988 *Data Communication, Computer Networks and OSI* 2nd edn (New York, NY: Addison-Wesley)

ADDITIONAL READING

Information Theory and Coding

Cappellini V (ed) 1985 *Data Compression and Error Control Techniques with Applications* (London: Academic)
Glorioso R M and Osorio F C C 1980 *Engineering Intelligent Systems: Concepts, Theory and Applications* (Bedford, MA: Digital)
Hamming R W 1980 *Coding and Information Theory* (Englewood Cliffs, NJ: Prentice-Hall)
Hill R 1986 *A First Course in Coding Theory* (Oxford: Oxford University Press)
Lafrance P 1990 *Fundamental Concepts in Communication* (Englewood Cliffs, NJ: Prentice-Hall)
Mansuripur M 1987 *Introduction to Information Theory* (Englewood Cliffs, NJ: Prentice-Hall)
Usher M J 1984 *Information Theory for Information Technologists* (Great Neck, NY: Macmillan)

Noise

Dupraz J 1986 *Probability, Signals, Noise* translated by A Howie (New York, NY: McGraw-Hill)
Schwartz M 1980 *Information Transmission, Modulation, and Noise: a Unified Approach to Communication Systems* (New York, NY: McGraw-Hill)
Winograd S and Cowan J D 1963 *Reliable Computation in the Presence of Noise* (Cambridge, MA: MIT Press)

Number Theory and Numerical Analysis

Rose H E 1988 *A Course in Number Theory* (New York, NY: Clarendon)
Rosen K H 1984 *Elementary Number Theory and its Applications* (Reading, MA: Addison-Wesley)
Schroeder M R 1984 *Number Theory in Science and Communication: with Applications in Cryptography, Physics, Biology, Digital Information, and Computing* (New York, NY: Springer)
Scott N R 1985 *Computer Number Systems and Arithmetic* (Englewood Cliffs, NJ: Prentice-Hall)
Spencer D D 1989 *Exploring Number Theory with Microcomputers* (Ormond Beach, FL: Camelot)
Spencer D D 1989 *Invitation to Number Theory with Pascal* (Ormond Beach, FL: Camelot)

Other Topics

Astrom K J and Wittenmark B 1990 *Computer-Controlled Systems: Theory and Design* (Englewood Cliffs, NJ: Prentice-Hall)
Cady J and Howarth B 1990 *Computer System Performance Management and Capacity Planning* (New York, NY: Prentice-Hall)
Reichgelt H 1991 *Knowledge Representation: an AI Perspective* (Norwood, NJ: Ablex)
Shannon C E 1949 *The Mathematical Theory of Communication* (Urbana, IL: University of Illinois Press)

3

Hardware Architecture of PC-based Instrument Systems

The integrated instrument systems discussed here use a personal computer (PC) as a basic component plus additional hardware and software arranged to meet the needs of a specific measurement and/or control problem. The elements needed to process experimental data are embodied in the computer:

- Hardware—physical elements associated with experimental control and processing.
- Software—programs that control the sequence of steps, operations or procedures carried out by the hardware.

The basic computer hardware is discussed in this chapter. Additional hardware considerations are found in chapter 6, particularly those elements associated with data acquisition. Software topics such as the computer's operating system as well as extensive discussion of applications programs are discussed throughout the remainder of the text. (While hardware considerations are important when specifying or designing an instrument system, software is becoming the increasingly dominant feature; the organization of this text mirrors the relative importance of the topics.)

Figure 3.1 shows the functional organization of the computer hardware and is a guide to the discussion below. The organization shown is neither unique nor exhaustive; it includes those hardware elements that are most important for automated instrument systems.

3.1 FUNCTIONAL COMPONENTS OF AN INSTRUMENT PLATFORM

A computer that is intended as a platform for data acquisition and process control requires four functional components:

Central processing unit (CPU). This is the logical heart of the computer and its principal function is to control execution of the instructions in an orderly manner. This further breaks down into a set of subordinate tasks: fetch

50 HARDWARE ARCHITECTURE

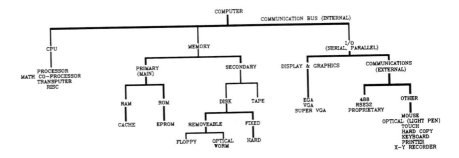

Figure 3.1 Functional organization of computer hardware for instrument systems.

instructions from memory, decode (decipher) them, and execute them—issuing commands to transfer data to and from memory or peripherals as required. Execution of an instruction requires the CPU to make logical decisions, compare logical entities, and perform simple arithmetic calculations.

Memory. In a physiological sense, memory represents a short- or long-term change in a physical property of matter caused by some sequence of events. Within the computer a more narrow view is required. Here it is interpreted as an apparatus in which information may be stored and retrieved at a later time. The memory stores three distinct entities: program instructions; partial (or intermediate) operational results; final results.

Input/output (I/O). If the computer is to be useful it must have some means by which to communicate with entities external to the PC itself. This is the functional domain of input/output or I/O. Two areas must be addressed: first, the hardware that moves the data, including the 'nature' of the data to be moved (e.g. human speech, visual information, electrical data); second, the strategy (or protocol) for moving the information. The computer is at one end of this communication path; however, the other source or destination may be either a machine (including another computer) or a human user.

System bus. The CPU, memory and I/O have a common path by which they communicate with each other; this is the *system bus* (or, simply, bus). Such buses include addresses (of data source and/or destination), the data to be moved, and finally a series of control lines normally referred to as *handshaking* lines for orderly transfer of the information.

3.1.1 The central processing unit (CPU)

The processing circuits contained within the computer use a binary number system for information representation. A fundamental characteristic or parameter of this number system is *the word size* which can be defined as *a group of bits considered as an entity and capable of being stored in one (storage) location* [1]. The word size is intimately related to the number of bits that can be processed in a given instruction time.

The instruction time includes the total time required to fetch (retrieve) the instruction from memory, to determine what the instruction is to do, and to execute the instruction. The time taken to complete an instruction depends on the type of instruction. A measure of computer speed is 'the number of instructions per second', which is an average over a mixture of various instruction types. Typically, a computer can complete millions of instructions per second; this is abbreviated as MIPS.

The word size affects the capacity (as well as the cost) of the computer. Common word sizes are 8, 16, and 32 bits with larger word sizes predominating. As the circuit density of integrated circuits increases in the future, word sizes within a PC can also expected to increase beyond 32 bits. (Currently, larger computers—so-called *main-frame* machines—already have considerably larger word sizes.)

To appreciate the effect of word size on the capabilities of the computer, refer to figure 3.2. In the most general case, instructions include: the action to be taken, the data source, and the data destination. (Typical actions are: move data; add data; subtract data. These will be discussed in chapter 4 in greater depth.) Not every instruction has this form but many do. If the word size is small (e.g. 8 bits) then a typical instruction may occupy as many as three sequential memory locations or words (figure 3.2(a)). (For example, the code for 'add data into the accumulator' appears as the first word; this is followed by two (eight-bit) words that, taken together, form the sixteen-bit address of the operand (data to be added).) As word size increases, the same instruction may occupy fewer words as shown in figures 3.2(b) and 3.2(c). (The structure found in figure 3.2(c) is characteristic of the *r*educed-*i*nstruction-*s*et *c*omputer (RISC).)

In order for the CPU to execute an instruction it must first access the memory to retrieve the code; this is referred to as a *fetch cycle*. The instruction is then executed; this is referred to as an *execute cycle*. The complete sequence is a *fetch–execute* cycle which is characteristic of the way in which the CPU operates.

For 'single-address' data structures (figure 3.2(a)), the CPU requires three fetch cycles to execute the generic instruction mentioned above completely; only two fetch cycles are required in the second case and a single fetch completely retrieves the instruction in the last case. Each fetch requires a small but finite time to complete. A fetch cycle can be considered a 'necessary evil'. Certainly a fetch is required to determine what is to be done, but additional fetch cycles for the same instruction usually slow operation of the computer. The so-called 'three-address' model represents a considerable speed advantage over the other alternatives. (It is to be noted that the gain in speed is offset by a more complicated and costly CPU, one that also consumes more power.)

In order to carry out the execution of instructions in an orderly manner and to implement completely the fetch–execute cycles that comprise the program, the following operational (CPU) elements are required:

- Means to *fetch* the next instruction to be executed.
- Some mechanism to *decode* (interpret) the instruction to be executed.

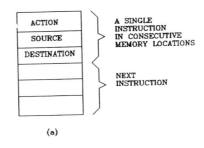

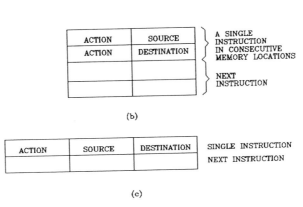

Figure 3.2 Word size and its importance for computer capability. (a) Small-word-size ('single-address') instruction format. (b) Intermediate-word-size ('two-address' instruction format). (c) Large-word-size ('three-address' instruction format).

- The means to *execute* the instruction including components to transfer data to/from other parts of the computer.

A functional block diagram of the CPU is shown in figure 3.3. The elements shown represent the composite architecture of several CPUs including features with advantages from each one. In a very real sense, the CPU can be considered a 'computer within a computer'; it is a 'computer in miniature'. Thus, it contains its own memory (LOCAL REGISTERS, INSTRUCTION FETCH QUEUE, and the INSTRUCTION DECODE), 'CPU' (CONTROLLER, ALU), 'I/O' (INTERFACE), and communication bus (INTERNAL BUS). The sequence of steps carried out by the CPU during a fetch cycle are summarized as follows:

- The address stored in the INSTRUCTION POINTER (IP) is sent out to the computer's primary memory (e.g. RAM). The number stored in the IP is the address of the next instruction to be executed.
- Increment the IP by one, which is necessary so that it references either the operand needed for the instruction or the next instruction if the present

FUNCTIONAL COMPONENTS OF AN INSTRUMENT PLATFORM   53

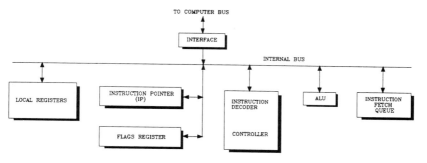

Figure 3.3 A functional diagram of a CPU.

instruction does not require an operand (e.g. negate the contents of the accumulator). It is possible that as a result of the current instruction this address may have to be adjusted (e.g. as the result of a 'GOTO' (or branch) instruction, but at this point in the sequence it is assumed that the next sequential memory location is required by the current instruction.

- The instruction stored in the address accessed by IP is returned to the INSTRUCTION DECODE component, which includes a *register* to retain this information. (A register is analogous to a location in the computer's main memory.)
- The output of the INSTRUCTION DECODE is an address that references a memory location within the CONTROLLER. This latter memory location is the start of a series of commands ('instructions') that the CONTROLLER must carry out in order to execute the instruction originally accessed in the computer's main memory.

The above sequence is carried out repeatedly during each fetch–execute cycle. The remainder of the CPU processing involves the other components shown in the figure. How these elements interact depends on the instruction to be completed. However, the purposes of each element can be summarized:

- The LOCAL REGISTERS are part of the CPU's local memory. Some CPUs contain only one such register, which is referred to (historically) as the *accumulator*. However, most modern processors contain many (e.g. 16). These registers store information (data) that will be used by the CPU during its calculations (e.g. the operands needed for the action to be carried out).
- The FLAGS REGISTER is a special-purpose register that stores certain information regarding the results of the calculations. For example, this register has provision for storing facts such as: 'the result of the operation was zero', or 'the result of the operation was negative', or 'the operation generated a carry', or 'an arithmetic overflow (results too big) was detected'. The register includes bit locations for each of these parameters.
- The ALU (or *a*rithmetic and *l*ogic *u*nit) is the computational heart of the CPU. Logic (gates) within the ALU can carry out binary arithmetic (ADD, SUBtract)

as well as logical operations (comparisons) such as combining operands using AND, OR, NOT (logical inversion) conjunctives.
- The INSTRUCTION FETCH QUEUE is another element that can be found in the modern CPU. This is yet another set of memory locations that store the instructions that follow the one currently being executed. The assumption (which is true for the most part) is that the instructions to be executed are found in sequential locations within the computer's memory. By retaining these instructions, the CPU fetch–execute cycle can be speeded up. This structure is known as *pipelining*—the instructions are 'in the pipe'. Of course, in some circumstances, the CPU is required to jump to a completely different portion of the memory in order to find the next instruction. In this case all instructions in the queue become invalid and must be replaced.

Historically, the arithmetic operations of multiplication and division were achieved by software (programming) techniques using, for example, repeated addition (for multiplication) and repeated subtraction (for division). Computer operations were slowed considerably.

A variety of algorithms for floating-point multiplication and floating-point division of real numbers (e.g. Booth's algorithm) have become standard and as a consequence could be 'converted to silicon'—transformed into a special-purpose integrated circuit. Such VLSI circuits are referred to as *floating-point co-processors*. These hardware circuits can complete a multiplication and/or division operation well within a single fetch–execute cycle. Thus instructions such as MULTiply, and DIVide were introduced into the CPU repertoire. Originally, the hardware needed to carry out these operations was included as separate circuits (chips) designed to work with the CPU; with increasing chip density, these circuits have been integrated directly into the CPU itself.

The popularity of high-level languages (see chapter 4) encourages development of processors with large numbers of instructions and methods of accessing operands ('addressing modes'). Consequently, the INSTRUCTION DECODER and CONTROLLER within the CPU have grown in size slowing the speed of operation of the CPU. However, a high proportion of the instructions that are actually used in applications programs are restricted to a relatively small subset of the CPU's repertoire. (CPUs with a rich complement of instructions are referred to as complex-*i*nstruction-*s*et *c*omputers or CISCs.) (The Motorola 68000 family of processors is one example of a CISC system.)

More efficient working of a CPU is achievable by restricting the instruction format to the 'three-addressing' format (figure 3.2(c)) and by restricting the number of instructions that the CPU can execute. The resulting architecture has become known as a reduced-instruction-set computer (RISC). Characteristics of such machines include the following:

- A relatively large amount of memory on the CPU itself; this greatly reduces the need to access primary memory, speeding overall operation.
- Transitions between parts of the program (e.g. subroutine implementations)

can be carried out efficiently. (This is, in part, a consequence of the on-chip memory.)
- RISC machines execute one instruction per fetch–execute cycle (on average)—a result of the fact that most instructions reference CPU registers as opposed to main memory locations.
- The CONTROLLER section of the CPU is designed to minimize complex sequences in order to complete an instruction.

In order to assess the speeds at which RISC machines execute programs, a new 'measure' of capacity must be considered other than the instructions per second scheme previously mentioned. This new measure (or benchmark) reflects the ability of a computer to complete real-world applications programs—the *throughput*.

Two commonly used units of measurement (today) are *dhrystones per second* and *whetstones per second*. The dhrystone is a synthetic benchmark based on the statistical characteristic of real programs. For the most commonly used version about half of the instructions are assignment statements—performing a calculation and storing the result in a memory location ('assigning' a value; see chapter 4). An additional third are control statements (e.g. IF ... THEN ...) and the remaining parts of the measurement program involve execution of separate program modules (e.g. subroutines). Normally, such benchmarks do not invoke the computer's operating system (see chapter 4) and thus are intended to determine the efficiency of the architecture (and high-level language compiler). Whetstones, on the other hand, are more representative of typical scientific programs, as they include integer and floating-point—real number—calculations, transcendental functions, subroutine calls and array indexing. Benchmarking results are shown for several CISC and RISC systems in table 3.1. The dhrystone unit is the number of times the benchmark can execute per second, and roughly 2000 dhrystones correspond to one MIPS. The number of whetstone instructions per second are shown for both single- and double-precision data.

From the table, we can conclude that the estimate of a performance improvement factor of 2 to 4 for a RISC is valid for the systems compared (e.g. SPARC or MIPS versus 68020, 68030, or 80386, i860 versus 486).

3.1.2 Memory

Memory is conveniently divided into two functional categories. Some information is needed at the same rate as the CPU executes instructions; such memory is considered to be *primary* storage. This is normally expensive and holds relatively limited amounts of information. With improvements in technology, high-speed memory has become less costly. Thus, the primary store normally includes a (relatively) small but very high-speed segment called a *cache memory* (and pronounced like the word cash).

Alternatively, *secondary*-storage systems hold large numbers of data but

Table 3.1 Comparison of CISC and RISC architectures.

Type	System	CPU	Speed (MHz)	Speed (10^3 Dhryst s^{-1})	Speed (10^3 Whetst s^{-1})	
					Single precision	Double precision
CISC	VAX 11/780		5	1.8	1.2	0.8
	SUN 3/200	68020/FPA	25	6.7	3.4	2.6
	HP 9000/345	68030	50	19.2	5.2	3.9
	SUN 386i	80386/387	25	8.8	1.9	1.5
	Compaq 486	486	25	25.9	9.8	9.5
RISC	SPARC 330	Cypress/TI	25	27.8	12.5	8.3
	MIPS M/2000	R3000/R3010	25	42.0	12.5	8.3
	Aviion DG6200	88000	25	35.1	23.0	12.5
	STAR 860	i860	33	94.6	25.6	20.0

access to such information is slow. The ratio of cost to (information) capacity for secondary storage is less than that for primary storage.

Primary (main) memory. A number of definitions and parameters are needed to describe the variety of possible memory structures within the primary memory. The descriptions that follow include some terms that differentiate primary from secondary memory.

Memory cell: this is the smallest unit of information storage within the computer's memory.

Word (size): a group of memory cells.

Address (physical address): the location of the word (or cell) within the space allocated for such memory purposes. The entire range of memory locations within the computer is referred to as the *memory space*. Within the computer, physical addresses are identified by a number; this number has a binary value. However, octal, hexadecimal or decimal equivalents are more convenient for human users. (When writing a (software) program physical addresses are assigned *names* or *identifiers* for designation purposes.)

Access time: the time taken to read from, or write to, a cell or word. Access time is composed of two parts: the time required for the computer's hardware to locate the cell; and the time it takes for the memory cell(s) to make the contents available to the computer or record the new value. An alternative designation for specifying read/write times (particularly semiconductor dynamic memories) is the so-called cycle *time*—the time, normally the smallest, that must be allotted between two successive read or write accesses.

Random access memory (RAM): memory for which the access time is effectively constant and independent of where, within the memory, the given cell (or word) is located.

Serial (or sequential) access memory: memory for which the access time is (strictly) dependent on the physical location of the cell or word. In such memory, access to the contents of the cell or word can vary widely depending on the amount of such memory.

Volatile memory: such memory loses its contents when the power source is removed. Memory that retains its information without external power is referred to as *non-volatile* memory.

Read-only memory (ROM): intended for applications where the contents of such memory can be read but cannot be altered (at speeds that are compatible with CPU requirements).

Static memory: the contents are retained until altered by overwriting or removal of power (if the memory is volatile).

Dynamic memory: such memory normally stores information in the form of a charge on the terminal capacitance of an FET (*field-effect transistor*). (This is further discussed later in the chapter.) As the charge will 'leak off' with time, it is necessary to *refresh* the information periodically if it is to be maintained.

Semiconductor memory. Each bit in the memory assumes one of two bistable conditions, which are retained until altered under computer (CPU) control. Bistable states may be achieved in several ways.

Static RAM (SRAM): two stable states are possible when the output of one semiconductor switch circuit is fed back to the input of a second switch (in addition to the reverse). Such circuits are designated as *static* memory cells and one such cell circuit is shown in figure 3.4(a).

The three-terminal semiconductor switch can be conveniently thought of as the analogue of a mechanical switch. (The technology currently in use to implement the semiconductor switching relies on a transistor architecture known as the *field-effect transistor* or FET [2]. The *gate* of the transistor corresponds (roughly) to the 'pushbutton' of the mechanical switch and works according to the following schedule:

Gate voltage	State of semiconductor switch	Pushbutton
At supply voltage	Closed	Depressed
At ground (0 V)	Open	Released

Consider the following situation for the elemental cell shown in figure 3.4(a):

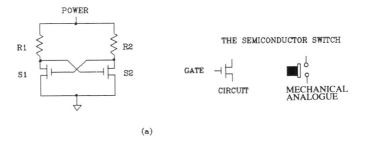

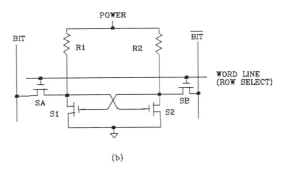

Figure 3.4 Semiconductor memory. (a) A basic cell. (b) Static cell architecture.

- Assume S1 is open—the gate of S1 is at ground.
- A path exists from the power supply through resistor R1 to the gate of S2.
- S2 is closed by virtue of its gate having been supplied with the supply voltage.
- The gate of S1 is at ground (0 V) because S2 is a short circuit.
- S1 is open because its gate is at ground potential; this confirms the initial assumption. The circuit will remain in this state (S1 open, S2 closed) indefinitely.

Starting with the opposite assumption (S1 closed) and using an argument similar to that given above a second stable state is achieved, namely S1 closed and S2 open. This circuit fulfils the requirements of a static memory cell with two stable (binary) states.

To enable the cell to be accessed for reading and writing purposes it must be modified and shown in figure 3.4(b); FETs SA and SB have been added. To access the cell for reading, the following should be noted:

- The computer's CPU (or other device that controls the primary memory) enables the WORD LINE by supplying it with a potential equivalent to the power supply.
- FETs SA and SB are closed (short circuited) because their gates are at power supply potential.
- Switches S1 and S2 are connected to the lines marked BIT and $\overline{\text{BIT}}$. These are

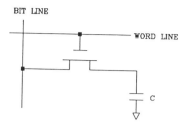

Figure 3.5 A DRAM cell.

in turn connected to the output of the memory, which typically transmits them to the computer bus where they are available to other parts of the computer (e.g. CPU or I/O).

The BIT and $\overline{\text{BIT}}$ lines carry the state of S1 and S2 to the memory output. S1 and S2 will be either at 0 V (ground) or at the power supply potential depending on their states (closed or open). As described above, they will always have complementary values.

Writing information into the cell is accomplished in a somewhat different manner:

- The device controlling the memory supplies the information (0 or 1) to the line marked BIT and its complement (1 or 0) to the line marked $\overline{\text{BIT}}$. (Because these lines were driven from the cell during a read access and are driven from outside the memory during a write, they must capable of 'two-way' operation as will be explained when discussing the typical organization of primary memory.)
- The WORD LINE is enabled.
- The values on BIT and $\overline{\text{BIT}}$ are transmitted to the switches S1 and S2.
- Since the values on BIT and $\overline{\text{BIT}}$ are complements of each other, the switches S1 and S2 will be forced to assume appropriate states according to the arguments presented above.
- The WORD LINE is disabled. The information stored in S1 and S2 will be retained until a new write operation alters the state of the cell, as SA and SB will be disabled.

In some memory modules the resistors (R1 and R2) are replaced by FET switches—not shown in the figure. A static memory cell (SRAM cell) may therefore have as many as six transistor switches. The number of such cells that can be included in a memory module (integrated circuit or chip) is thus limited by the relatively large number of FETs per cell.

<u>Dynamic RAM (DRAM)</u>: the dynamic cell (DRAM) requires a single FET to store information—not counting the refresh circuitry—see figure 3.5 [3]. The reading and writing requirements are similar for those of the SRAM:

- For a read access, the CPU enables the WORD LINE. The charge stored on the capacitor C is discharged onto the BIT LINE. This is sensed by an amplifier (not shown). If the capacitor is charged (to the power supply voltage) a logical 1 will be reported to the memory module output. If the capacitor is discharged before the read operation, the sense amplifier will generate a logical 0. (Note that during a read operation the capacitor will be discharged; this is called a *destructive read* and will require the capacitor to be recharged (refreshed) to retain/restore the original data.)

- To write data, the CPU puts the data onto the BIT LINE. Next the WORD LINE is enabled; this closes S1. Capacitor C will be charged to the supply voltage if a logical 1 is to be stored. Alternatively C will be discharged to ground (0 V) if a logical 0 is to be written.

During a read operation, the data stored on capacitor C will be destroyed. Additionally, capacitor C will gradually—typically 2 ms to 4 ms—lose its charge. Thus, additionally circuitry refreshes each cell in order if these are not being accessed by the CPU. Cells are automatically refreshed during a read operation. Refresh requirements tend to slow the operation of the DRAM memory module somewhat but the vast increase in the number of cells that can be fabricated in a given module greatly outweighs the speed reduction limitation.

Other forms of semiconductor memory. There are numerous circumstances in which non-volatility is advantageous. For example, many programs included in the computer's operating system are not likely to change very often, yet it is important to include them as part of primary memory for its speed advantages. Certain forms of semiconductor memory are suitable for such environments. Several technologies can be used for such purposes (see figure 3.6) and are briefly described:

Read-only memory (ROM): the user specifies the information for each word of the memory. The fabrication process ('factory installed') includes a step that fixes ('forever') the bit values. These are also referred to as *mask programmable* which mirrors the manufacturing step in which the data are fixed.

Field-programmable read-only memory (PROM): the user receives the memory module (in integrated circuit form) with all bits set to a default value such as logical 0. Using a special instrument, the user changes any cell value by 'blasting' (destroying) a tiny link (fuse), which changes the value of the bit.

Programmable logic arrays (PLAs): these are similar to ROMs as regards the technology. However, in the case of PLAs, fewer words can (or are required to) be addressed. These also are available in field-programmable form.

Erasable programmable ROMs (EPROMs): fixing a memory array indefinitely can sometimes be a disadvantage, such as when systems are under development. For such instances an erasable ROM can be used. Electrical signals applied to a cell cause semipermanent changes to the FET (10 to 100 years). To erase the array, the module is exposed to ultraviolet light. Programming and erasing may be repeated hundreds of times before the array becomes unusable.

Electrically erasable PROMs (EEPROMS): these are similar to EPROMS but can be erased electrically without the need for ultraviolet light. A variety

FUNCTIONAL COMPONENTS OF AN INSTRUMENT PLATFORM 61

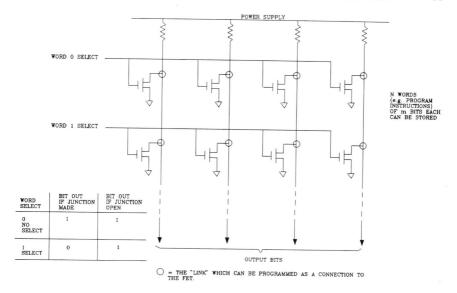

Figure 3.6 One form of a ROM array.

with somewhat different technology is known as *electronically alterable ROM (EAROMs)*.

Typical organization of primary memory. There are a number of physical ways in which to group memory cells. (This is not necessarily the same as the 'logical' organization in which the memory space is divided into regions called the *memory map*. Groups of physical addresses may be set aside for such purposes as program storage, subroutine storage as well as application programs like an editor or compiler, both of which will be discussed in later chapters.) Figure 3.7 shows two common configurations for primary memory organization. These are referred to as *bit-wide* and *byte-wide* architectures. For example, if the memory is to include 64K bytes—with a byte equivalent to eight bits—it may organized in (at least) two ways: 64K modules (eight of them) each with one bit; 8K modules (eight of them) with one byte per physical address location. Further, the CPU must be able to access specific locations within the memory space. For this, address lines and control lines must be added to complete the picture; this is shown in figure 3.8.

Address lines determine which physical location within the memory array is being accessed. These are decoded so that one of the 2^n words—consisting of one or more bits per word—is accessed using the n address lines—that is the purpose of the DECODER. This one word is then enabled ('opened') for reading or writing. A typical memory module uses three control lines to accomplish reading and/or writing. The control lines are:

Chip select (CS): enables entire module (not explicitly shown).

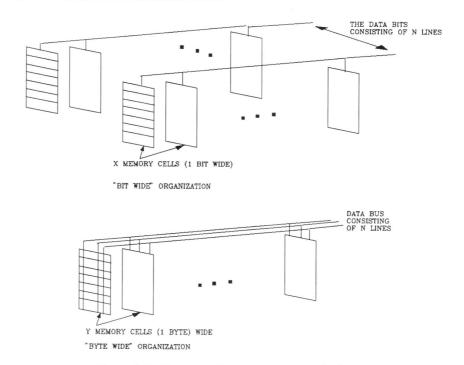

Figure 3.7 Common primary memory organizations.

Read/write (R/W): determines which access is required (read or write).

Output enable (OE): allows data from memory to pass to the computer's data bus. This feature supports multiple data storage arrays and larger memories (not explicitly shown). Table 3.2 shows what happens for each state of the control signals with the exception of OE. Figure 3.9 includes a *flow diagram* describing the sequence of steps that support orderly access to a memory module.

Figure 3.8 also includes the means for two-way operation of the BIT LINEs. The read/write signal (R/W) inherently determines the direction of information flow to/from the DATA BUS. If this signal is a logical 1 then the read access is implied; a write operation is implied if it is a logical 0. The triangular symbols included in figure 3.8 identify gates that permit data to flow in one direction. During a read operation, data are permitted to flow from the DATA BUS to the BIT LINE. When a write access is signalled, information flow is in the opposite direction.

Secondary memory (storage). This type of storage can be contrasted with main memory; it is characterized by high data (storage) capacity, slow data-access times, data retained 'indefinitely' (non-volatility), and low cost per bit of information stored. Such memory includes disk memory (in several forms), tape storage systems, and incremental data recorders.

FUNCTIONAL COMPONENTS OF AN INSTRUMENT PLATFORM 63

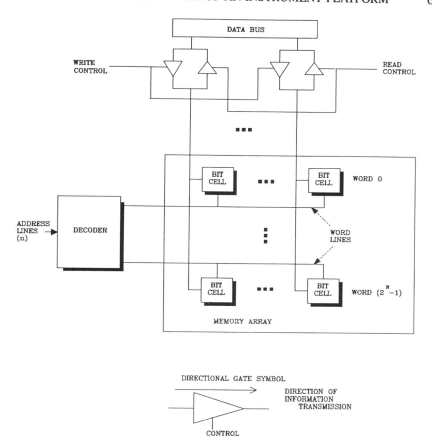

Figure 3.8 Memory module organization.

The storage schemes shown in figure 3.1 rely almost exclusively on magnetic storage techniques—of those shown, only the WORM (write-once–read-many) secondary storage depends on optical technology. There are a variety of magnetic recording implementations; these differ primarily in the physical arrangement of the head (the mechanism that transfers the information to and from the media) and the media.

Magnetic storage principles. The principles are depicted in figure 3.10, where functionally (figure 3.10(a)) the information is in electrical form, either a current (when writing data to the secondary store) or a voltage (when retrieving or reading data from the storage system). This information is ultimately retained as appropriate magnetic changes in the storage medium. Two important laws govern the way in which conversions take place:

- *An electric current gives rise to a magnetic field in its vicinity; the field is*

64 HARDWARE ARCHITECTURE

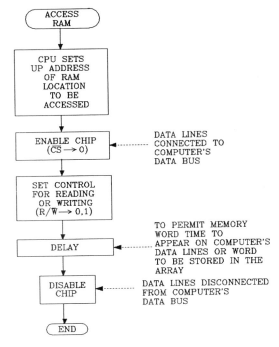

Figure 3.9 A simplified flow diagram for a RAM read or write operation.

Table 3.2 The various outcomes.

$\overline{CS}$	$R/\overline{W}$	What happens
0	0	Data are stored (a write operation)
0	1	Data are transmitted (a read operation)
1	0	No operation
1	1	No operation

proportional to the current producing it, with proportionality being a function of the surrounding material.
- *A changing magnetic field gives rise to, or induces, a voltage in a wire in its vicinity—the voltage may result from a changing field, or a fixed field and moving wire, or both.*

A physical schematic diagram is shown in figure 3.10(b). Information is recorded according to the following sequence of events:

- Binary information is present by virtue of an electric current flowing in the wire. (The direction of the current in the wire may be determined by the binary data (0 or 1) to be stored.)

FUNCTIONAL COMPONENTS OF AN INSTRUMENT PLATFORM 65

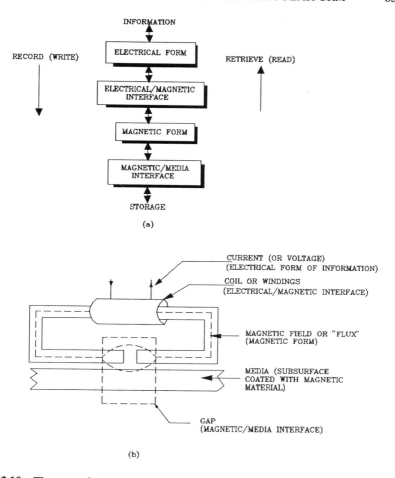

Figure 3.10 The operating principle of magnetic storage systems. (a) A functional scheme. (b) A physical schematic diagram.

- According to the first principle (above) this gives rise to a magnetic field in the medium around it, in this case a core of magnetic material.
- The magnetic field (in the form of 'flux') remains almost entirely within the core except at a small gap. At this point the flux crosses the air gap and in doing so extends into the magnetic media nearby—either in physical contact or extremely close.
- Either the head or, more commonly, the media are moved to a new position to record an additional bit of information. On some occasions both the head and media move.

Retrieving or reading data may be understood by reversing the steps. In this case the relative motion of the media moving past the head produces a changing

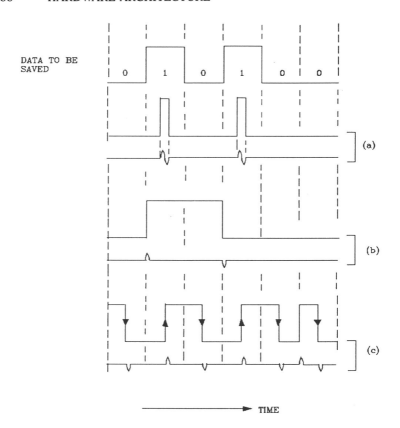

Figure 3.11 Traditional magnetic encoding schemes. (a) 'Return-to-zero'. (b) 'Non-return-to-zero'. (c) Phase ('Manchester').

field in the gap. This produces a change in the flux which appears as an induced voltage in the coil according to the second law cited above. The voltage is subsequently detected and converted into either a binary one or zero.

Common writing (encoding)/retrieval (decoding) schemes. Figure 3.11 shows several common schemes that are used to record and/or retrieve information from magnetic media. (Keep in mind the laws cited above to understand how encoding/decoding is accomplished. Included in the figure are the more 'traditional' schemes. Figure 3.12 shows two standards that can be found where higher recording densities (bits per inch) are used.) In the example depicted the data set to be recorded—the binary sequence 010100—is the same for each scheme. The first record (in each part of the figure) is the writing current while the second record in each case is the voltage that is generated during the retrieval process. The various encoding processes are summarized in table 3.3.

These methods may be compared according to a number of important

FUNCTIONAL COMPONENTS OF AN INSTRUMENT PLATFORM

Table 3.3 Summary of magnetic encoding schemes

Method	Description
RZ (return-to-zero)	Logical 1s and 0s stored as two different magnetic field values.
NRZ (non-return-to-zero)	For each logical 1 magnetic flux is reversed. When retrieving data, each flux reversal signals a logical 1.
Manchester	For each bit to be stored: low-to-high (head) current transition occurs for logical 1 and high-to-low transition occurs for logical 0.
Single-density—SD (FM or frequency modulation)	At the start of each bit the flux is reversed. If the bit is a logical 1 the flux is reversed again at the midpoint of the bit cycle.
Double-density—DD (MFM or modified frequency modulation)	If the bit is a logical 1 the flux is reversed (as in FM). If two successive 0s are to be recorded, flux is reversed at start of second logical 0.

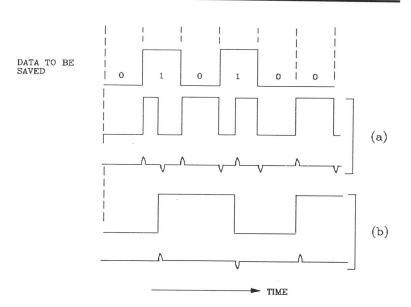

Figure 3.12 Higher-density magnetic encoding schemes. (a) Frequency modulation (FM). (b) Modified frequency modulation (MFM).

characteristics or criteria:

Efficiency: the number of stored bits per flux reversal—one flux reversal per bit corresponds to an efficiency of 100%.

68 HARDWARE ARCHITECTURE

Table 3.4 Comparison of magnetic recording methods

Method	Efficiency	BW	Clocking	Complex	Noise
RZ	50%	Low frequency needed	Not self-clocking	Medium	Poor
NRZ	100%	Low frequency needed	Not self-clocking	Medium	Fair
Manchester	50%	Good; no low frequency	Self-clocking	High (IC exists)	Good noise immunity
SD	50%	Good; no low frequency	Self-clocking	High	Good noise immunity
DD	100%	Good; no low frequency	Self-clocking	High	Good noise immunity

Bandwidth (BW): the range of frequencies that must be reproduced in order to record and recover the symbols. If the flux remains constant for long sequences on the media then the system must be able to deal with very low frequencies (approaching 'DC' or direct current); this is a more demanding case, as is the situation that includes many flux reversals creating the need to process high frequencies correctly.

Self-clocking: a self-clocking scheme includes means for automatically sensing the start of each bit cycle such as the SD encoding technique (table 3.4) where a flux reversal is made at the start of every bit independently of whether it is a 1 or a 0.

Complexity: a measure of the circuitry (as well as its accuracy) required to store and retrieve the data.

Noise: how well the scheme deals with unwanted and/or extraneous signals. Lost signals are referred to as *drop-outs*, while signals extraneously introduced are called *drop-ins*. Signals picked up from adjacent regions (tracks—see the section on hardware architectures below) are referred to as *cross-talk*.

Table 3.4 summarizes each recording scheme according to the characteristics cited above.

Magnetic storage hardware. There are a variety of head and media arrangements for magnetic data recording [4]. These are designed to have one or more advantages with regard to: speed, (small) size, power consumption, reliability (immunity to noise, and environmental robustness), and cost. (The basic functional arrangement is shown in figure 3.10.) Hardware recording schemes include thin-film heads as well as vertical recording systems. We will restrict our discussions to the architectural elements of magnetic recording

FUNCTIONAL COMPONENTS OF AN INSTRUMENT PLATFORM   69

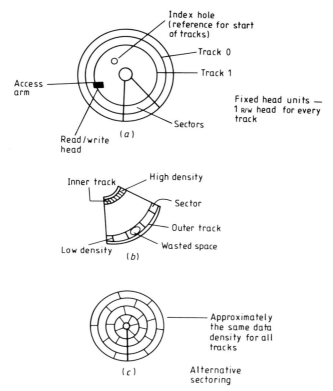

Figure 3.13 Functional organization of a secondary-storage (disk) system. (a) Overall architecture. (b) A problem associated with sectoring. (c) One solution to the sectoring problem.

systems, as these are the most important to understand as they relate to integrated instrument systems. Secondary-storage systems using magnetic recording generally employ either disk or tape media. These will form the basis for the discussion that follows.

Magnetic disk technology. Figure 3.13 shows important elements of disk-based storage systems. In some disk systems ('floppy' disks) the recording/playback head is in contact with the media, while in others ('Winchester' systems) the head is positioned from 0.5 to 2.5 μm above the surface of the disk when writing or reading information (1 μm = 10^{-6} m).

Data on the disk are recorded in concentric circle regions called *tracks*. The disk is further divided into sectors, which are pie-shaped areas on the disk; these are numbered and can hold a specified number of characters. A more detailed relationship between tracks and sectors in shown in figure 3.13(b). A disk limitation is immediately evident; because all tracks within a sector must contain the same number of bits, tracks closer to the inner edge of the sector

Table 3.5 Comparison of various secondary-storage technologies.

Type	Speed of rotation	Capacity	Access time
Floppy	360 RPM	180 Kbyte–1.44 Mbyte	83 ms
Hard (Winchester)	2400–4700 RPM	20–1000 Mbyte	15–60 ms
Tape[a]	32–640 cm s^{-1}	Many billions of bytes	Up to 15 minutes
Streaming tape	N/A—see text	100 Mbyte	N/A—see text
Cassette	76 cm s^{-1}	40 Mbyte	N/A—see text
Optical (WORM)	200–500 RPM	500 Mbyte	100–200 ms

[a] Normally associated with mini- and main-frame computers, not PCs.

have a higher bit density than those at the outer edge. This not only wastes space on the disk, it also makes more demands on the read/write head and drive by virtue of the increased density. (The head must be able to increase its rate of detection of magnetic alternations as it traverses the media.) To equalize the bit density over the surface of the media a sector organization similar to the one shown in figure 3.13(c) may be used. A summary of various disk technologies is included in table 3.5. A brief description of the more popular technologies follows.

Floppy (diskette) secondary-storage systems. These take their name from the fact that the early recording media consisted of thin plastic disks coated with magnetic material. Each disk was enclosed in a protective envelope. It was thin enough to bend easily, as contrasted with other disk systems, which had very rigid surfaces. Today, the most common 'floppy disks' are small (8.9 cm or 3.5″), and are not at all flexible, but are still referred to as 'floppy' primarily because of their similarity to the older systems in both data organization (format), and methods for writing and retrieving information. The media look similar to the schematic diagram shown in figure 3.13(a); the index hole is used to mark the start of the first sector on the track. Such disks are called *soft-sectored*; they must be formatted prior to use. Such formatting consists in (magnetically) labelling the track and sector number of each sector. Normally the data are organized on the floppy according to a standard known as IBM 3740. However, when the recording format, disk size, track spacing, and number of disk sides used (there are single- and double-sided disks) are taken into account there are some 32 valid options. To this must be added standards imposed by the computer's operating system. (The potential advantage offered by the floppy of being able to move data between computers is offset by the need for physical and software compatibility between systems.)

The access arm positions the read/write head over the track where data are to be stored or retrieved. In evaluating the performance of such mechanisms several parameters may be used: seek time (the time required for the head to reach the appropriate track); latency (rotational delay or time required to reach the correct sector within the track); head switching time (time to choose correct read/write head for multiple-head or multiple recording-surface systems); and transfer time (time for actual movement of data from disk to main memory). The mechanism used to move the read/write head and provide for the rotation of the media as well as the circuitry to write or read data are called the drive. The access arm and actuator in many floppy drives consists of a stepping motor (whose shaft is rotated through a fixed number of degrees each time that it is energized), and gear (lead screw) which advances in a linear fashion for each advance of the stepper. The read/write head is attached to this lead screw and is this moved from track to track as the stepper motor rotates.

Hard- (fixed-) disk storage systems. While many of the concepts associated with floppy-disk systems are applicable to hard-disk systems, the design of the read/write head leads to a number of important differences. Because of the precise nature of the mechanism, hard drives are often sealed to prevent dust and other contaminants from reaching the disk. The design of the hard drive is intended to increase recording density and hence data capacity. Modern hard drives intended for use with PCs can store more than 1 Gbyte (1 Gbyte = 10^9 bytes) of data, about as much as 835 floppy disks (400 000 pages of textual material). In order to achieve such capacity, hard drives may use more than one recording surface and the actuator/access configuration is different from those found in floppy systems. Figure 3.14 is a schematic representation of such systems. The outer (top and bottom) surfaces of the multi-platter system are not used. Corresponding tracks on each recording surface form a cylinder. Such multi-platter arrangements not only increase capacity, they also reduce seek time. This is accomplished by techniques such as the clustering of recorded data where data that are to be accessed together are located on the same or adjacent cylinders. In addition, the most frequently needed data may be placed near the middle cylinder. By recording data in two places, disk latency may be reduced; no more than half a rotation may be required to access the data. (This, of course, reduces overall capacity.)

New magnetic storage systems. Recently, an innovative high-density storage system employing a two-axis, high-acceleration linear actuator has been reported in the literature [5]. Figure 3.15 is a schematic representation of the system. Instead of using a rotating disk and radially translated head mechanism, the proposed system employs a fixed recording medium and a read/write head capable of rectilinear (two-axis) motion.

A dual-track system uses one for the clock and a second for the data (with NRZ recording). Potential advantages for this system include:

- No rotational and seek latencies associated with rotating systems.
- Less weight; smaller; less power consumption—no need for large, bulky motors.

Using stepping actuator technology (similar to what is used in X–Y plotters) the dual head can be translated directly to the desired location with minimal

72 HARDWARE ARCHITECTURE

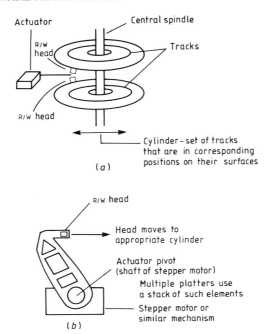

Figure 3.14 Hard-drive storage systems. (a) A schematic diagram of the multiple disk organization. (b) A sketch of the actuator system.

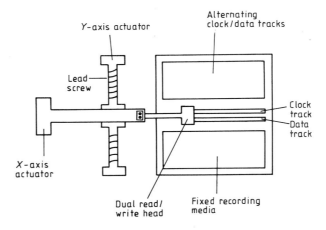

Figure 3.15 A sketch of a rectilinear magnetic storage system.

delay by actuating X and Y simultaneously. The system can be configured with the head in contact with the media (as floppy-disk drives) or designed with an air-bearing head structure (like hard-disk drives). Using magnetoresistive sensing, the playback signal would be independent of scanning speed—as contrasted with the inductive sensors previously discussed—greatly reducing compensation (amplitude equalization) during a read access. Projected storage densities are (impressive): 1 Gbyte in^{-2} ($\simeq$ 0.155 Gbyte cm^{-2}) and reading/writing data capacities of 5 Mbyte s^{-1}.

Magnetic tape systems. This is a non-volatile storage medium, which finds application primarily in archiving, backing up and transferring data between installations. Except in a form of this system called streaming tape—see below—it has been used principally in main-frame environments. Schematic representations of the drive and tape format are shown in figure 3.16. Such systems are very high-performance versions of (consumer) reel-to-reel tape recorders. (The differences make such systems considerably more expensive.) In order to reduce data access time, the system must be able to move tape past the R/W heads at great speed and to start and stop the tape rapidly. Since the tape spools are relatively large they have considerable inertia and starting and stopping these in a short time is difficult. The tape, which has much less inertia, 'flies' off the spool when the latter is stopped suddenly. The transport mechanisms shown in figure 3.16(a) are designed to provide for tape equilibrium. Two systems are shown; in the first, a vacuum is used to sustain a high tape speed across the R/W heads with rapid start/stop characteristics; in the second a (very low-inertia) 'flying' shuttle is used to accomplish the same end. Figure 3.16(b) shows some elements of the tape format. The tape is divided into nine tracks and separate R/W heads are used on each track. Each column ('frame') stores one byte (eight bits) plus one bit for parity error checking. The tape is divided into several regions: the start of a tape contains information about the tape itself (e.g. how the data are recorded); this is followed by the actual data, divided into blocks, each block containing one or more 'records' of information; finally the tape contains a trailer control label (including additional error-checking information) and a special marking that signals the end of the file (EOF mark). A common reel size contains 2400 feet of tape and may contain a hundred million characters.

The modern PC uses the Winchester technology to implement the principal secondary-storage system with system capacities in excess of 200 Mbyte. But this is a fixed medium and in order to transfer files and to preserve the contents of the secondary memory some other scheme is required. Transfer of files (e.g. programs, data) between PCs is conveniently done by the floppy disk; however, its relatively low capacity limits its use as a means for archiving information. A so-called *streaming tape* is suited for back-up. This contains an inexpensive transport mechanism which may require a second to start or stop; this is to be compared with a tape transport which starts and stops within 1 ms. In order to get around this disparity, some scheme is employed to 'buffer' the information being transferred between the PC and the streaming tape. Two methods are in common use:

- In one method the tape are allowed to overshoot the end of the region where data are being recorded. A region of blank tape is set aside to act as an

74 HARDWARE ARCHITECTURE

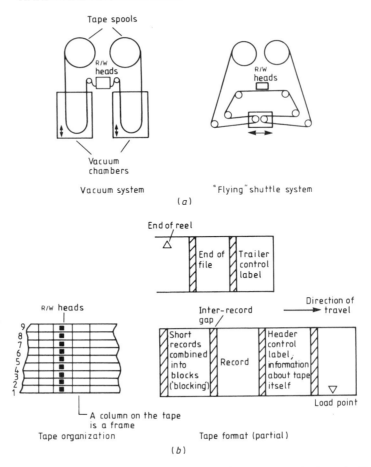

Figure 3.16 Tape storage systems. (a) Two types of transport mechanism. (b) Elements of the tape format.

inter-record gap between recorded regions. The tape is then rewound to a point before the inter-record gap. When it is restarted it has time to reach the correct speed before recording again.
- A second method makes use of a semiconductor (memory) buffer. Data are first read into this buffer and then transferred to the PC at high speed. The streaming tape may then pass from one record to the next without loss of data.

Optical storage. The optical disk is increasingly being used for secondary storage, particularly where large amounts of data are involved and such information is retained as part of a data-base. Optical storage systems are referred to as CD-ROM (where CD stands for compact disk) or WORM drives. The storage capacity may be equivalent to several hundred thousand pages of text

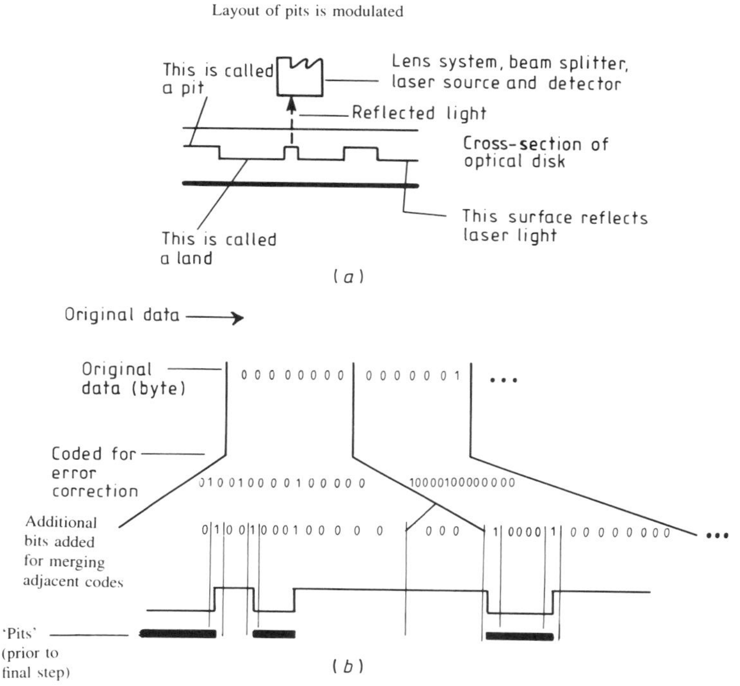

FUNCTIONAL COMPONENTS OF AN INSTRUMENT PLATFORM 75

Figure 3.17 Operating principles of optical storage systems. (a) A schematic diagram. (b) Data encoding and pit/land modulation (a fragment only).

or 500 floppies holding 1.44 Mbyte each. Improvements in laser technologies have made such high-density storage devices feasible. (Very precise and high-accuracy systems are required since dimensioning errors as small as 0.8×10^{-6} m ($0.8\ \mu$m) can result in data errors.)

The operating principles of optical storage systems are depicted in figure 3.17. CD information is ultimately stored in the modulated lengths of regions on the optical media called pits. Laser light is reflected from both the pits and the intervening regions called lands. However, the light from the pits is scattered because reflected light from the surrounding lands cancels reflections from the pits. This happens because the disk is fabricated with the pit/land depth equal to a quarter wavelength of the incident laser light. The total difference in optical path length is then equal to half a wavelength producing destructive interference between the surfaces (pit and land).

Reflected light from lands is bright while light received from pits is dim. Within the CD technology transitions in light intensity—bright to dim or the reverse—are interpreted as logical one-bits (1s) and constant reflection levels are treated as strings of logical zeros (0s). Each byte of data to be recorded must therefore be encoded as strings of 0s separated by logical 1s. A more immediate question is that of how to determine the number of 0s represented by a given length of pit or land. This is accomplished by a clock which is derived from the (actual) data on the CD—optical encoding is 'self-clocking'. While the laser beam is very narrow it does have a small finite width which limits the ultimate bit resolution—the ability to deposit adjacent pits and lands. Currently, this dimension coupled with the linear speed at which bits pass under the detector corresponds to a bit clock rate of 4.3218 MHz.

The encoding steps that transform experimental data into coded segments (lengths of pits and lands) on the CD are shown in figure 3.17(b). (While two distinct operations are shown, additional transformations are needed to reduce errors further; this is discussed below.) The original byte (8 bits) is encoded into a 14-bit code. The original byte has 256 valid states (2^8) while the 14-bit code has 16 384 permissible states, of which only 256 are 'legitimate'. Such coding schemes support both detection and correction of errors, which can be introduced during the recording process (see chapter 2). A second operation is added to eliminate possible encoding violations that might occur between adjacent bytes. (The coding scheme currently used for optical storage systems prohibits adjacent 1-bits; they must be separated by at least two 0s. Additionally, the longest string of 0s is limited to ten.) The *merging* bits carry no information; they are merely code separators.

Another encoding operation (not shown) further reduces the possibility of error by combining (encoding) 24 bytes of data into 32 bytes using a scheme known as Reed–Solomon. This added layer of coding, coupled with the techniques discussed above, combines to produce a system that is highly immune to error. CD-ROMs have undetected-error rates of less than 1 in 10^{13} bits (10 trillion). Even when errors are bunched together ('error bursts'), up to 450 successive bytes (of the original data) that contain errors can be properly decoded.

Rather than having a series of tracks as found in magnetic storage systems, the CD is arranged much like a phonograph record—a continuous spiral. The information (modulated pits and lands) must pass under the detector at a constant (linear) rate—otherwise the number of 0s and 1s will be misinterpreted. Therefore, the angular speed of the disk varies from outer to inner edge as the number of pits and lands (per rotation) varies between the edges. The CD speed varies between 200 and 500 RPM. Varying speed, coupled with a complex 'head' that is required to house the light source, the detector, and the laser positioning system, leads to a relatively slow access time—of the order of 200 ms. While this is slow compared to that of hard-disk drives (~ 20 ms) the CD can store the equivalent of 500 Mbyte of data on a single disk less than $5''$ ($\simeq 12.7$ cm) in diameter.

To maximize data densities, adjacent 'tracks' are separated by dimensions of the order of the width of the laser beam (1.66 μm). Therefore, small variations in the movement of the read head will cause severe error. Two head positions must be considered: the height of the head above the surface of the disk; and the horizontal position of the head relative to the edge of the disk. Light entering the optical head is split into two parts; four photodiodes—paired into groups of two—report differences in received light. Unless the head is properly focused (proper height above the CD) and it tracks properly (laser beam striking the centre of the pit), appropriate differences in photodiode signals are not equal to 0. This 'error' signal is used to position the detector lens in the vertical and/or horizontal position such that the error signal is reduced to 0.

WORMs. Some optical storage systems will write data once on the medium. This information will then be available to be read many times with minimum deterioration. Such systems are referred to as *write-once–read-many* (or mostly) devices or WORMs. There are a variety of writing techniques:

- Using organic dyes that are vaporized by the write pulse.
- Surface ablation of non-reflecting layers above a reflecting background to create a pit.
- Melting a region of plastic coating to create a pit.

Read–write optical storage. Erasable CD systems are found among the newer optical storage systems. One technology for supporting read and write operations employs an active material (tellurium iron cobalt) that can change the polarization of reflected laser light. The magnetization of the material determines the direction of the polarization of the reflected light. To write data, a burst of high-energy laser light heats the film (locally) changing its magnetic properties and, in combination with an electromagnet under the disk, changes the direction of magnetization once the material has cooled. A 'weak' laser will read the disk by detecting its polarization. To erase a bit the area previously magnetized in accordance with the data is restored to its initial magnetization in a manner similar to that of the writing protocol. Developments in erasable optical technology can be expected to continue in view of the potential for high data capacities.

3.1.3 Input/output (I/O)

The principal human interface to the PC is the keyboard and cathode-ray tube (CRT) monitor. These are referred to as the *terminal* although they are completely separate components. These devices perform critical functional purposes; they convert 'real-world' information (alpha-numeric symbols and/or icons) into machine-readable form (binary) as well as the reverse. To these devices must be added a number of other interface resources: the mouse; light pens; touch-screens; printers; strip-chart recorders; and $X-Y$ recorders. A rapidly emerging facility that can be found on the modern PC-based instrument is the speech synthesizer.

Keyboard. User keystrokes are converted into a standard code which can be recognized by programs executing within the PC. The keyboard includes the hardware necessary for the detection of keystrokes, and (permanent) software

78 HARDWARE ARCHITECTURE

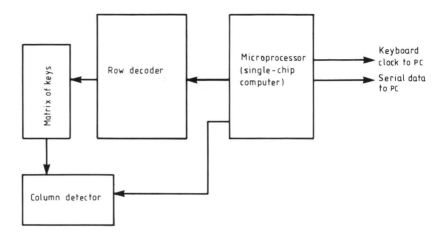

Figure 3.18 Hardware architecture of the PC keyboard.

within the PC ultimately converts this information into a standard ISO/ASCII-encoded equivalent. Figure 3.18 is a block diagram of a typical keyboard; it includes its own microprocessor for carrying out the detection. The keys themselves are arranged as a matrix; keystrokes result in a connection between a row line and a column line. The processor provides signals that can determine which key has been depressed. It produces a series of 'interrogation' signals on each of the horizontal lines. If no key is depressed then the interrogation returns a negative response. If one of the keys is actuated then it remains to determine which column (of the row in question) has been energized. The processor further interrogates the COLUMN DETECTOR circuitry to determine the exact switch. This information is transmitted to the PC in serial fashion; included is a second (clock) signal, which permits the PC to detect each bit of the transmitted data. The PC uses this information to generate the ASCII equivalent of the key that was depressed. In addition to the code itself, it is important to know for how long the key has been held down. One protocol that is in general use is transmitting the character code when the key is first depressed (the 'make') and a second character when the key is released (the 'break'). Normally the second character is similar to the key that is depressed but with one bit altered so that the PC 'understands' that this terminates the stroke. Furthermore, the microprocessor within the keyboard provides for other contingencies. For example, when a key is depressed it will 'bounce' repeatedly, so the onset (or termination) of the keystroke is not 'clean'. This can cause significant error if not corrected. The microprocessor includes a timing delay so that the keys are interrogated only after they have had time to settle—this is called *debouncing*. Finally, the keyboard encoder may send out a stream of characters to the PC if the key is held down longer than approximately 0.5 s. This repeat character capability is useful in a variety of circumstances (e.g. to use a keystroke to move a *cursor*,

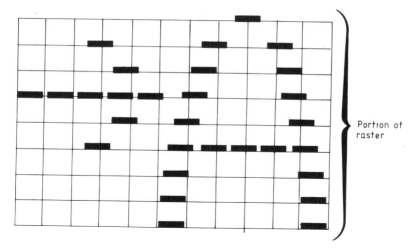

Figure 3.19 Raster image generation.

or 'marker icon', on the display). The number of repeat characters transmitted per second is often under program or software control.

The PC accepts this serial character and using the associated clock transforms the serial information into parallel form. This is accomplished using a hardware element designed for such purposes—a *serial-to-parallel converter*. It is then available to be converted into ASCII form; this is carried out by firmware ('permanent software') within the PC.

CRT monitor. The display that the user sees evolved from television (TV) technology where mass production has minimized the cost of this part of a computer system. The image is created by a beam of electrons that strike the surface of the CRT; the face of the monitor is coated with a phosphor that glows when struck by the electrons [6]. The image relies on an underlying pattern of lines created by deflecting the beam using electrostatic or magnetic deflection of the beam; the resulting pattern in called a raster. The raster pattern can be intensity modulated, so any arbitrary image can be produced (see figure 3.19). Such images can include colour if the CRT is coated with colour phosphors; colour monitors may be referred to as RGB (red–green–blue)—after the phosphors on the CRT. For PC applications the specifications of the CRT greatly exceed those of an ordinary TV receiver. The CRTs must be able to handle intensity variations of 20 MHz (20×10^6 or 20 MHz) in order to reproduce the fine details ordinarily not important for TV viewing (with a bandwidth of 4.5 MHz).

The raster pattern is viewed as a series of positions on the face of the CRT whose intensity and/or colour can be modulated. Each (dot) location is referred to as a picture element or *pixel*. With technological advances, the number of pixels that can be displayed has increased as well as the number of colours

80   HARDWARE ARCHITECTURE

Table 3.6 A summary of adapter capabilities. (Monitor resolution must be compatible with number of pixels.)

Name	Number of pixels (per row × per column)	Colour
Monochrome Display Adapter (MDA)	720 × 350	Black/white (also with amber or green phosphors); no graphics capability
Hercules Adapter	720 × 348	Black/white but with graphics capability
Color Graphics Adapter (CGA)	640 × 200 320 × 200	2 colours 4 colours
Extended Graphics Adapter (EGA)	640 × 350	16 colours
Video Graphics Adapter (VGA)	640 × 480	16 colours
Super Video Graphics Adapter (Super VGA)	1024 × 768	16 out of 256 colours (selectable)

that are available to the user. Table 3.6 lists some of the capabilities that are commercially available. (The Super VGA is currently considered to be 'state-of-the-art'.)

In modern computer-based instruments, the monitor is viewed as 'another instrument', which may be used as either a display device or as a tool for storing an image. Since the monitor is a fundamental component of the system, the hardware used to control it is included within the PC; only the electronics needed to modulate the beam and generate the raster are included within the monitor itself. The hardware that provides the signals to the CRT is contained in an *adapter* (card) which interfaces directly with the system bus. (See figure 3.20, which includes a simplified block diagram of a CRT controller.) These adaptor boards function as stand-alone computers dedicated to serving the CRT monitor. The processor is a specially designed integrated circuit (e.g. the 6845 CRT controller). An important part of this adapter is the dual-port memory. This allows the PC's CPU or the CRT controller to access the cells. Because this memory is within the address space of the CPU, the organization of the CRT adapter is said to be *memory mapped*. The VIDEO MEMORY (RAM) stores information about the image to be displayed. In general, each byte is divided into two regions: part of each word contains the data to be displayed and a second part contains additional information, namely some attribute of the information to be displayed. The following is a representative list of data and attributes:

Data:
- A single bit that indicates whether or not the pixel should be intensified. Such an arrangement is called *bit mapped*. One bit is retained for each pixel

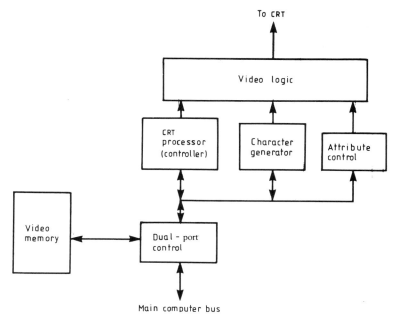

Figure 3.20 A simplified video adapter block diagram.

position on the CRT. The size of the memory must exceed the number of pixels that can be displayed (e.g. a 640×200 display requires 128 000 cells which corresponds to 16 kbyte of RAM).
- A number of bits that indicate not only whether or not the pixel is to be intensified, but also its colour (e.g. a 4-bit code could select 1 of 16 colours—including black, which would not show up on a black background).
- An ASCII code specifying which alphanumeric character is to appear on the CRT. The ASCII codes appear in sequence within the memory and correspond to one of the (sequential) positions on the CRT.

Attributes:
- Foreground intensity.
- Underline.
- Normal.
- Reverse video.
- Highlight.
- Blink.
- Combinations of the above: for example, specifying blinking and highlighting.

The graphics-controller processor continually cycles through the information stored in the VIDEO MEMORY. If the information to be displayed is a character, the ASCII code for the character is sent to the CHARACTER GENERATOR.

This is a ROM device which stores the pixel pattern that is to be displayed. The CHARACTER GENERATOR transmits the pixels, one row at a time, for each line of text to be displayed. The adapter must also be designed with logic that coordinates accesses to the VIDEO MEMORY from the CRT PROCESSOR and the CPU. If uncoordinated, the display will appear to jump. (This was a shortcoming in early versions of the PC/adapter systems.) The CRT PROCESSOR also supplies synchronization signals in the form of horizontal and vertical synchronization (synch) signals to the VIDEO LOGIC component, which combines the information into a form that is compatible with the CRT monitor. (There are a great variety of CRT monitors with a wide range of characteristics and parameters; a recent addition to the array of monitors is the flat-screen machine with improved ergonomic characteristics.)

Printers. One of the simplest forms of hard-copy device is the printer, which has two basic forms: impact and non-impact. Such electromechanical machines are generally low in cost, and reliable. (They are an outgrowth of the manual typewriter, which has been refined over the many years of its history.) However, they suffer from one of the following limitations: (i) low resolution; (ii) limited ability to reproduce graphical images and/or textual fonts. They come in a number of different forms with varying capabilities, cost and intended applications. The various printer types are now briefly described:

Dot matrix (impact). Characters formed in a way similar to that in the CRT. The dots are formed by wires (needles) which press an inked ribbon onto the paper. Solenoids control the wires. The needles are propelled by the solenoids (impact) into the ribbon and hence to the paper.

Cylinder/ball (impact). Preformed characters are embossed on either a cylinder or a golf-ball-shaped printing head. The head is rotated to position properly the character to be printed. (For the cylinder, a combination of rotation and vertical positioning is used because it has several rows of characters around the periphery; for the ball, rotation and tilting are used to position the character.) The ribbon and paper are positioned in front of the head. When the character is positioned properly, a hammer strikes the cylinder (or the golf-ball is moved by a pivot and cam) such that head strikes the ribbon/paper combination against a platen which is part of the system that moves the paper.

Daisy-wheel (impact). Instead of a ball, this print head looks like a daisy. It consists of a disc with slender spokes (petals) arranged around the periphery. At the end of each spoke is an embossed character. The disc rotates in the vertical plane at high speed and when the correct letter reaches the ribbon a solenoid hammer (impact) forces the character against the ribbon/paper/platen combination.

Line-printer (impact). An entire line of text is printed in one 'print cycle'. A drum contains embossed characters around its circumference at each potential printing position, often 132 positions with 64 possible characters at each position. The drum rotates, and when the correct character (for a given position) is aligned with the paper, a hammer located behind the paper strikes a ribbon thus generating the required symbol. Such printers: (i) may operate at 500 lines per minute (or faster in some cases) and produce generally low-quality type; (ii) are useful for high-volume environments; and (iii) are expensive to buy, operate and maintain. Another form of such printers is the so-called belt printer. This replaces the drum with a belt of embossed characters; it is slower but less costly.

Ink-jet printer (non-impact). The operating principle is similar to that of the CRT described above. In place of a stream of electrons (from the 'electron gun'), this printer generates a stream of fine ink particles. Two sets of electrodes can deflect this stream in the vertical and horizontal planes. The deflected drops strike the paper forming an image of the character (or graphic) to be produced.

Thermal printer (non-impact). Using specially coated paper, an array of wires—the dot matrix—heats the surface of the paper. The paper turns black (or blue in some instances) generating the desired character or graphic image. This is a 'quiet' printer but requires an inventory of the special paper.

Arc printer (non-impact). This employs another method requiring special paper. In this case the paper is coated with a thin film of aluminium. A high potential between the matrix of wires and the aluminium causes an electric (current) arc to be generated. This arc vaporizes the aluminium revealing the black paper backing. On energizing the proper wires, characters and images are formed.

Laser printer (non-impact). This is an important hard-copy device because it can reproduce sharp images—it has very high resolution. The reproduction capability is the result of the use of an extremely narrow beam created by a low-power semiconductor laser. Its basis of operation is charged-particle transfer which can be seen in figure 3.21. Certain organic chemicals are 'light sensitive'—their electrical properties change when they are exposed to light. Briefly stated, the charge-transfer process proceeds according to the following steps:

- A drum whose surface is coated with a light-sensitive chemical is charged to a high electrostatic potential. (The coating is a natural insulator and thus holds the charge.)
- White regions of the image to be generated reflect light onto the drum; these regions become conductive and lose their charge.
- The rotating drum passes close to the *toner*, which is a fine black powder.
- The particles of the toner are attracted to the charged regions on the drum. The image is now on the drum as regions with and without toner.
- The paper to which the image is to be transferred has been given a charge higher than that of the drum. The drum comes into contact with the paper.
- The drum transfers the toner to the paper creating an image.
- The paper is heated to 'fix' the toner—this makes the image permanent.

To accomplish this sequence, the elements of the block diagram (figure 3.21(b)) function as follows:

84 HARDWARE ARCHITECTURE

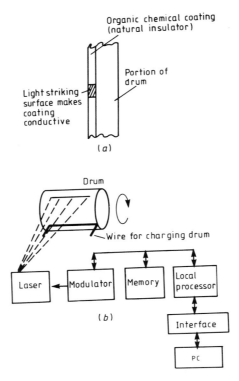

Figure 3.21 Laser printing. (a) The operating principle. (b) A simplified block diagram of a laser printer.

- A low-power semiconductor laser produces a very fine spot of light. (This is the basis for obtaining a high-resolution image.)
- The laser light is modulated (altered) by the image to be produced (e.g. by the current driving the laser). The light from the laser illuminates the drum using any of the following: a mechanical system (a series of mirrors); a CRT; or a liquid crystal whose opacity changes with the image information.
- The image is transferred to the rotating drum one (high-resolution) line at a time.

The image is often stored in the printer's memory. For example, a resolution of 300 dots per inch (in both the X- and Y-directions) requires approximately 1 Mbyte of memory to retain an 8×11 image (black and white). (Each dot occupies one memory cell.)

Liquid crystal display (LCD). Certain organic substances have elongated molecules that can modify their orientation under the influence of an electrostatic field. This state is called *liquid crystal(line)* (a few degrees above the melting point of the solid state). In its liquid crystal state the molecule exhibits unusual optical properties, in particular optical polarization. If polarized light (waves) pass through a second polarized filter, light will only pass through the combination if their polarizations match. For a twisted nematic liquid crystal, polarization of the molecules rotates as an electrostatic field is applied; the amount of rotation is 90° for applied voltages of less than 3 V. The electrostatic field must be of the alternating-current (AC) type or electrolysis of the material will result and the crystal will be destroyed.

As flat-panel display components, LCDs come in two varieties. On making one side of such displays into a reflecting surface, light will either be blocked or returned to the viewer depending on the relative polarization (which, of course, depends on the electrostatic field driving each segment). This technology is difficult to view and has been largely replaced by a so-called 'back-lit' design. A fluorescent panel located behind the LCD provides the initial illumination. Shades of grey may be reproduced by changing (modulating) the proportion of time for which each cell element is off or on. Screens made up of LCD elements have found wide use in small, low-power computers called *laptops*.

3.1.4 Communication

The elements of the instrument platform are linked together by means of a functional channel known as the bus or system bus. (This idea can be extended to describe the interconnection of larger functional units such as the computer and the external devices to which it may be connected, including other computers. In this case it may be referred to as a *local-area network* (LAN). When such devices are separated by large distances (e.g. connected by satellite), the resulting architecture is known as a *wide-area network* or WAN. Such arrangements are not characteristic of instrument systems and will not be discussed in this text.) Buses, whether they are computer buses, LANs or WANs, include three distinct functional elements:

- Identification of the source or destination of the information—*address lines*.
- Elements that carry the information—*data lines*.
- Means to provide for orderly transfer between source and destination—*control lines*.

For the physical elements noted above, some standard must exist regarding the transfer of the information. Normally, the CPU controls the orderly transfer of information between computer components (memory and I/O). This is accomplished by the following (simplified) sequence of events:

- The CPU places the source or destination address on the address lines.
- A directional control signal is provided next: 'read' if information is being requested by the CPU, 'write' if the CPU is going to transmit information.
- The actual data are placed on the bus by the source.
- A time delay is introduced to allow the receiver to accept the information.
- The transaction is terminated.

Because a number of components are connected to the same electrical path, some mechanism must be provided to ensure that two or more components do not attempt to access any of the wires comprising the bus for the same purpose (transmit or receive) at the same time. The specialized circuits are referred to as *buffers*; in particular, these circuits may be totally disconnected from the bus and for this reason they are referred to as *tri-state* buffers. The name follows from the fact that their outputs may assume voltage levels corresponding to two logical states (0 or 1) and a third state in which their outputs are totally disconnected from the bus lines. The modern bus architecture is a direct consequence of the development of such circuits.

Other control signals include so-called *handshaking* lines. These are meant to increase the reliability of the data transfer. For example, if for some reason the source in a read operation is unable to provide the requisite data, the receiver will not issue an *acknowledge* signal. Should the receiver (the CPU) fail to receive the data it may take evasive action. (One simple action that may be invoked is to inform the user that a *time-out error* has occurred; the computer (program) may then terminate and await appropriate intervention by the user.)

The CPU may not always control the transfer of information on the computer bus. It is possible, for example, for an I/O element to transfer a large block of information directly to the memory without intervention by the CPU. Such arrangements are called *d*irect *m*emory *a*ccess or DMA transfers and are discussed in chapter 5 in the context of data acquisition.

3.2 MULTIPLE-PROCESSOR SYSTEMS

Early computers were designed according to the architecture shown in figure 3.22(a); it is known as the 'Von Neumann' scheme to honour the contributions made by him to the development of the modern computer. One of the limitations of this design is the requirement for all data to pass through the CPU; this significantly reduces the machine throughput (capacity). It has become known as the 'Von Neumann bottleneck'.

To improve the computer's ability to process data, machine architecture developed along the lines shown in figure 3.22(b). It was facilitated by the development of *dual-port* memory—with memory cells that could be accessed from two separate sources. Processors of this type could then be coupled together via a 'high-speed' bus; this greatly increased throughput. Each processor (P) with its own I/O could be assigned a subtask with cooperation between processors supervised by one processor variously referred to as the 'master' or 'supervisor' [7, 8]. The design has become known as the *multiprocessor* architecture (figure 3.22(c)). For example, one processor could be programmed to gather data, a second processor designed to display the results, and yet a third processor arranged to archive the information on secondary storage and/or provide tabulated or graphic output. In another example, each processor

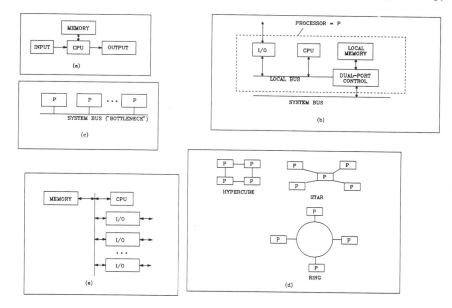

Figure 3.22 Multiprocessor architectures. (a) The 'Von Neumann bottleneck'. (b) A newer structure. (c) Multiprocessor systems. (d) Additional examples of multiple-processor architectures. (e) The transputer.

could control a separate external instrument associated with the automation project. With improvements in the capacities of the individual processors, the multiprocessor bus once again became a limiting factor in the speed at which data could be processed. Other arrangements have been formed and are shown in figure 3.22(d); they include *ring, star*, and *hypercube* designs. Communication between processors (which have evolved into interconnections of PCs) is based upon a 'message'-passing system. (When data are to be sent, they are packed into the message.) In the case of the ring, a *token* (message) with a destination address (the processor) is placed on the communication bus by the originator. The token is passed from one processor to the next until one of the processors takes it off the bus by recognizing its own address. In the case of the star, each message is sent to a central processor where it is rerouted to an appropriate destination. (This approximates a basic component of the modern telephone system.)

A variety of standards regarding the communication paths themselves have evolved. These are meant to deal with such problems as, 'what to do when two processors try to send messages at the same time', in an orderly manner. The communications networks are known as either LANs or WANs depending on the extent of the system [9].

A more recent architecture for dealing with laboratory environments is shown in figure 3.22(e). Laboratory applications often require a considerable number of I/O facilities (e.g. the need to control a variety of valves, pumps, instruments in process control or automated analysis). These can be supported by the *transputer* design which includes 'modest' CPU and memory capabilities but with considerable I/O. Additionally, transputers can be used to implement the hypercube architectures.

3.3 ARCHITECTURE OF INSTRUMENTS FOR AUTOMATED ENVIRONMENTS

The architecture of computer-based instruments intended for automated environments (including laboratories) functionally separates those elements (or modules) of the system that measure or control data. Such special-purpose modules are to be distinguished from the general-purpose computer elements (including CPU, memory, and I/O). Examples of these 'instrumental' modules include: oscilloscopes, signal generators (of various types), (volt)meters, and data-acquisition systems. Because of this modular construction, the user must be attentive to both the specifications of these instrument modules and the way in which they communicate with the general-purpose elements of the computer. At times, such modules may generate data (and send the data to the PC) while at other times they may receive data from the PC. In the former instance they are (data) *sources* and in the latter circumstances they are (data) *consumers*. The communication architectures are the focus of the discussion at this point.

Communication between instrument modules and the PC requires three components: an address to identify source and/or destination; the data to be transmitted or moved; and orderly control of the transfers. These requirements duplicate those of the PC system bus. A variety of bus architectures for communication between the PC and the instrument modules have become generally accepted standards and will be described with the aim of providing a basic understanding.

3.3.1 The RS-232 convention

Many computer-to-computer and computer-to-instrument (communication) systems are based on previously existing communication systems—the telephone network to name but one (see [10]). This architecture has evolved into the RS-232 protocol and is configured as a serial communication system (see figure 3.23).

Serial communication is an economical information interchange scheme when system capacity (or system throughput) requirements are low. RS-232 does not include a separate line for a clock signal and therefore several questions are raised: 'How does the device receiving the data know when a message has begun?'; 'How does the receiver know when the information has ended?'; 'How

ARCHITECTURE FOR AUTOMATED ENVIRONMENTS

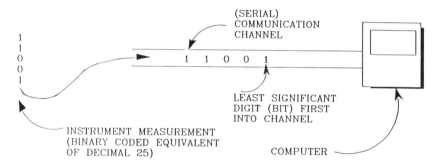

Figure 3.23 Serial interconnection between a computer and a remote instrument.

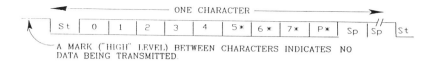

Figure 3.24 A format for serial communication.

does the receiver detect when one bit in the code has ended and the next begun?'; 'How does the transmitter know whether or not the receiver is free to receive a message?'. To answer these questions, a commonly accepted format exists and this is depicted in figure 3.24. S_t is the low-level 'start' bit which signals the receiver to listen. The intervals marked '0', '1', ..., '7' are bits that comprise the code for the character (or data). An asterisk indicates that inclusion is optional in the transmission. Thus the coded characters may include 5-, 6-, 7-, or 8-bit positions. 'P' signifies that an error-detecting bit—the *parity bit*—may be appended to the transmission. Each of the intervals depicted in the figure defines a *baud period*. Specification of this parameter is framed in terms of its inverse and is called the *baud rate*. The trailing bits designated S_p (*stop bits*) terminate the transmission and are equivalent to 1, $1\frac{1}{2}$, or 2 (high-level) baud periods.

The convention outlined above is an asynchronous protocol; there is no clock to detect the end of one baud period and the start of the next. However, when the receiver detects the start of a transmission (via the start bit) it 'starts' a clock of its own. Even if this clock is not strictly in synchronism with the clock that was used to generate the transmission at the source (producer), it will retain enough stability in the time needed to complete the transmission to ensure that the baud intervals remain in synchronism. (The frequency of this clock may be many times that specified by the baud rate—perhaps 64 times—which ultimately leads to better synchronization.)

The RS-232 convention for information communication was designed to transmit data over long distances without error. There are two popular forms for such transmission. The first is the *current-loop* system which relies on two different values of current to distinguish between binary values. It is intended for those applications that have to operate over long distances but that do not have stringent data throughput requirements. For those instances where separations are shorter but need higher capacity, the *voltage-level* protocol is employed. Two distinct voltage levels correspond to the binary signals. This is the most commonly used variety.

In addition to the two kinds of information transmission method just mentioned, there are three kinds of communication pathway architecture. These are depicted in figure 3.25 and are: the simplex method (one way only, using one transmitter); the half-duplex (sharing the path); and the full-duplex (two-way) structures. Each of the communication pathways noted in the figure requires 'coordination' (in addition to the clock synchronization) between the transmitting station and the receiving station. (For example, even before the transmission is initiated, it must be determined whether or not the computer is 'ready' to receive the information.) Table 3.7 includes the signals associated with RS-232 transmissions including the pin number on the connecter that is specified in the standard.

In order to maintain an orderly RS-232 communication path several control lines (in addition to the data lines) must be added. These lines support full-duplex communication schemes, among other purposes. These control lines are now briefly defined:

Request-to-send (RTS): indicates that one device wishes to transmit data.

Clear-to-send (CTS): indicates that a device is ready to receive data.

Data set ready (DSR): indicates that the device that is to receive the data is ready (e.g. it is switched on and in its normal functioning mode rather than a self-test mode, for example).

Data terminal ready (DTR): this line is analogous to maintaining a phone connection between the transmitting and receiving devices. Disabling it is akin to 'hanging up the phone'.

For a proper RS-232 communication channel, these signals may be arranged to 'fool' the transmitter and receiver into accepting data. A very popular, and simple interconnection for achieving this is shown in figure 3.26. This arrangement does not provide for complete handshaking; thus careful (software) control is required for proper operation. (Notice that the RTS and CTS signals are shorted within each device; each one is 'fooled' into believing that it is free to send data without regard to the status of the other device.)

3.3.2 The IEEE 488 convention

While the RS-232 protocol has an advantage in its relative simplicity, it is limited in its capacity; data rates are limited to $19\,200\,\text{bit}\,\text{s}^{-1}$. In addition, the PC requires a different port for each device. By contrast, the IEEE 488 standard (and its variations) is a network-oriented architecture, which greatly increases

SIMPLEX (ONE TRANSMITTER):

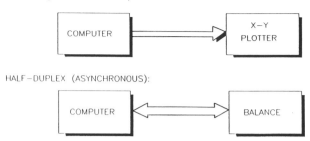

HALF-DUPLEX (ASYNCHRONOUS):

THE SINGLE CHANNEL MUST BE TURNED AROUND AT APPROPRIATE TIMES.
EXAMPLE:
1. COMPUTER SENDS COMMAND OVER CHANNEL TO BALANCE TO TAKE A READING.
    (COMPUTER IS THE TRANSMITTER)
2. COMPUTER SENSES CHANNEL TO SEE IF BALANCE HAS DATA READY.
3. BALANCE SENDS DATA OVER CHANNEL TO COMPUTER (BALANCE IS THE
    TRANSMITTER).

FULL DUPLEX (SIMULTANEOUS, TWO-WAY TRANSMISSION)

Figure 3.25 Serial communication pathways.

the number of devices that can communicate as well as the data rates that can be achieved [11]. It is both a communication and a control convention which was originally developed by the Hewlett–Packard Corporation; it was called the *General Purpose Interface Bus* or GPIB. This subsequently evolved into the recognized IEEE 488 standard.

Devices that are connected to this bus fall into three categories; for a given module the classification may change from time to time:

Listener. This corresponds to the (data) consumer cited above; an example of such a device is a pulse injector that operates in conjunction with an instrument that measures the mobility of charged carriers in a substance.

Talker. This corresponds to a data producer (described above) and one example is a photographic scanning system output from an analytical ultracentrifuge.

Controller. Normally, but not exclusively, the PC. The Controller can be compared to the central switching office of the telephone system. It monitors the IEEE (GPIB) network and when it notices that a party—one of the devices

Table 3.7 RS-232 signal descriptions. Notes: signals marked with * are used for handshaking; RS-232 is the standard set by the Electronic Industries Association (EIA); binary 1 is −3 to −20 V, or 20 mA current; binary 0 is +3 to +20 V, or no current; the connector is a 25-pin 'D-Shell', also called the DB-25; DTE = data terminal equipment; DCE = data communication equipment.

RS-232 signal	Input/output	Pin numbers Terminal/DTE	Modem/DCE
Signal ground	o	1	1
Data (transmit)	I	2	3
Data (receive)	o	3	2
Request to send (RTS)*	I	4	5
Clear to send (CTS)*	I	5	4
Data set ready (DSR)*		6	20
Chassis ground		7	7
Carrier detect (CD)*		8 (IN)	8 (IN)
Data terminal ready (DTR)*	o	20	6

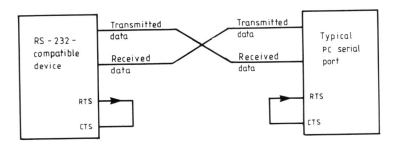

Figure 3.26 A simple interconnection for RS-232 communication.

connected to the net—wants to call or send a data message, it connects the caller (Talker) to the receiver (Listener). The call may be to a 'party line' which includes more than one Listener.

Some devices may inherently be Listener/Talkers, with their mode dependent on their status as producers or consumers. Examples include: a nuclear magnetic resonance (NMR) instrument, and a mass spectrometry system that synchronizes data production with a particular parametric setting (e.g. field current).

The Controller usually addresses or enables a Talker and one or more Listeners before the Talker can send its message to the Listener(s). After the 'message' has been sent, the Controller may 'unaddress' (disable) both the producer (source) of data and the consumer(s) (destination(s)) of the data. Figure 3.27 shows a typical GPIB/488 application including a table that describes how each device may change its category over time.

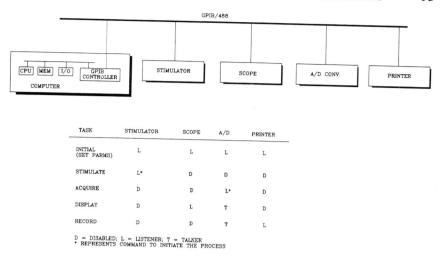

Figure 3.27 An example of an IEEE 488 instrument architecture.

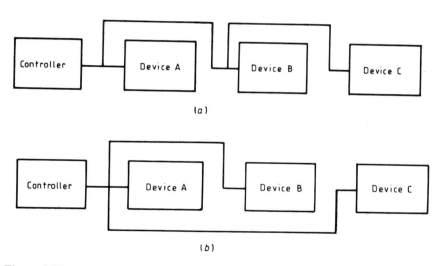

Figure 3.28 Interconnection examples for the GPIB (IEEE 488) network. (a) Linear. (b) Star.

The stand-alone devices on the network envisioned by GPIB can be interconnected in several ways as depicted in figure 3.28. The first IEEE standard in this category, IEEE 488.1, outlined hardware specifications as well as characterizing the nature of 'messages' that could be sent over the network. The message architecture is shown in figure 3.29.

Fifteen devices may be connected to the communication channel (not all

HARDWARE ARCHITECTURE

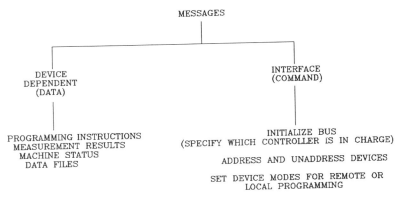

Figure 3.29 Message types in the IEEE 488.1 protocol.

Table 3.8 IEEE 488 handshaking signals.

Signal mnemonic	Signal name	Purpose
NRFD	'Not-ready-for-data'	Indicates when a device is ready or not ready to receive a message.
NDAC	'Not-data-accepted'	Indicates when a device has or has not accepted a message.
DAV	'Data valid'	Tells when signals on the data lines are valid.

talking at once) and data rates of up to 1 Mbyte s^{-1} can be achieved primarily because the protocol employs a *parallel* architecture. This data capacity includes all necessary handshaking. The bus consists of sixteen wires grouped as follows: eight lines (one byte) for data; three lines for orderly data transfer (handshaking); five lines for device management (establishing Talkers, Listeners, etc).

Additional system limitations include the following:

- The maximum separation between two devices is 4 m.
- The average separation over the network must be less than 2 m.
- The cable length is limited to a maximum of 20 m.
- At least two thirds of the devices—fifteen maximum—must be turned on (operating).

(The network can be extended, however, using specialized hardware devices known as *bus extenders*.)

Two categories of control signal provide for the uncontaminated transfer of messages between devices on the bus. These are summarized in tables 3.8 and 3.9 and include:

- *Handshaking lines*: three-wire interlocked signals that guarantee that

Table 3.9 IEEE 488 interface management lines.

Signal mnemonic	Signal name	Signal source	Purpose
ATN	'Attention'	Controller	1: data lines have commands; 0: Talker can send data messages.
IFC	'Interface clear'	Controller-in-charge	System Controller drives this line; initializes bus and becomes CIC.
REN	'Remote enable'	Controller	Places devices in remote or local program mode.
SRQ	'Service request'	Any device	Requests service from Controller.
EOI	'End or identify'	Talker or Controller	If Controller, tells devices to identify their response when polling (interrogating). If Talker, marks end of a message string.

information on the data lines is sent and received without transmission error.
- *Interface management lines*: these five lines manage the flow of information across the interface.

The first IEEE 488 standard [12] specified the GPIB hardware that was to be used; it did not specify command and data formats, status reporting, common commands, controller requirements, and message-exchange protocols. IEEE 488.2 was drafted in 1987 [13] and addressed these issues. As a result, a number of standard protocols were established; these are summarized in table 3.10.

In addition, IEEE 488.2 specifies a series of control sequences by defining the IEEE 488.1 messages that must be sent, as well as their order if more than one message is to be sent. These 'low-level' sequences are used to implement the protocols described in table 3.10. (They are not included here.)

3.3.3 New GPIB standards

If a 'standard instrument architecture' is envisioned then the protocols defined under IEEE 488.2 could be used to implement highly 'abstract' commands to those instruments that are compatible with the GPIB standards. Such commands have a number of advantages: errors are minimized; instrument design is

Table 3.10 A summary of IEEE 488.2 protocols.

Mnemonic	Name	Description
RESET	'Reset system'	Initialize GPIB (establish controller-in-charge); clear all devices to known state.
FINDRQS	'Find device requesting service'	Controller senses SRQ line for service requests; devices are polled in prioritized order with important devices queried first.
ALLSPOLL	'Poll devices (serial)'	Polls devices in serial order; returns status byte of each device.
PASSCTL	'Pass control'	Pass control of bus to other device(s).
REQUESTCTL	'Request control'	Device can request control of the bus.
FINDLSTN	'Find Listeners'	Monitor bus lines to locate devices that are present on the bus. Controller issues a Listener address, then monitors NDAC line to see whether or not device is at address; returns a list of addresses for all located devices.
SETADD	'Set address'	
TESTSYS	'System self-test'	Each device runs self-test; returns status ('ready to run' or 'problems').

made easier because the designer's intentions can be clearly interpreted; and development time is reduced because software does not have to be recreated for each instrument. In 1990, a group of instrument manufacturers agreed to the *Standard Commands for Programmable Instruments* (SCPI). By judiciously choosing a generic instrument architecture, they saw that it would be possible to simplify instrument compatibility with differing environments. Figure 3.30 depicts the model that they used. Table 3.11 describes the function of each block in the diagram. Not every functional component is required by each instrument model. By using the standards, high-level commands can be readily generated without burdening the designer with myriad programming details, which are 'hidden' and follow directly from IEEE 488.2 standard protocols and controller sequences. A simple example will suffice:

```
:MEASure:VOLTage:AC?20,0.001
```

which translates as follows. The first colon signifies the start of the command; MEASure signifies that a measurement is to be taken; VOLTage specifies the generic entity to be sampled; AC indicates that an alternating-current value is required; '?' signifies that this is the value to be returned; '20' specifies the range at which to set the instrument; '0.001' sets the resolution of the instrument to

Table 3.11 A summary of the SCPI functional model.

Model	Element	Purpose	Examples
Measurement function	Signal routing	Connect signal to instrument's internal functions	
	INPut	Condition input signal	Filter, bias, attenuate
	SENSe	Convert signals into internal data	Control range, resolution, gate time
	CALCulate	Convert acquired data into more useful format	Convert units, rise and fall times, frequency parameters
	FORMat	Convert to form compatible with bus	
Signal generation	FORMat	Convert to form compatible with internal functions	
	CALCulate	Convert data into generator compatible form	Convert units, change domains
	SOURce	Generate signal based on parameters	Modulate amplitude, power, current, voltage, frequency
	OUTput	Condition signal after generation	Filter, bias, attenuate
	Signal routing	Connect to external world	
TRIGger		Synchronize with external events or other instruments	
MEMory		Store data inside instrument	
DISplay		Control presentation	

1 mV. This 'simple' command would be suitable for a digital voltmeter that is compatible with the SCPI/IEEE 488.2 standards.

3.4 THE COMPLETE 'COMPUTER-ON-A-CHIP' AND PORTABLE INSTRUMENTATION

The ability to manufacture high-density integrated circuits has led to development of the *computer-on-a-chip*. Such devices are particularly useful for dedicated applications such as measurements or process control at a remote or

98 HARDWARE ARCHITECTURE

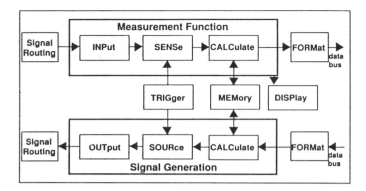

Figure 3.30 An SCPI instrument model.

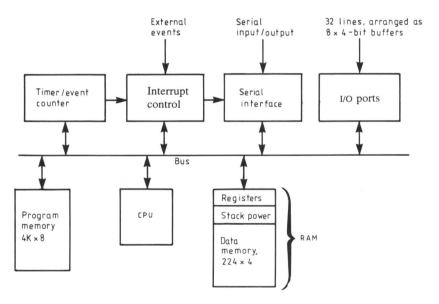

Figure 3.31 Architecture of a single-chip computer.

inhospitable site, or for portable instrumentation. They may be battery operated and draw very little power. A (simplified) block diagram of a representative circuit (NEC μPD7500 family) is shown in figure 3.31 [14]. A number of variations of this architecture can be found within the family of computers adapted for different potential applications. (In one version, the program memory is configured as an EPROM; this can be used during development.) For single-chip computers, certain design goals and compromises are introduced in order to achieve several advantages. For example, some of the advantages include:

• Has extremely low power consumption: ranging from $10\,\mu$A when the

computer is 'put to sleep' (standby mode), up to 300 μA when operating at its highest clock rate.
- Includes features requiring many additional circuits as found in the PC: timer/event counter; vectored, prioritized interrupts; extensive I/O capability (32 lines); serial port.
- Capable of operation from a single battery, variable from as low as 2.5 V up to 6.0 V; using a lithium cell as the power source an extremely small instrument can be developed.

However, some limitations may have to be considered:
- The data word size is four bits although the program word size is eight bits.
- The system clock is normally at 200 kHz which may result in slow execution of programs, and then placing of limitations on some real-time applications.
- Limited memory is set aside for data.

In spite of these limitations, there are many instances when such a 'complete' device is extremely useful. For example, a behavioural shaping instrument employing just such a chip has recently been described [15]. A block diagram of this is shown in figure 3.32. The program accepts information regarding some aspect of animal (human) behaviour, compares this against a goal, and returns a reinforcement—in the form of an audible tone, for example—which encourages the user to an improved level of performance. Periodically, the user's record is examined and the goal adjusted so that overall performance is gradually improved. (The instrument shown has been used to manage idiopathic scoliosis [16].) The instrument could also communicate with a PC, so the user's performance over a period of many days could be recorded and further analysed.

With a similar architecture, such an instrument could be arranged to record information that did not require frequent sampling. Periodically, the information could be 'off-loaded' onto a PC. By linking two four-bit data memory words together, larger word sizes could be implemented. Counting, for example, could be extended from four bits to eight bits by using the overflow from the lower-order significant digit to increment the higher-order significant digit; such software is known as double-, triple-, quadruple- (etc) precision arithmetic. In addition, a one-chip computer would lend itself well to applications requiring measurements in the field (e.g. chemical measurements, radiation measurements).

3.5 CHOOSING A PC PLATFORM

There is no well defined method for deciding on which PC platform is appropriate for laboratory environments. The user must use practical, rule-of-thumb, or heuristic considerations where past experience is an important consideration. However, there are a series of broad guidelines that can help to select a particular PC from among the many choices available to the individual.

100   HARDWARE ARCHITECTURE

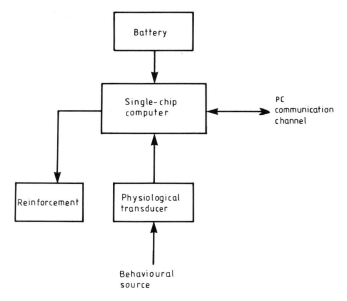

Figure 3.32 Block diagram of a behavioural shaping instrument employing a single-chip computer.

A foremost consideration is the intended purpose of the computer. In this regard the user must decide whether the PC will be used primarily in a dedicated ('turnkey') application or whether it must be able to support multiple tasks ('general purpose') such as a combination of experimental control, data acquisition, data analysis, and report generation.

3.5.1 Dedicated applications

The computer-controlled automated substance analysis system using a robotic arm (shown in figure 1.8) is a typical example in which the PC is used to instrument a limited class of experiments. The analytic sequence (program) and its associated instruments may change between tasks, but the system requirements remain bounded by the most demanding operation. In cases such as these, the user must determine the following:

- The capacity (or throughput) demanded by the most demanding task. This parameter, in turn, impacts on the speed (clock rate) of the PC.
- The number of experimental data that the system must store. This is needed to determine, in part, the amount and nature of the main memory (e.g. the size of the cache, if needed), as well as the size of the secondary store.
- The software architecture and complexity of the programs to be executed. For example, platforms that must support *concurrent* architectures will require more memory than those platforms that do not support such applications.

(Concurrent concepts are discussed in chapter 4—see section 4.1. Concurrent operation would increase throughput if the computer was generating a report for one analytic test while a second test was being conducted 'at the same time'.)

- The I/O hardware that is needed to control or communicate with the external instruments.
- The display (monitor) to be used, depending on its mode of use. If textual information is to be displayed then a low-resolution monitor may suffice. If graphs must be displayed then monitors with higher resolution could be required.
- Possible additional hard-copy (printing) facilities required by the PC platform. Here again, the specification of this resource would depend on the nature of the information to be recorded (e.g. graphics generally require higher resolution). Questions such as that of the reliability of such equipment must be addressed. (For example, if the printer is to be used rarely, then a less costly printer is indicated.)

As a general rule, *for dedicated applications the user should acquire only those resources that will be fully utilized*. If, for example, the PC includes 8 Mbyte ('8 Meg') of main memory but the intended application never requires more than 4 Meg of main memory then the additional memory represents a wasted expenditure. In other words, only pay for what you will need and/or use.

3.5.2 General-purpose systems

When choosing a PC platform for multipurpose applications or for those cases where the user cannot anticipate all of the hardware and software requirements, the general rule to be followed should have the following character [17]:

Acquire as much PC power as the budget will allow.

As a starting point, the user should consider a minimum general-purpose system to consist of:

- A 32-bit processor with built-in or added floating-point capability.
- 8 Mbyte or more of RAM.
- 250 Mbyte or more of hard-disk storage.
- High-performance graphics and display.
- Built-in networking capability.

The key criteria for choosing such systems from the many competing models include five key issues: long-term availability of support, technology, applications, cost, and institutional needs. These factors are now briefly described.

- **Support**. If the architecture and system that the user chooses prove to be short lived, he or she will be forced to repeat the process of acquiring resources,

102 HARDWARE ARCHITECTURE

Table 3.12 A comparison of factors that influence processor viability. (Numbers indicate qualitative ratings with 1 highest and 6 lowest.)

Factor	Alpha	MIPS	PA-RISC	Pentium	Power PC	RS/6000	SPARC
Market Share	5	4	4	1	2	4	3
Third-party support	5	4	4	1	2	3	2
Open architecture	6	4	4	6	1	4	1
Future product investment	6	4	6	1	4	6	4

requiring new investments in cost and (more importantly) time. Some of the factors that help to sustain a product line include: market share; support for products from sources other than the manufacturer (third-party support); open architecture; manufacturer's investment; and competitive intensity (future product investment). A sampling of processors and their corresponding ratings for these factors are shown in table 3.12.

- **Technology**. A computer must be able to execute the user's applications. These vary from user to user and a processor's capability is often judged by the time that it takes to run certain 'standard' programs called benchmarks. Some of these benchmarks are summarized below:

MIPS. This is the best known and represents the number of MIPSs, that a processor can complete. Its companion is called *Mflops* (for millions of floating-point operations per second). (These have come to be considered of minimum value.)

Whetstones. This measures floating-point performance.

Dhrystones. This measures integer performance.

SPECmarks (from the Standard Performance Evaluation Corporation, administered by the National Computer Graphics Association, Fairfax, VA, USA). This is a suite of programs designed to perform various mathematical functions including integer and floating-point operations. (There are versions of SPECmarks that combine both types of operation as well as separate versions for floating-point and integer operations.)

Khornerstone (from Workstation Laboratories). This measures CPU/integer and floating-point performance.

Graphstone (from Workstation Laboratories). This tests low-level elements basic to most high-level two-dimensional graphics functions. (This would include the graphics library efficiency as well as processor capability and graphic hardware capabilities.)

Table 3.13 Performance ratings of various processors. (1 = highest; 5 = lowest; same integer indicates a tie.)

Benchmark	Alpha	MIPS	PA-RISC	Pentium	Power PC	RS/6000	SPARC
Khorner-stones	1	3	2	3	2	2	4
Dhry-stones	1	2	3	4	2	2	5
Graph-stones	3	3	2	5	1	1	4

Table 3.13 summarizes some performance ratings for various processors. It should be pointed out that benchmark performance is dependent on a number of factors including: clock rate; memory size (including the amount of cache memory); graphics hardware and library; disk speeds; and the operating system being used. In particular, benchmarks are particularly sensitive to the compiler and its settings. (Some compilers do not take full advantage of the architecture of the processor on which a program is to run because the compiler was designed for the processor's predecessor.)

Performance of the computer's hard disk can be judged by the sustainable rate at which data can be transferred between the main memory and the disk.

- **Applications**. Users should consider their most important application when evaluating competing systems. In particular, programs that they may wish to run in the future are important factors. Generally, a generic system would need to support a word-processing package as well as data-bases, spreadsheets, communications/e-mail, presentation graphics and statistical packages. Minimum systems usually come packaged ('bundled') with: cache memory (often on the processor chip but with an external cache for extra performance); a large secondary store; a generous amount of RAM (8 Meg is currently common); high graphics resolution; and I/O interfaces of various kinds (e.g. a printer port and two serial ports). A CD-ROM is currently an extra feature in most systems, but it is very likely that these will be used to distribute software, documentation, and data-bases in the near future. Expansibility and ease of upgrading must be taken into account, and in this regard the system's power supply should be able to support hardware additions to the computer.
- **Cost**. The buyer must consider competing prices carefully. Be certain to determine the price for the minimum system including all of its components. (Typically, a supplier will quote an attractive price but fail to reveal that a monitor is not included and must be purchased separately.)

Software upgrades must be included in the budget, if only as a cost of system maintenance. The software company's record and policies for upgrades should be noted. If one includes such costs as the entry system, cost of peripherals, installed software costs, discounting policies, processor price-to-performance ratio, and graphics price-to-performance ratio then we can assign a score to the various processors mentioned thus far as follows:

Alpha	6.7
MIPS	7.7
PA-RISC	6.8
Pentium	8.2
Power PC	8.4
SPARC	7.2

(The higher the score, the better the rating. Note that these scores represent an overall average. Ratings may vary in any one category (e.g. price-to-performance ratios greatly favour RISC architectures) and this may be an overriding factor when making a decision.)

- **Institutional needs**. Compatibility of the system with the overall goals of the institution must be considered when acquiring a computer intended for generalized applications. In particular, a common architectural trend is the client–server organization. This involves a distributed computer system in which the performance of tasks by one or more available hardware and/or software entities in the system can be performed at the request of another entity. These entities communicate over a local- (or wide-) area network (LAN or WAN) and individual nodes (stations) in the system must support the hardware and software needs of the network protocol.

REFERENCES

[1] *Webster's Dictionary of Computer Terms* 1988 3rd edn (Englewood Cliffs, NJ: Prentice-Hall)

[2] Webb R J 1988 *Digital Technology with MOS Integrated Circuits* (New York, NY: McGraw-Hill)

[3] Floyd T L 1986 *Digital Fundamentals* 3rd edn (New York, NY: Merrill)

[4] Clements A 1993 *Principles of Computer Hardware* 2nd edn (Kent: PWS)

[5] Katti R R 1992 Compact translating-head magnetic memories *NASA Tech. Briefs* (May)

[6] Sargent M and Shoemaker R L 1986 *The IBM PC from the Inside Out* 2nd edn (New York, NY: Addison-Wesley)

[7] Silverman G 1984 Automation in the biomedical laboratory *IEEE Trans. Biomed. Eng.* centennial issue (December)

[8] Silverman G and Harrison M 1982 Microprocessors: a biomedical toolkit *Proc. 4th Annual Conf. of the IEEE Engineering in Medicine and Biology Society (Philadelphia, PA, 1982)* (New York, NY: IEEE)

[9] Tanenbaum A S 1989 *Computer Networks* 2nd edn (Englewood Cliffs, NJ: Prentice-Hall)

[10] Wolf S and Smith R F M 1990 *Student Reference Manual for Electronic Instrumentation Laboratories* (Englewood Cliffs, NJ: Prentice-Hall)
[11] *National Instruments IEEE 488 and VXIbus Control, Data Acquisition, and Analysis* 1992 National Instruments
[12] *ANSI/IEEE 488.1 1987, IEEE Standard Digital Interface for Programmable Instrumentation* 1987 (New York, NY: IEEE)
[13] *ANSI/IEEE 488.2 1987, Codes, Formats, Protocols and Common Commands for Use with IEEE 488.1* 1987 (New York, NY: IEEE)
[14] The μCOM75 family CMOS 4-bit microcomputer *User's Manual* NEC Electronics
[15] Silverman G and Dworkin B R 1991 Instrumental behaviour shaping automata *Digital Biosignal Processing* ed R Weitkunat (Amsterdam: Elsevier)
[16] Silverman G and Dworkin B 1992 Automatic operant conditioning system especially for scoliosis *US Patent* 5,082,002 (January)
[17] Juliussen E 1994 Focus report: workstations and PCs *IEEE Spectrum* (April)

4
Software for Instrument Systems

Early computers required the user to enter a series of binary numbers directly into the memory (via the CPU) using switches mounted on the front panel of the machine. The operator could execute the short program represented by these binary numbers to read a larger program into the computer's memory. The short program was called a *bootstrap loader* which implies its function—to '*bootstrap*' the computer into operation. (If, for some reason, the loader or larger program became corrupted or contaminated during operation, the entire time-consuming operation had to be repeated.) With the development of ROMs, the sequence described above has become automatic; the user now turns power on to initiate this 'transparent' sequence.

The programs that are automatically loaded into memory provide for communication between a user and the computer; control of the keyboard and the monitor are among the numerous tasks that are supervised by these software modules or processes commonly referred to as *drivers*. These 'primitive' programs have been used to develop still more powerful programs that manage primary and secondary memory, as well as control communication with other devices (e.g. printers). Taken in totality, the complex of programs, drivers and additional software procedures are referred to as the computer's *operating system* (OS). (One formal definition of the OS is: 'A set of programs and routines that guide a computer in the performance of its tasks, assist the programs (and programmers) with certain supporting functions, and increase the usefulness of the computer's hardware' [1].

This chapter provides a functional description of the OS as well as the nature of programming, and various programming languages, including those closely aligned with the hardware (*assembly* languages), as well as more abstract languages (*high-level languages*, HLLs). In turn HLLs lead to programs that provide the user with advanced (specialized) capabilities (e.g. the word-processing application program that was used to develop this text). While such applications can perform powerful operations, the capabilities are generally limited to one sphere of activity.

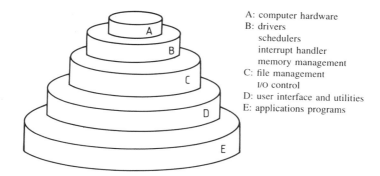

Figure 4.1 Layered organization of an OS.

4.1 COMPUTER OPERATING SYSTEMS

The services provided by the OS belong to two categories:

- Operational control of the hardware.
- Supervision of the allocation of hardware (and software) facilities to user programs.

Traditionally, the OS has been viewed as a layered structure with the more primitive code (*kernel*) at the centre and more abstract layers built upon these procedures. One popular formulation of the OS structure is shown in figure 4.1 [2–4]. To the user this arrangement appears as an abstract, *logical*, entity, to which the term *virtual* machine is applied.

At the very lowest level is the machine's hardware. Elements of the hardware are supported by the most primitive of procedures, which control basic tasks such as scheduling jobs, responding to interrupts, and managing the memory; this layer is referred to as the kernel (or nucleus) of the OS. It relies heavily on routines that reside in a part of the computer's primary memory that contains ROMs; the code within is referred to as firmware because of its permanence. For example, the detailed instructions for handling communication with the keyboard are found in these ROMs. Taken together, these routines form the *basic input/output system* or BIOS. To this are added the following layers (given together with a brief description of what is included):

- *Layer C.* Using the drivers, and memory management software, OS procedures are designed to support the connection between logical references to I/O devices (e.g. the printer) and the underlying I/O hardware. A particular characteristic of this layer is *device independence*. (An application program may require graphical data to be displayed on the monitor; software in this layer should function properly whether the CRT is high resolution or low resolution.) In addition, a most important (user) function defined in this layer is the *file system*. Computer-based instrument systems commonly include a secondary-storage system—typically a fixed disk—which is divided into variably sized logical groups called *files*, each consisting of a collection of data with a common (logical) *name*. The file management (software) resource may refer to these files without regard to their size or physical location on the disk—such details being supervised by the disk driver together with the disk controller hardware, which can translate logical address accesses into physical addresses and control signals. A most important function of the OS allows users to *create*, *delete* (*erase*), *rename*, *copy*, *append*, and/or modify files and the data within. To manage this information, the OS maintains a *directory* of such files, which itself can be modified when necessary.

- *Layer D*. The user interface coordinates user commands with the rest of the OS. The software includes a user *command interpreter* often referred to as a *shell* that permits a user to enter commands and data into the system via the keyboard (in cooperation with the monitor). Of growing importance is the *graphic user interface* (GUI) which supplants the *Job Control Language* (JCL), with a combination of icons (graphical images). These icons move in correspondence with movement of a manipulator mechanism (mouse, trackball, light pen, etc) allowing the user to 'point' at the desired task(s). (Such an arrangement may be called a WIMP environment—standing for Windows, icon, mouse, pointer—which permits highly unsophisticated users to master the computer quickly.)
- *Layer E*. The applications layer, as this is called, refers to highly abstract programs and packages that can execute successfully within the OS. Such programs include HLLs, spreadsheets, word-processing packages, simulators, data-bases, and proprietary software for experimental measurement, control and analysis of results.

4.1.1 The disk operating system (DOS)

A variety of OSs have emerged (commercially) within the 'layered' framework outlined above, each targeted to a specific community of interests. For computer-based instruments one of the most popular is the 'single-user' OS (e.g. CP/M, OS9). A popular OS that this text will address to some extent from the user's point of view is the disk operating system developed by Microsoft—known by the name MS-DOS [5].

DOS has been adapted for use with powerful *multitasking* user interfaces—see Windows in chapter 12. Multitasking is a term applied to the way in which procedures are processed within the computer. In particular—see figure 4.2—a program (or job) may be decomposed into a series of procedures that need to be executed by the computer and its resources. Some of the tasks require considerable time to complete, particularly those associated with transfer of data between the CPU and the secondary storage system, or with peripheral devices such as external memory stores (buffers) within an A/D or D/A converter element. While such transfer is taking place, the CPU may be idling; this results

COMPUTER OPERATING SYSTEMS 109

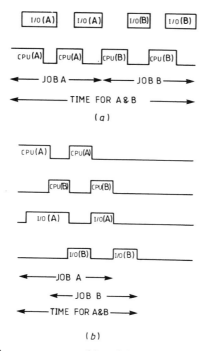

Figure 4.2 Multitasking oss are capable of higher throughput than job-at-a-time organizations. (a) A timing diagram for a job-at-a-time system. (b) The same job structure executed in a multitasking environment.

in a most inefficient use of the computer's capacity. Figure 4.2 details the difference between resource allocation under a 'one-at-a-time' strategy and a multitasking organization. Two jobs are depicted (A and B); two resources are shown, the CPU and some unnamed I/O device, perhaps the (hard) disk. The improved performance as measured by the ability to complete the two jobs—the *throughput* —is clearly demonstrated for the multitasking alternative. In the figure, both jobs belong to the single user who is in control of the machine.

MS-DOS allows a user to 'stack' a series of tasks in the form of OS commands. Using the computer and the OS in this fashion is known as *batch* operation and the file that contains the commands to be executed is known as a *batch-file*. Alternatively, the user may enter the commands one at a time, interactively, via the keyboard and monitor—taken together these hardware elements form the *console device*. Commonly, the user does not require MS-DOS to complete these commands in a specified time interval. OSs that respond to an event or job within a well defined time are referred to as *real-time operating systems* (RTOSs); they are extremely important for applications where information data rates demand rapid OS response. When real-time specifications are important, the user may write a special-purpose program in one of the

110   SOFTWARE FOR INSTRUMENT SYSTEMS

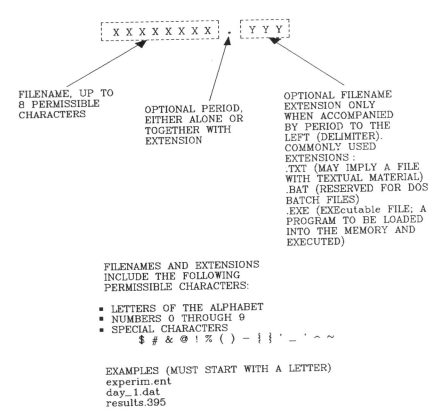

Figure 4.3 The file naming convention.

popular languages (HLL or Assembly). While RTOSs exist for large, multiuser installations, that is not the case for computer-based instrument environments. Some isolated instances of RTOSs have been reported but are not widely available [6, 7].

From the user's point of view, the file system and maintenance of his or her files is extremely important. One of the principal MS-DOS software tools (utilities) supervises, manages, and controls the file system. As noted above, a file is considered to be a group of data (e.g. alpha-numeric characters or binary-coded numbers representing experimental results) that is assigned a single name—the *filename*. The syntax (format) of an MS-DOS filename is shown in figure 4.3. The files are normally stored in a part of the secondary-storage facilities.

Within MS-DOS, file systems may be viewed in a way that is analogous to a document-based system with the following relationships:

Conventional file system entity	DOS 'equivalent'
Filing cabinet	Volume
File draw	Directory
File folder	File

An important difference between a conventional document system and MS-DOS follows from the possibility that a directory within MS-DOS may include (reference to) another directory that is subordinate to the first directory and referred to as a *subdirectory*. A subdirectory is considered to be the *child* of the *parent* directory in which it is first referred to. This organization results in a *hierarchical* arrangement of files. A sample of such a system is shown in figure 4.4. In order to access a file for execution, or refer to a particular file within an MS-DOS command, its path within the file system must be specified. The path may start with the particular secondary-storage element on which to find the file—the drive indicated by a letter followed by a colon (:) (e.g. a:,

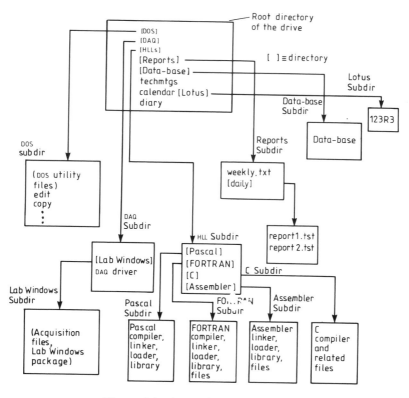

Figure 4.4 A sample directory system.

b:, c:, d:, e:). Within figure 4.4, the path to the file named 'report1.tst' would be specified as:

```
c:\reports\daily\report1.tst
```

assuming that it is located on drive c:. The symbol '\' (backslash) is a reference to the root or starting point of the particular directory or subdirectory. Directories and subdirectories may be created or erased using MS-DOS commands (or by pointing to appropriate symbols if using a GUI shell such as DOSSHELL or Windows). A summary of MS-DOS commands is included in table 4.1; this is by no means an exhaustive tabulation but includes those commands that are most appropriate for computer-based instrument systems. (There are a number of versions of DOS and some of the commands included in the table were not implemented in early versions of the OS.)

The allocation of computer storage in the PC is aimed at maximizing processing efficiency. To support this goal the OS includes various routines for converting logical references to memory locations into physical addresses. For example, a physical address may refer to a location in secondary storage as opposed to primary memory. The software logic that makes such determinations and 'translations' possible comprises the memory management module of the OS. There are several broad circumstances in which *memory management* comes into play—see figure 4.5:

- A program is needed only on rare occasions. The DOS program that permits disks to be formatted does not need to reside in main memory because it can be loaded in the relatively rare instances in which it is used; thus it resides in secondary storage.
- A program contains a main portion and a number of additional modules called *overlays*. This permits a number of different programs of this type to reside in main memory at the same time, making multitasking more efficient. Overlays may be loaded into main memory as needed.
- Writing programs in a manner that is consistent with overlays puts constraints on the programmer. A more general solution is possible when primary memory is to be shared by several tasks being executed concurrently. Main memory may be divided into fixed increments called pages. Procedures and/or programs that are actually running are loaded into main memory. A program may occupy several pages (in secondary memory) but only one page at a time is needed in main memory. When the CPU detects that a logical (program) address belongs to one of the pages residing in secondary storage it exchanges or swaps (out) one of the pages in main memory and loads the required page. A somewhat more general memory management scheme based on page swapping is known as *virtual memory*.

In summary, memory management within the OS is responsible for placing, fetching, and removing data, pages, or segments into or out of the main memory of the computer system.

Table 4.1 A summary of MS-DOS commands (partial).

Command (mnemonic)	Purpose	Example
APPEND	Instructs DOS to search additional paths when looking for a file. (See PATH command for another command of this type.)	append=a:/experim.ent DOS will look in drive a: if it does not find the file experim.ent in the current directory.
ASSIGN	Redirects references from one drive to another.	assign a=c A reference to drive a: will be redirected to drive c:.
ATTRIB	Allows a file to be marked for reading only (protected), or for archiving purposes.	attrib +r experim.ent Marks the file experim.ent for reading only; cannot be overwritten.
BACKUP	Allows user to make back-up copies of disk files—creates an archive of the file(s) in question.	backup c:experim.ent a: Creates a backup version of the file experim.ent on drive a:.
BUFFERS	DOS allocates memory space for disk/file transfers.	buffers=20 Some software applications packages require user to set aside disk buffers in memory.
CALL	A batch command that executes one batch file from within another.	call nextone The batchfile nextone will be executed; when completed, control returns to the batch-file that invoked nextone.
CD (CHDIR)	Changes current directory, or displays current directory's path.	cd a:\datadir Changes current directory to the directory designated datadir on drive a:.
CHKDSK	Reports on status of directories and subdirectories on a disk.	chkdsk/f With the /f option, DOS will correct certain types of disk error if they are encountered.
CLS	Clears screen and moves cursor to upper-left-hand corner of monitor.	cls
COMP	Compares contents of one file with those of another; reports mismatches.	comp file1.txt file2.txt
COPY	Copies an existing file; combines two or more into a single file; transfers data between peripheral devices and files.	copy c:file1.txt prn: Copies the file named file1.txt on drive c: to the printer (prn:)—in other words, it prints the contents of the file.

Table 4.1 (continued)

Command (mnemonic)	Purpose	Example
DATE	Displays current date; used to change the date.	`date`
DEL	Deletes (erases) one or more files from a disk—alternative form for erase.	`del junk`
DEVICE	Used to install a device driver; can only be used within a file called `config.sys`	`device=mouse` Adds the mouse driver.
DIR	Lists directory entries (some or all files and subdirectories).	`dir/w` Lists all files in a multicolumn arrangement, so that it fits on a single screen (with no scrolling).
DISKCOMP	Compares the contents of two floppy diskettes.	`diskcomp a: b:` Compares diskettes in drives a: and b:; if a mismatch exists, the track and side where mismatches occur will be displayed.
DISKCOPY	Used to copy contents of one floppy diskette onto another.	`diskcopy a: b:` Copies one track at a time from a: to b:, all pre-existing data on b: ('target diskette') will be overwritten.
ERASE	Erases one or more files from a disk.	`erase c:\expdata\test` Erases the file 'test' in the root of the subdirectory named expdata.
FILES	Sets aside memory for handling files; can only be used in the file called `config.sys`.	`files=20` Some applications programs/packages require the user to set aside space for 20 file handlers.
FIND	Searches for a given string of alpha-numeric characters—the text a string—in a file or files.	`find/n 'sample' c:\expdata\test` locates any lines containing 'sample' in the file named test in the subdirectory named expdata
FORMAT	Makes disks ready to work with DOS.	`format a: /s/v` Formats the disk (floppy) in drive a:; adds system files (/s option) which allows the floppy to 'boot the system'; allows the user to name the volume (/v option).

Table 4.1 (continued)

Command (mnemonic)	Purpose	Example
GRAPHICS	Permits the user to print whatever is on the screen when the shift key and the 'Prt Scrn' key are depressed simultaneously.	graphics
KEYB	Loads a keyboard device driver that supports non-US keyboards.	keybd fr Loads a driver for a French keyboard.
MEM	DOS displays how much memory is being used.	mem DOS will report how much memory is used and how much is available in the system.
MKDIR	Creates a subdirectory.	mkdir c:\expdata\trials Creates a new subdirectory in the subdirectory named expdata; this subdirectory is named trials.
MODE	May perform any of the following functions: setting mode of operation of parallel printer; acting as colour display adaptor, acting as protocol for asynchronous communications port, and various (program) coding designations.	mode com1: 9600,e,8,1,, The asynchronous communications port com1: is set to communicate at 9600 (baud), with even parity (e), having eight-code bits per word, with one stop bit; the repeated comma instructs DOS not to retry communication when the line is busy (use 'p' if retry is desired).
MORE	Outputs 23 lines of data at a time.	more < bigfile Allows one screenful of bigfile at a time.
PATH	Instructs DOS as to the directories and subdirectories to search to find executable files.	path=c:\first;c:\second;c:\third DOS will first search the indicated subdirectory, i.e. (first) in the c: drive—then subdirectory second, and finally subdirectory third.
PRINT	Prints a list of file(s).	print fileone filetwo file3 DOS will print the files indicated while concurrently performing other tasks.
PROMPT	Sets the prompt that the user sees when DOS is ready to accept commands.	prompt pg This form of the DOS command will produce the current directory followed by the > symbol.

Table 4.1 (continued)

Command (mnemonic)	Purpose	Example
RECOVER	Recovers data from files or an entire disk that has bad sectors or a damaged file directory.	`recover abadfile` Only data in the undamaged sectors are recovered; this is a limited DOS facility with more powerful commercial products being available (PC-Tools, Norton Utilities, Mace Utilities).
RENAME	Renames a file.	`rename new old` The file with the name 'new' is renamed to the name 'old'.
REPLACE	Replaces or adds files.	`replace a:source b:` Replaces a file named source on drive b: with the file source on drive a:.
RESTORE	Restores one or more files from one disk to another.	`restore a:backup1` Restores the file backup1 on drive a: to the current drive (normally c:) if it was stored with the back-up command at some time in the past.
RMDIR	Deletes a subdirectory.	`rmdir temp` The subdirectory temp will be removed, but only after all files in the subdirectory have been erased.
TIME	Displays the current time and allows the user to change it if desired.	`time`
TREE	Displays directory paths.	`tree a: /f` All directories and subdirectories in drive a: will be listed; the /f option will also list the files.
TYPE	Displays the contents of a file on the screen.	`type a:testresu.lts`
VER	Displays the version of DOS that is installed on the system.	`ver`
VOL	Displays the name of the volume (disk) if one was assigned.	`vol a:`

COMPUTER OPERATING SYSTEMS

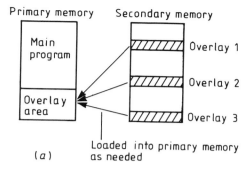

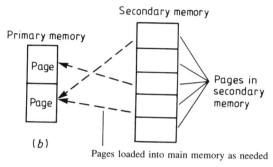

Figure 4.5 Architectures supported by memory management routines in the os. (a) Overlay. (b) Paging.

There are a number of ways to create files within DOS. For example, the DOS copy command may be used to create a file as follows:

```
c:\>     copy con: myfile
         This is an example for creating a DOS file.
         ^z
```

The symbol 'c:\>' is often seen on the monitor; it is the prompt that informs the user that the computer is ready to receive DOS ('job control language') commands. The user types the copy command as shown; the phrase 'con:' advises DOS that the source of the data to be copied is the console device, namely the combination of the keyboard and monitor. The data destination is specified by the next phrase—'myfile.' This is the name of the file (in the directory c:\), which will be created if it does not already exist. If it does exist it will be overwritten. (Each line in the example above must be followed by a carriage return.) All (alpha-numeric) text following the command line (This is...file.) will be stored in the file. The last line (^z) is formed by depressing the keyboard key marked 'Ctrl' and the key 'z' at the same time; it is referred to by the name 'control z'. It is needed when using the copy command to signal the end of the file—it is the *end-of-file mark*. It too must be terminated by a

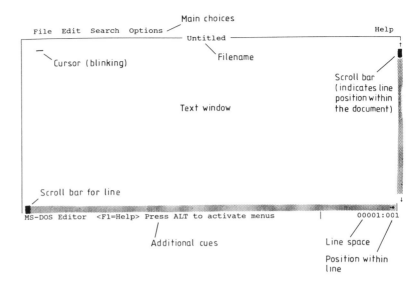

Figure 4.6 A sketch of the MS-DOS edit screen display.

carriage return. The user should once again see the DOS prompt. This is not the preferred method for creating files. In the first instance, one cannot edit existing files in this way; they will be overwritten. A much better method makes use of a program (as distinct from the DOS 'command') designed specifically for such purposes. A great variety of such programs exist; they include *editor* programs of various descriptions as well as word-processing packages.

An editor program provides the user with a number of useful features. (It is not as powerful as a word-processing program, which includes many 'publishing' capabilities; however, an editor is convenient and permits the user to create programs (e.g. Pascal, C, etc)). Using such an editor program permits a user to perform corrections on, insertions into, modifications to, or deletions from, a program or data file. An editor program that is included in MS-DOS is invoked by typing edit (and validated with a carriage return) when the prompt is visible. The 'user-friendly' screen shown in figure 4.6 will appear in response to the command. The user may enter textual material at the position indicated by the cursor. The cursor position can be moved by using the keyboard arrow keys (e.g. keys marked →, ←, ↑, ↓). This editor, like many programs, is menu driven. On depressing the key marked 'Alt'—the alternate key—the menu choices become accessible. For example, once the Alt key is depressed followed by the key marked ↓, additional menu choices will be revealed as shown in figure 4.7. Using the arrow keys an appropriate choice is highlighted. The carriage return key will validate (select) the choice. Help is available by depressing the key marked F1 (which is referred to as the *function key* F1).

COMPUTER OPERATING SYSTEMS 119

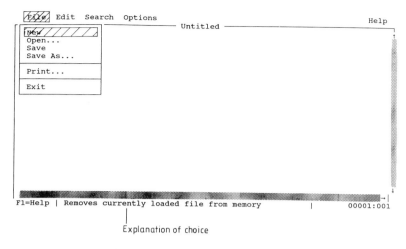

Figure 4.7 Editing menu choices.

4.1.2 An introduction to Windows

The DOS operating system has an installed base of users exceeding tens of millions. It provides features that most users need, particularly for single-user PC architectures [8]. As people with less intensive training have begun using PCs for instrument and control applications, it has become increasingly necessary to provide a simple, intuitive way in which to control the computer.

DOS is criticized in two ways: firstly, it is difficult to use; secondly, it runs only one program at a time. To overcome these limitations the GUI was introduced; the Apple Macintosh was one of the first computers to support such an architecture. Where computers employ GUIs, users are presented with a system of graphic symbols—pictures if you wish—called icons. These symbols are used to make selections. Along with the icons, a system of menus make it easy for users to identify and select available programs, commands, and options.

PC installations now rely on the Windows operating environment where a mouse (manipulator) moves a corresponding icon on the monitor and permits the user to choose actions to be carried out by the computer. Also of importance is the ability of Windows to manage more than one program at a time.

The heart of Windows is the *Program Manager*—a software element—which permits the user to run other Windows applications. Other important components of the Windows display includes the following:

Desktop. The area on the screen where the Program Manager runs.

Work area. The border of an active application window (like the Program Manager).

Window frame. The four edges that define the border of a window.

Title bar. This displays the name of the window and indicates whether or not the window is active by its colour.

Menu bar. This is displayed below the title bar and contains a number of commands that can be selected when the window is active.

Control menu box. This always appears at the upper left-hand corner of a window (or other interactive display called a *dialogue box*) and lists a set of commands for controlling the window (e.g. Close (the window and end its activation)).

Size (button). This can reduce the entire window to a single icon. It can also be used to increase the size of the window so that it fills the entire monitor screen.

Group window icons. In the Program Manager's window is a set of icons ('children') called group window icons. When one of these is activated (opened), a new group of applications are displayed and may be subsequently invoked.

Program icons. Within a group window one finds icons representing specific programs (e.g. a Calculator). The user can activate (run) such programs by 'clicking' on a button located on the mouse manipulator.

Figures 12.1 to 12.3 provide examples of the features described above. A more detailed discussion of Windows and its use in instrument environments is found in chapter 12.

4.2 THE NATURE OF PROGRAMMING

A *program* is a detailed and explicit set of directions for accomplishing some purpose, the set being expressed in some language suitable for input to the computer, or in machine language [1].

Natural languages, such as English, include considerable redundancy. (For example, within a sentence one might encounter the word 'qeen'; it is likely that the reader would recognize this as an error and replace it with 'queen', later to determine whether or not the substitution is consistent with the rest of the sentence.) While natural-language programming remains a goal of research, most computer languages require (in general) considerable structure and syntactical formalisms. Linguistic constraints imposed by programming languages are not particularly onerous in view of the fact that a program must be able to serve two conceptually simple purposes:

- To specify what calculations and operations are to be performed (on the data).
- To specify the order or sequence in which calculations are to be performed.

(While these are simple fundamentals making programming 'easy' on the face of it, programs can become exceedingly complicated as the number of operations—as measured by the number of lines of code—increases. It is not unusual for large programs to exceed 100 000 lines of code although programs of this size are unusual in computer-based instrument applications.)

Operations that the computer can perform include: data transfer or movement, arithmetic, and information comparison. Unless otherwise instructed, the

THE NATURE OF PROGRAMMING 121

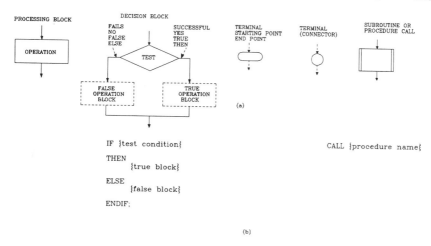

Figure 4.8 The basic building blocks of software control organization. (a) A flow diagram. (b) Pseudocode equivalents (where applicable).

computer executes one instruction at a time in the order in which the instructions are encountered. The power of the computer comes from its ability to change this strict sequence. When developing a program the user normally designs a logical solution to his or her problem—the *algorithm*. Language statements will support the underlying algorithm and, except in the most trivial examples, include a number of *control structures* for altering the strict sequential order. There are six commonly recognized *primitive* control structures and an interrelated sequence of these structures can be used to define the algorithm. (These primitive structures are themselves composed of several *basic* elements.)

Two methods have been used to implement algorithmic designs; one is graphical and the second is linguistic [9].

The graphical method for defining a program is embodied in a *flow diagram*. The basic building blocks of *flow diagram* are shown in figure 4.8 (see figure 4.8(a)) and these include:

- *Processing block*: an operation to be performed is specified within the processing block.
- *Decision block*: a test, normally a comparison, is specified; if the test is passed, the program continues on the branch named (variously) *yes*, *true*, or *then*; if the test fails, program control is transferred to the branch named (variously) *no*, *false*, or *else*.
- *Procedure or subroutine*: this represents a distinct body of code that is specified within another flow diagram. Upon entry into such a block, control is transferred to the (program) name specified within the procedure block. Upon completion of the code referred to, the program continues on the branch leaving the procedure block.

- *Connector*: this specifies a program continuation, and is especially useful for large flow diagrams, which may be spread over several pages. Terminals with the same designator (e.g. 'x') represent common continuation branches in a flow diagram.
- *Terminal*: this specifies the start or end of a program and/or procedure.

All languages include syntactical elements (statements) to support these blocks. Because the syntax varies from one language to another, a generic form of these statements known as pseudocode is used by many people in place of the flow diagram graphics. Algorithms are then described using this pseudocode rather than a flow diagram. The *pseudocode* equivalents of the flow diagram basic building blocks are shown in figure 4.8(b) and include:

- The pseudocode for a decision block:

```
IF {condition}
    THEN
         {true block pseudocode statements}
    ELSE
         {false block pseudocode statements}
ENDIF;
```

The '{condition}' phrase will be replaced by the test to be made in any given application; {true block ... statements} and {false block ... statements} will also be replaced by appropriate actions including the possibility for no action (absence of the THEN or ELSE branch), or another decision block (*nested decision blocks*).

- The pseudocode for a procedure block is:

```
CALL {procedure entity or name};
```

A specific object (name) will replace {procedure ... name} in a given application.

The other flow diagram block elements do not have pseudocode equivalents; however, the primitive control elements that are built from these basic elements do have pseudocode representation as described below.

The basic building blocks may be combined into a set of primitive structures as shown in figure 4.9; also shown are the pseudocode equivalents. The simplest organization consists of sequential processing blocks; this is representative of the computer executing instructions one at a time, in the order in which they are encountered without deviation. The remaining primitives are all concerned with repetition in one form or another. There are three types of repetition:

- *DO LOOPs*: a DO loop may be employed when a group of instructions is to be repeated a fixed number of times. The loop is characterized by three parameters that determine the control of the loop counter: the {start} value is the initial setting of this counter; the {stop} value determines when the loop

THE NATURE OF PROGRAMMING

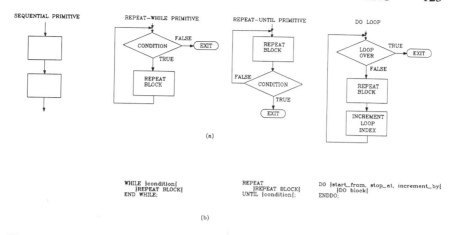

Figure 4.9 Primitive software control structures built from the basic elements. (a) A flow diagram. (b) Pseudocode equivalents (where applicable).

is terminated—the loop terminates when the number of repetitions *exceeds* the {stop} value; for each repetition of the statements within the loop, the loop counter is incremented by an amount denoted by {increment}. The *loop block* includes the instructions to be completed; the block can consist of additional basic or primitive structures (e.g. another DO loop which results in a nested DO loop organization). Initially, control is transferred to the start of the loop; values for {start}, {stop}, and increment are determined by instructions prior to entry—this is *loop initialization*. On occasion, the initial value of {start} may exceed the value of {stop}; in such instances the loop is not to be repeated at all. In most cases the loop will be repeated one or more times. When it has been executed the requisite number of times the loop is terminated—*exit*. (Within the loop block there may be additional programmatic tests which may result in an *abnormal* termination of the loop. An example of such tests is checking for an error such as the presence of a negative result when none is expected.)

- *DO WHILE*: this is the first of two structures in which the number of repetitions of the *repeat block* may not be known in advance of its execution. For this architecture a condition is tested at the beginning of the structure. If the condition is validated (*true* branch) control passes to the repeat block. This block contains other basic or primitive structures one of which normally alters the test condition. (If it did not, the loop would never terminate.) The condition is tested again and, if it fails, the loop is terminated.
- *DO UNTIL*: this is the second form of repetition where a modification within the loop determines whether or not the loop is to continue. It differs from the DO WHILE primitive in one important respect; the repeat block is executed *at least* once—condition is tested *after* processing of the repeat block. In other

respects it is similar to the DO WHILE primitive.

A given computer language may not include the entire repertoire of basic and primitive structures outlined above, but each includes the possibility to 'emulate' the ones that it does not support directly.

There are different 'styles' of program development:

- Top-down development: some users prefer to start from a 'global' view of the problem solution and provide procedures to support the algorithm in increasing detail; each required process block is further decomposed into appropriate structures.
- Bottom-up development: other users prefer to implement procedure details first, and then combine these into a coherent program that supports the algorithm.

(Some programmers use a combination of top-down and bottom-up approaches, detailing code that supports more delicate aspects of the problem early in the development cycle in order to establish feasibility.)

An example using pseudocode. Various animal behaviours, including alcohol consumption, are to be studied, perhaps to gain insights into addictive behaviour [10]. (The pseudocode to be described will support any number of animal studies (e.g. obesity research).) Of interest are such things as: how much exercise (spontaneous locomotive behaviour) do these animals get; what is the nature of their caloric intake (e.g. calories from alcohol or from solid food); how much water is ingested. The instrumentation set-up is to include resources for a large population of animals (e.g. 128 stations). Each station includes a solid food feeder, several tubes filled with water and/or water/alcohol mixtures, an exercise wheel that can be freely accessed by the animal—for purposes of this discussion mice are assumed to be the animals under investigation. Each station is equipped with transducers that convert licking behaviour into electrical pulses that can be detected by the computer—see chapter 6 for a discussion of transducers and chapter 3 for functional hardware for collecting the data.

Experimental parameters include the following: the number of animals in the given experiment; how often each station is to be sampled; the starting time of the experiment; the stopping time of the experiment, possibly some indication of experimental groups that represent alternatives (hypotheses) being studied.

The pseudocode is shown in figure 4.10. (All programming languages have some method for including comments within the code; in the case of pseudocode, any textual material preceded by an asterisk (*) will signal the start of a *comment*—the comment is considered to end at the next carriage return. The *end* statement is equivalent to the terminal element in a flow diagram.) The pseudocode is organized using the top-down programming style. The pseudocode is virtually self-documenting but additional comments and explanations for each module follow:

- *Main program*: the programming problem subsumes three tasks: the user must enter the parameters associated with the experiment (call to procedure

```
* Start of the main program
      CALL parameter_input;
      CALL data_acquisition;
      CALL report_generator;
END main;


* PROCEDURE parameter_input
      REPEAT
            prompt_user_for_parameter;
            read_value;
            check_for_errors; *wrong number
      UNTIL (all_parameters_read);


* PROCEDURE data_acquisition
      WHILE (current_time <> stop_time and current_time > start_time)
            IF (time_to_sample)
                  THEN
                        DO (station_1, last_station, increment_1)
                              CALL collect_counts;
                        ENDDO;
            ENDIF;
      END WHILE;
END data_acquisition;


*PROCEDURE report_generator
      print header; *experiment name, data, parameter values
      DO (station_1, last_station, increment_1)
            print station_number;
            print group_type;
            DO (start_time, stop_time, sample_time)
                  compute calories_of_solid_food;
                  compute water_intake;
                  compute alcohol_intake;
                  compute exercise;
                  print calories_of_solid_food;
                  print water_intake;
                  print alcohol_intake;
                  print exercise;
            ENDDO;
      ENDDO;
END report_generator;

* Physical end;
```

Figure 4.10 Pseudocode for a behavioural study.

parameter_input); continued collection of data from the stations during the experimental trial (call to procedure data_acquisition); generate a printed tabulation ('hard copy') of the results (call to procedure report_generator).

- *Procedure* parameter_input: logically the module calls for the user to enter repeatedly the parameters of the experiment. (The pseudocode might even specify all the parameters to be entered but such detail would overly complicate the example.) Entering parametric data requires the computer to prompt the user (e.g. 'Please enter the number of stations'); the computer reads the value that is entered; the computer checks the validity of the value. This last process block is included to make a point: a good program will often include checks on data that are entered from the keyboard. For example, suppose the datum to be entered is a real number and the user enters alpha-numeric data by mistake, the computer should recognize this error and prompt the user for an appropriate value.

- *Procedure* `data_acquisition`: the computer continues to collect data from all the stations (DO (`station_1, last_station, increment_1`)) from the start of the experiment until the end of experiment at times specified by `time_to_sample`. The symbols within the WHILE loop test condition (<> and >) stand for 'not equal to' and 'greater than' respectively. The procedure, call `collect_counts`, is included as a functional (logical) requirement, but corresponding code has been omitted for simplicity.
- *Procedure* `report_generator`: the final procedure supports the tabulation of results and provides a printed or hard copy of the data that have been collected. A heading is printed summarizing general information (experimental name, date, parameter values, etc). For each station the data that have been collected are first converted into appropriate units (e.g. 'licks on the solid-food dispenser' is converted to 'calories'—the caloric properties of the food as well as the sample taken with each lick being known, of course). This is followed by a print statement for each of the collected variables.

Many programming details have been omitted; however, the pseudocode provides the programmer with a logical framework in which to specify syntactical statements within the programming language. It is an important aid for anyone who must understand the program details and for anyone who is called upon to modify the program at a later date.

4.2.1 High-level languages (HLLs)

The basic, and primitive, control and process structures described above can be formalized within a set of English-like statements that comprise a high-level language (HLL). (Each statement ultimately corresponds to several machine code instructions and HLLs make programming considerably easier.) By some estimates there are thousands of such languages when various versions of a given language are included; some of these languages are not widely used. All of them have fundamental goals:

- Determine which data are involved in the calculations.
- Allow a user to specify what operations are to be carried out.
- Allow a user to structure the order in which the operations are to be performed.

Generic facilities for achieving these purposes were described above. At this point we shall briefly examine one such language, namely *Pascal*. (This is not intended as an exhaustive treatment of the rules of Pascal; rather its purpose is to relate some of the organization of a program written in Pascal to the generic structures previously discussed.)

THE NATURE OF PROGRAMMING   127

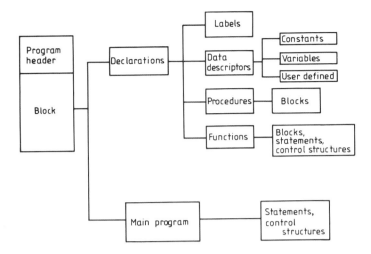

Figure 4.11 Organization of a Pascal program.

Examination of a program written in Pascal will reveal a number of well defined sections as shown in figure 4.11. This figure shows how such a program is decomposed into its constituent parts (e.g. blocks consist (in part) of declarations, which themselves may include labels (declarations), data descriptors, procedures, and functions, which are each further divided). Note particularly that within the region set aside for declarations one is likely to find the specifications for procedures (as well as entities called functions) that will be required and these in turn will include blocks; thus the organization is 'recursive'—inherently repetitive.

Within a program, names (*identifiers*) are assigned to quantities of interest. These quantities may include integers, real numbers, characters, and data of our own definition (e.g. we can define days_of_the_week to be quantities of interest that take the 'value' {monday, tuesday, ..., sunday}). Pascal is one of the HLLs that requires the programmer to associate the name of the quantity of interest with an appropriate type of data; it is considered to be a strongly typed language. That is one of the main purposes of the declaration section of the program.

Identifiers within a Pascal program may be considered to be constants. For example, the name 'pi' is a good substitute for the number whose value is 3.141 59...; it makes programs easier to read and follow (logically). Some identifiers within a program are obviously going to change during the course of execution; such entities are designated variables and will be identified by the keyword VAR in Pascal programs. (A keyword is an identifier that is set aside for use by the compiler; it should not be used by a programmer to define constants or variables.) Other aspects of the language will become apparent on examining a simple program; this is shown in figure 4.12.

128 SOFTWARE FOR INSTRUMENT SYSTEMS

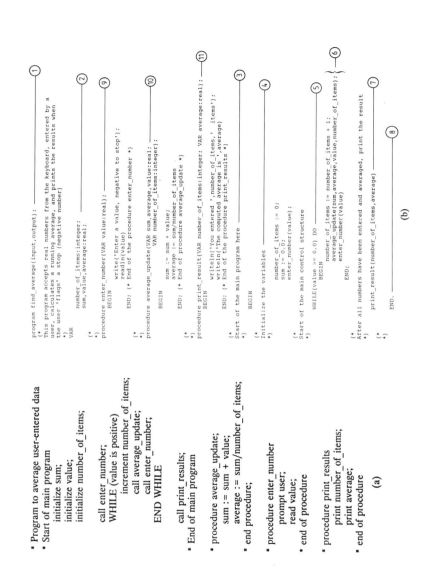

Figure 4.12 A Pascal program to average a series of user-entered real numbers. (a) Pseudocode representation. (b) The Pascal source code.

The outline of the program can be followed from the pseudocode description (figure 4.12(a)). A simple arithmetic average is calculated by adding the numbers to be averaged and dividing by the number of items that are being averaged. The number of items (to be averaged), as well as the sum of such numbers, must be set to 0 initially. (The computer should be directed to establish initial values for quantities of interest and not left to make its own choices (defaults); otherwise unexpected—usually 'unpleasant'—results follow.)

A WHILE control primitive is used to direct the operations: the computer calls the procedure that will permit the user to enter one of the numbers to be averaged. If the value that has been entered is positive, it may be factored into the average. If the value is positive—zero being treated as a positive number—the procedure average_update is invoked after the number of items (to be averaged) has been incremented (by one). (Prior to the start of the WHILE primitive, the enter_number procedure is invoked once to allow the user to enter the first number to be averaged.)

The WHILE primitive is terminated if the value that has been entered is negative and program control is transferred to the statement following END WHILE. At this point the results are tabulated for the user—print_results is invoked, and when it concludes the program terminates. Using a negative number to terminate the process is a 'trick' that programmers use to *flag* a termination event. (If negative numbers are to be included in the average, then some value that is to be excluded—'impossible to occur'—is chosen as the flag.)

The procedure average_update calculates a running average by incrementing the sum and dividing the new value of the sum by the number of items. The procedure print_results reports the number of items that have been averaged together with the final average value.

The actual Pascal program is included in figure 4.12(b); the numbers shown in the figure are not part of the program; they are references to the remarks shown below. (In Pascal, all textual material that follows the symbols '(*' and up to and including the symbols '*)' is considered to constitute a comment, and is ignored by the computer. In addition to important annotations, judicious placement of such symbols can be used to create space between segments of the program—'white space'—which makes the program more legible.)

1. This is the program header; it includes the name of the program (find_average) as well as certain resources that the program will require. In this case the program will need to communicate with the standard *input* device (the keyboard) and the standard *output* device (the video monitor). If the program requires additional files (from the secondary-storage system), those facilities will also appear in the header statement.

2. The statements following the keyword VAR are part of the program's declarations. One of the purposes served by statements within the declarations segment is to provide information about various data to be encountered in the program; this is descriptive material rather than processing directives (instructions). Within the program the computer can expect to find identifiers—

variables whose values will change from time to time—whose names are sum, value, and average. These are to be treated as real numbers; should they be used in an inappropriate manner within a statement—syntactically incorrect—the computer will detect the error, thus avoiding incorrect results before the program is ever executed. In a similar manner, the identifier number_of_items is an integer quantity.

3. A comment signals the start of the main portion of the program towards the end of the figure. (Procedures used within the program are defined in the declaration segment of a Pascal program; they must appear before they are invoked. The procedures will be described below.) All statements bracketed by the keywords BEGIN and END form a block, in this case bracketing the main portion of the program.

4. The first portion of the block defines initial values for some of the key variables in the program. For example, the first instruction is

```
number_of_items := 0;
```

This is a most important Pascal statement; it is referred to as an *assignment statement*. Expressions on the right-hand side of an assignment statement are evaluated and the result is 'assigned' to the variable (identifier) on the left-hand side of the symbol ':='. (Professor Niklaus Wirth who introduced Pascal in 1971 deliberately chose ':=' as the assignment operator; older languages had used '=' for assignment, and to avoid confusion with arithmetic equality he used another symbol. Assignment is not arithmetic identity, else a programming statement such as (item = item + 1) could be used to 'prove' that 0 = 1!). The symbol (;) signals the terminator of the instruction—the *delimiter*. All instructions must end with this delimiter except the one preceding an END statement. A second assignment statement initializes the value of the variable *sum*. Next the procedure enter_number is invoked. This allows the user to enter the first number to be averaged. Calls to a procedure should include all data that are to be transferred—to the procedure as well as 'returned' from the procedure; these are found within the parentheses. In the present instance the variable *value* representing the real number to be included in the average (and supplied by the user) will be returned from the procedure.

5. This points to the start of the principal control sequence within the main segment of the program. In this case a WHILE primitive is used—see the discussions in section 4.1. In Pascal, the syntax for this software architecture includes the keywords WHILE which precedes the condition to flag termination as well as DO which defines the start of the loop itself. A block (BEGIN...END;) delineates all statements included within the loop. The loop will be repeated as long as the variable *value* is greater than, or equal to 0.0 (value >= 0.0); notice that a real number (0.0) is used in this case because *value* has been declared to be a real quantity and 0.0 is a constant that is consistent with such designation.

6. As part of the main segment of the program, the number of items is incremented by one and the procedure average_update in invoked to calculate

the latest value of the average. (There are other logical possibilities for this program. For example, all the data can be stored within memory in a structure called an *array* and the average computed only after all numbers have been entered. One 'advantage' in the present design is the reduced need for memory.)

7. The user signals the end of the list of numbers by entering a negative value; this terminates the WHILE loop and transfers control to the statement identified as 7 within the figure. This statement invokes the procedure that will tabulate the results. In order to display the data the procedure requires the information to be reproduced. This is included as `number_of_items` and `average` within the parentheses.

8. The last physical statement in a Pascal program must be END terminated by a period (.).

9. Two new Pascal statements are exemplified in the procedure `enter_number`. They are both intimately involved in user interaction. The first, `write('Enter ... stop');`, will reproduce the message included between the apostrophe marks on the monitor. (Had the keyword `writeln` used in place of `write`, the computer would have supplied a line feed, carriage return sequence to the monitor; if additional output statements were included, they would thus appear on the next line.) The second Pascal statement is '`readln(value)`'; this allows the user to enter the number to be averaged and have it assigned to the variable whose name is value. Errors made while typing ('typos') may be corrected by using the *Back space* key prior to validation; the Enter key generates a carriage return, line feed signal that informs the computer that a value has been entered.

10. The procedure `average_update` is a straightforward calculation of the 'running' average of the numbers entered to that point in the program.

11. Procedure `print_result` displays the values of the variables `number_of_items` and `average`. These results will only appear on the monitor. Program modifications would be required if the user wanted a printed or hard copy of the data. One possibility is to direct the results to a file in secondary storage, and when control is returned to the OS upon the termination of the program, obtain a hard copy using OS commands.

Only a small subset of the repertoire of Pascal facilities, statements, and structures has been included in this example. A more extensive treatment of this popular HLL can be found in many other texts [11]. A short list of Pascal statement types includes: Assignment, Procedure call, If–Then, Case (transfer to one of several branches), While, Repeat-until, for, with, and goto. Syntactical details as well as functional descriptions of these can be found in the aforementioned reference.

4.2.2 The program translation operation (interpretation, compilation)

With the exception of machine code, all other languages employ linguistic and syntactical rules to specify the instructions to be carried out by the computer.

The resulting statements must therefore be translated into machine code before they can be executed properly. The translators, programs themselves, must reside within the computer, at least until the translation is complete, and in some cases while the program is being executed. The original program is referred to as the *source* (code) no matter what the translation scheme. The source code is created using an editor (program) similar to one that was discussed above.

The two most common forms of translation are *interpretation* and *compilation*; the corresponding programs are referred to as *interpreters* and *compilers*. They are described as follows:

Interpreter: in this approach each statement of the source code is translated into machine-executable form as it is encountered in the program, one line at a time. This happens each time the program is to be executed.

Compiler: a compiler accepts source code and converts this into a machine language form before it is actually executed. Initially, the compiler may generate an intermediate form of machine code known as an object program. Some code within the object program may refer to programs that are located elsewhere in secondary storage—in particular to certain standard procedures (or routines), which are found in a *library* of such utilities. The second stage of the translation occurs when unresolved references to subroutines or library routines in the object code are satisfied. This step is carried out by another program called a *linker* or *linkage editor*. There are thus two steps that must be completed without error before machine code is ready for execution: *compile* and *link*. The output of the linker program is machine-executable code, which can be *loaded* into an appropriate (main-) memory location and run (executed).

Compiled languages have an ultimate speed advantage over interpretive ones. The reason for this follows from the fact that translation takes place once—assuming that there are no errors—before the program is ever executed. Thereafter, whenever the program is to be executed, the machine code is ready to run and further translation is unnecessary. In contrast, when using an interpretive language, translation of each line in the source takes time; the code cannot be executed until this process is complete. The translation time thus reduces the efficiency and speed at which the program executes. Each time that the program is to be run, the translation must be repeated.

Based on this assessment, the reader might question why anyone would choose an interpretive language to develop a program; there are several potential answers. When speed of operation is not a critical factor, some programmers choose an interpretive language particularly if it has an 'undemanding' syntax and structure; 'non-elegant' programs may then be developed rather quickly. The interpretive languages often have a variety of features that provide them with considerable convenience during program development. For example, the source code can be modified interactively; errors can be corrected 'on the spot' without the need to run a separate editor program as would be the case for a compiled language. In addition, interpretive programs can be executed one step at a time (*single-step* operation); between steps, results can be checked for validity. (In compiled languages temporary print statements may be included to check intermediate results and these statements removed when the program has been completed.) Some programmers find such features convenient for *debugging* purposes—finding and eliminating errors—and may then choose to use interpretive languages. In addition, such languages may be useful for testing the feasibility of an algorithm; once the algorithm has been refined it may be rewritten using a compiled version of the same, or another, language to regain the speed advantage.

4.2.3 Assembly language programming

Early in the history of computer development researchers saw the need to replace the cumbersome, tedious, and error-prone process of having to enter binary numbers into the computer in order to execute a program. One of the first software developments was to replace the binary numbers with mnemonic equivalents and then to have the computer itself convert such symbols into binary numbers. (Of course a program for such translation had to be written in machine code.) The language that replaced machine code was *assembly language* (AL). It is often referred to as a low level language to distinguish it from the HLLs described above [12].

An example will help to explain some differences between ALs and HLLs. A simple (Pascal) assignment statement might appear as

```
c := a + b
```

The AL equivalents of this statement might look like the following:

```
lda    first
adda   second
sta    result
```

The first of these statements *moves* data from a memory location named 'first' (corresponding to the location of the variable 'a' in the HLL fragment) into the CPU's accumulator; the next instruction *adds* the contents of the memory location designated as 'second' (the memory location retaining the value of the variable 'b' in the HLL statement) to the same accumulator; finally, the contents of the CPU's accumulator are transferred ('stored') in the location named 'result' (the location corresponding to 'c' in the Pascal fragment). The AL fragment shown above depends heavily on the particular processor that is being used. (In particular, a single AL statement might be possible for a 32-bit machine. In addition, the mnemonics vary across assembler programs.) The AL fragment is not as abstract as the HLL equivalent; consequently, following the logic of an AL program is considerably more difficult. However, when speed of program execution is essential, an AL format is necessary. (AL code may be used in some time-critical sections of a HLL program and HLLs often include facilities for incorporating AL code.)

AL statements have a variety of formats; while there are no universally accepted standards, many gravitate toward the following organization of fields within a source program (within the software community a field is a specified set of bit locations in a computer word, used for a particular category of data):

| LABEL | OPCODE | OPERAND | COMMENT |
| FIELD | FIELD | FIELD | FIELD |

LABEL FIELD: an optional field; it might represent a memory location where data are to be stored or a point in the program where control might be transferred.

OPCODE FIELD: a line of code must include an OPCODE field; it specifies the instruction (operation(al) code) to be executed.

OPERAND FIELD: the contents of this field are determined in part by the OPCODE. There may be no entry, or two mnemonics separated by commas or other operational symbols (e.g. +). Such symbols eventually are converted to the data that are the 'targets' of the OPCODE.

COMMENT FIELD: this is ignored by the computer (the assembler program that translates the AL source code) but is essential for good documentation practice. (Not every line of code should be followed by a comment, but important logical elements should be identified as such.)

Some code within an AL program is 'informative'; these are *assembler directives* and specify information about the assembly process. Some directives include:

Origin: specifies at which memory location to start assembling (e.g. ORG 1000H means start the assembly at memory location 1000 (hexadecimal)).

Equate: Allows the programmer to use a mnemonic in place of a constant (e.g. count EQU 25H).

Set aside (reserve) a specified number of bytes within the program (perhaps to store a series of data).

Set aside (reserve) a specified number of words—the same as the above, but sets aside word locations rather than byte locations.

In a manner similar to that for HLLs, the sequence for successfully developing an AL program starts with generation of source code using one of the many editors that are available. The resulting text (source program) is the 'input' to the assembler (translator) program, which produces an *object* (code) module. Basically, object code is the machine language equivalent needed to execute the program on a particular processor. For each line of code an address, and a completely specified operational code (opcode) including identification of the data to be used or manipulated, must be specified. When this is the case, the code is said to be completely *resolved*. On occasion, such code cannot be completely resolved because there may be a reference in the code to some other section of code that has not yet been encountered (e.g. a reference to a memory location

Table 4.2 Summary of assembly language operations

Type	Mnemonic	Meaning
Data transfer	MOV	Move data between parts of the computer.
	PUSH, POP	Move between the CPU and a special portion of the memory called the *stack*.
	IN, OUT	Move (transmit) data between the CPU and the external environment.
Arithmetic	ADD, SUB	Add or subtract.
	INC, DEC	Add or subtract one to/from data.
	NEG	Negate the data.
	MUL, DIV	Multiply or divide.
Bit manipulation	AND, OR, NOT	Carry out the logical operations of intersection (AND), union (OR), or inversion (NOT).
	SHL, SHR	Shift the data 1-bit to the left or right.
Program transfer	CALL	Execute a predefined set of instructions (subroutine or procedure call).
	RETURN	Transfer control to the instruction following the last call instruction.
	JE, JG, JL	Transfer control of the program to a new location if comparison of binary quantities produces a true result: JG—jump if first operand is greater than second operand; JL—jump if first operand is less than second operand; JE—jump if two operands are equal.

after the location in question). Such *forward* references may be resolved once the source code has passed through the assembler once; a second pass through the assembler is necessary and the assemblers that work this way are called *two-pass* assemblers.

Once the object code has been completed the program is ready to be *loaded* into the computer and executed. Up to this point the operations (editing, assembling) can be performed on *any* computer (on which the assembler is installed), not necessarily on the processor for which the AL program is destined. The final code may now be loaded into the machine for which it is intended—the *target* machine. *Crossassembly* is the name that describes this case. The program for achieving the placement of the code within the computer is called the *loader*; it must reside on the target computer.

The target machine usually includes a *debugger* program that helps to resolve logical programming errors and/or verify results. A debugger includes facilities such as: executing the program one instruction at a time; examining registers (in the CPU) as well as locations in main memory; installing temporary breaks (*break points*) to permit examination of intermediate results.

Assembly language instructions (opcodes) themselves fall into four major

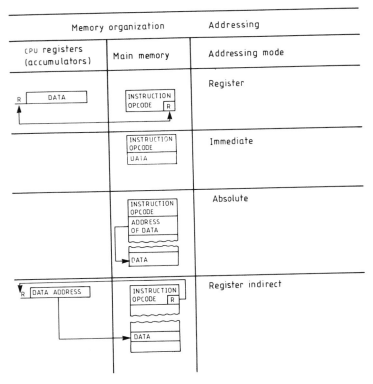

Figure 4.13 Common assembly language addressing modes.

classifications, including those that: transfer data (moving data); perform arithmetic; manipulate bits within a memory location; and transfer control of the program. A sampling of such instructions together with representative mnemonics is shown in table 4.2.

In addition to the OPCODE, the 'target' data—the data to be used in the operation—must be specified. These data may be defined in a number of ways characterized as the *addressing mode* or 'mode of address'. Addressing modes are intimately related to the type of processor being used. For example, figure 4.13 shows a representative compilation of addressing modes; it is not exhaustive but includes generic forms from a variety of processors, including corresponding organization of the computer's memory and registers within the CPU. Each is now briefly described:

Register: an instruction within the main memory (where the program being executed is located) includes as part of its format a reference to one of the general-purpose registers within the CPU. That register contains the target data—the operand.

Immediate: in this addressing mode the operand is included within the

```
pseudocode
Initialize: output data, delay time counter,
port address;
Enable the port bit (send 1 to port);
while (delay time counter < > 0)
     decrement delay time counter;
end while;
Reset port bits (send 0 to port);
end of pseudocode;
```

```
         ;start of assembly language code segment
         mov  cx,n      ;store a number "n" in CPU register cx
         mov  dx,port   ;store address of port
                        ;in CPU register dx
         mov  al, 1     ;store a logical 1 in CPU register al
         out  dx,al     ;turn on the bit by moving contents
                        ;of CPU register al (1) to the address
                        ;specified in CPU register dx (the
                        ;addressof the port.
loop:    dec  cx        ;decrement the delay time counter,
                        ;count down.
         jnz  loop      ;wait until delay counter is 0;
         dec  al        ;subtract 1 from CPU register al,
                        ;making it 0.
         out  dx,al     ;reset the port bit by sending a 0
                        ;(contents of CPU register al) to the
                        ;address of port (contents of dx).
         ;end of assembly language segment;
```

Figure 4.14 Pseudocode and assembly language segments for sending a control signal to a peripheral.

program itself; it follows directly after the instruction (OPCODE) in main memory.

Absolute: the address of the target data follows directly after the OPCODE itself within the program section of main memory. Some processors only provide for one byte after the OPCODE limiting the address to a byte (eight bits) and consequently the data must be within 256 locations of the instruction itself. Newer processors include provision for 16-, 32- and even 64-bit (multiword) data addresses.

Register indirect: part of the instruction OPCODE includes a reference to one of the general-purpose registers within the CPU. This general purpose register (R) includes an address; the operand (target datum) is found at that address.

The code fragment shown in figure 4.14 shows how assembly language source code might appear in a program. This segment turns on, or enables, one of the computer's output (port) bits for a specified period of time after which it is disabled or reset. (In this way it can be used as a control for some peripheral device such as a pump, stimulator, electronic instrument (e.g. oscilloscope) or other external device.) Included in the figure are comments to explain the purpose of each instruction as well as a pseudocode equivalent. Registers within the CPU retain important data associated with the code; locations within main memory can also be used but addressing modes are more complex and the program executes more slowly:

To retain important data associated with the code CPU register 'cx' is initialized with a number 'n', which has to be determined by the programmer; its value will determine the length of time that the output port (bit) will remain enabled (at logical 1). A key portion of the code decrements the number stored in this register. The computer requires a small but finite time to decrement the register each time that it is executed ('dec cx'). This time multiplied by the number stored within the register determines the length of the time delay.

The CPU register 'dx' contains the address of the port to be accessed; whenever the instruction 'out dx,al' is executed, the contents of the CPU accumulator 'al' are moved to the port whose address is specified by the contents of 'dx'. The first time that this happens, 'al' will contain a logical '1'; this enables the least significant bit of the port because the binary equivalent of a '1' is 0000001 for the byte stored in 'al'. The next time that the same instruction is executed the value in 'al' will be 00000000. As a consequence of these two instructions and the 'delay loop' introduced between their execution, the least significant bit of the output port will:

1. Take on a logical value of 1.
2. Remain at that value for a specified period of time.
3. Return to a logical value of 0.

4.3 APPLICATIONS PROGRAMMING LANGUAGES AND PACKAGES

While assembly language programs are important in many applications, especially where speed of execution is critical, the majority of users are concerned with carrying out functional operations of an abstract ('high-level') nature. In this regard they are interested in applications programs such as 'spreadsheets' (discussed in chapter 7), or commercially available data-acquisition programs where only parameters (e.g. 'sampling rate') need to be supplied; the computer already contains all the necessary instructions to receive, convert, and store the data and the program details are considered to be 'invisible' or 'transparent'. Data-acquisition and data-processing software packages are extensively covered in subsequent chapters.

One of the more significant problems faced by a user, particularly when dealing with a computer-based instrument with very large capacities, is the orderly maintenance of information. The goal of such *data-bases* is to maintain a non-redundant collection of properly interrelated data items that is available to one or more users. A very popular form of such packages is the *relational database* [13]. Examples of commercially available data-bases include dBASE III Plus^{TM}, dBASE IV^{TM}, dBXL^{TM}, Clipper, FoxBase^R Plus, FoxPro^R, and Quicksilver^{TM} [14].

The relational data-base model is based on the concept of a relation whose physical representation is a *table*—the name 'relational' follows from the mathematical notion of a relation that can be expressed in the form of a table.

APPLICATIONS PROGRAMMING LANGUAGES AND PACKAGES

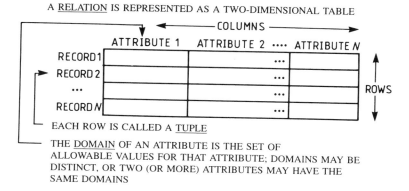

Figure 4.15 The table is a fundamental element of a relational data-base.

A relation is represented as a two-dimensional table as shown in figure 4.15. Several characteristics of the table should be noted:

- Each cell in the table contains only one value.
- Each column has a distinctive name—the name of the attribute (e.g. 'data rate').
- Values in a column all come from the same domain—they are all values of the corresponding attribute.
- The order of the columns is not important.
- Each *tuple* (row) is distinct—no duplicates.
- The tuple order is not important.

User interaction with a data-base may be accomplished in two ways: the first involves a set of commands that permits the user to create tables, enter information, and retrieve information; the second is incorporating the commands into a customized application, often menu driven, and eliminating the need for remembering commands (at the expense of total flexibility). The various kinds of *data manipulation languages* (DMLs) are depicted in figure 4.16. An SQL (structured query language) that permits a user to create a table—that is, to define its structure in terms of records and attributes—is given below:

```
create table exp
    (exp#      char(5)         not null,
     type      varchar(20)     not null,
     drate     integer         not null,
     nchans    integer         not null,
     fscale    decimal(5,1)    not null,
     primary key (exp#) );
```

In this example, a table, named 'exp' (a mnemonic for 'experiment') has been created; it consists of the following attributes for each tuple:

140 SOFTWARE FOR INSTRUMENT SYSTEMS

RELATIONAL DATA MANIPULATION LANGUAGES (DML)

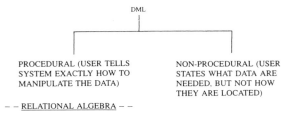

OTHER TYPES OF DML (NOT QUALIFIED AS PROCEDURAL OR NON-PROCEDURAL)

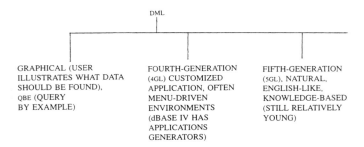

Figure 4.16 Classifications of relational data manipulation languages (DML).

- exp#—experimental number; this is a unique identifier, which will be assigned to each experiment to be included in the table.
- type—the type of experiment to be recorded; 'varchar' notes the fact that this cell will contain a varying string of up to 20 characters.
- drate—will retain the data rate that was used in the experiment and will be represented as an integer quantity.
- nchans—another integer quantity indicating the number of channels that were recorded.
- fscale—the full-scale setting of the data-acquisition element; decimal(5,1) indicates that there will be five digits (and sign), with an assumed decimal point one digit from the right-hand end of the number.
- primary key—this refers to the fact that the attribute named 'exp#' will be a unique identifier, having the following properties: no two tuples have the same attribute (value); the key is a 'minimal' identifier—if it is not included in the attributes identifying a tuple then tuples will no longer be uniquely determined.

APPLICATIONS PROGRAMMING LANGUAGES AND PACKAGES 141

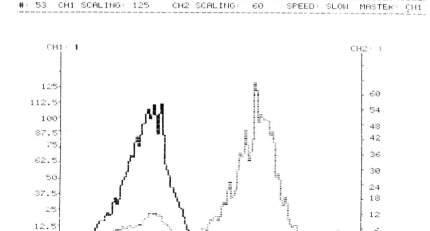

Figure 4.17 The integrated EMG in the time-domain obtained from a flexion–extension movement in abducted arm from a normal subject.

The table could be updated interactively or might be part of an applications program that also permits the user to make entries and changes, summarize

Figure 4.18 Tabulated data for figure 4.17.

data, 'filter' data, or generate reports. Data-bases find utility in a wide variety of environments; one example is described below.

An EMG example. Human cells are the basic source of all bioelectric potentials. These cells consist of an ionic conductor separated from the surrounding environment by a semipermeable or selectively permeable cell membrane [15]. This arrangement results in a difference in potential across the cell wall or membrane. Bioelectricity is studied from within the cell as well as remotely (skin surface) via the electrolytic current flow from a cell or group of cells. In electrophysiology it is common to penetrate the cell in order to study its internal potential, particularly that in response to some electrical or mechanical stimulation. In addition, we also make measurements external to a group of cells while these cells are supplying electrolytic current flow. Typically, these signals are detected at a site relatively remote from the underlying cells.

Electrolytic current flow within the fibres of a human muscle can be generated by an appropriate electrical stimulation or as a result of some voluntary behaviour. The electric field produced by the flow of current can be detected by suitable electrodes. The electric potential can be picked up either by a needle electrode (inserted into the muscle) or by electrodes placed on the surface of the subject or patient over the muscle being studied. These signals are then amplified (and preprocessed by filtering) and displayed or stored in a variety of ways by a computer-based instrument.

Electromyographic (EMG) signals recorded from appropriate sites on the skin reflect underlying muscle activity. They can be used to measure and quantify the extent of deficit as a result of insult to brain [16]. Such recordings also have the potential for use in rehabilitation [17]. The basic electrical activity detected in any muscle is the single-motor-unit action potential, which generally appears in a recording as a series of pulses, with repetition rate ('density') being proportional to muscle activity.

A better representation of behaviour may be achieved by rectifying and integrating these signals, producing a signal that reflects a subject's 'effort'. The resulting unit of recording is the *microvolt second*—the unprocessed signal has units of μV of activity and finding the area under the curve (integration with respect to time in seconds) results in units of $\mu V\,s$. A typical recording from the forearm (agonist and antagonist) of a normal individual during flexion–extension is shown in figure 4.17. (Figure 4.18 is a tabulation of these same data. For comparison's sake an abnormal response is shown in figure 4.19; notice the presence of serious 'co-contraction'—resulting in spastic, uncoordinated movement—towards the end of the record.) A data-base of such records was deemed necessary for clinical as well as experimental documentation. The organization of the data-base to retain such information is shown in figure 4.20. Three interrelated tables form the basis of the system:

Master table: attributes include the patient's identification (ID—Social Security number or similar identification number when the SSN is not appropriate), the name, participant's age, sex, date of the insult or deficit

APPLICATIONS PROGRAMMING LANGUAGES AND PACKAGES

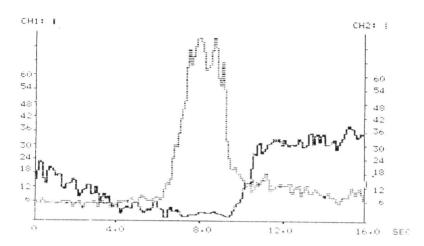

Figure 4.19 The integrated EMG in the time domain obtained from flexion–extension in an abducted arm from a patient with a right hemiplegia.

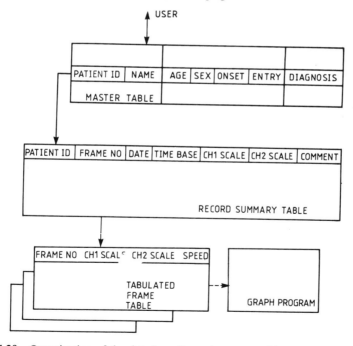

Figure 4.20 Organization of the data-base for maintenance of integrated EMG records.

('onset'), date of entry into the treatment programme (or experiment), 'entry', and diagnosis. The SSN is considered to be the primary key (as two different patients could have the same name).

Record summary table: contains a summary of all records within the database. Each tuple includes the patient ID, a frame number that identifies the particular record on a given date (the same frame number may occur on different dates), the date on which the record was taken, the time base, scale settings for the two channels, and a comment field.

Tabulated frame table: each frame occupies a separate table with a tabulation similar to the one shown in figure 4.18.

The entire data-base program is 'menu driven'; by using directional keys ('arrow keys') a user may scroll through records, and select a given record (or group of records) for review, or to obtain a printed report.

REFERENCES

[1] Parker S P (ed) 1989 *McGraw-Hill Dictionary of Scientific and Technical Terms* 4th edn (New York, NY: McGraw-Hill)
[2] Zarrella J 1982 *Microprocessor Operating Systems* (Suisan City, CA) Microcomputer Applications
[3] Shaw A C 1974 *The Logical Design of Operating Systems* (Englewood Cliffs, NJ: Prentice-Hall)
[4] Clements A 1993 *Principles of Computer Hardware* 2nd edn (Boston, MA: PWS-Kent)
[5] Simrin S 1989 *The Waite Group's MS-DOS Bible* 3rd edn (New York, NY: Sams)
[6] Silverman G 1984 Design discipline for computer-based instruments *Proc. 6th Annual Conf. of the IEEE Engineering in Medicine and Biology Society (Los Angeles, CA: 1984)* (New York, NY: IEEE)
[7] Silverman G and Harrison M 1982 Microprocessors: a biomedical toolkit *Proc. 4th Annual Conf. of the IEEE Engineering in Medicine and Biology Society (Philadelphia, PA, 1982)* (New York, NY: IEEE)
[8] Weiskamp K and Aguiar S 1992 *Windows 3.1* (New York: Wiley)
[9] Schick W and Silverman G 1993 *Introduction to Engineering Computation Using Fortran 90* (New York, NY: Wiley)
[10] Dole V P, Ho A and Gentry R T 1994 An improved technique for monitoring the drinking behaviour of mice *Physiol. Behavior* **30** 971
[11] Silverman G and Turkiew D 1989 *Computers and Computer Languages* (New York, NY: McGraw-Hill)
[12] Wakerly J F 1981 *Microcomputer Architecture and Programming* (New York, NY: Wiley)
[13] Date C J 1992 *Introduction to Database Systems* 5th edn (New York, NY: Addison Wesley)
[14] Jones E 1990 *The dBASE Language Reference* (Osborne, FL: McGraw-Hill)
[15] Strong P 1971 *Biophysical Measurements* (Beaverton, OR: Tektronix)
[16] Song G 1987 The analysis of the integrated electromyogram using the discrete Fourier transform *Master's Thesis*, Fairleigh Dickinson University, Teaneck, NJ
[17] Brudny J, Korein J, Grynbaum B B, Belandres P and Gianutsos J 1979 Helping hemiparetics to help themselves; sensory feedback therapy *J. Am. Med. Assoc.* **241** 814

5
Data-processing Considerations

5.1 COMPUTER-BASED INSTRUMENT CAPACITIES

The performance of a computer-based instrument system should be matched, as closely as possible, to the broad range of its environmental requirements [1]. For example, a 'general-purpose' laboratory computer may be required for: performing numeric computations (e.g. fluid flows); performing simulations; real-time applications; large-memory, input/output problems (data-bases); artificial intelligence. Some of the components of a computer designed to meet all these needs may not be used in each application. Consequently, the system may be highly inefficient in spite of its high performance. A series of specialized machines may be more cost effective than a single, high-performance system.

Many computer-based instrument system applications require *real-time* performance—that is, completion of assigned task(s) in a specified amount of time. The term 'real time' can be further qualified by a number of other descriptors including the operation, processing, programming and system (i.e. real-time programming). All refer to those instances in which the computer-based instrument that controls an ongoing process can deliver it outputs (or control its inputs) not later than the time when these are needed for effective control (e.g. chemical process control). In other words some time constraint is placed upon the computer, its hardware and software. A number of computer architectures have been cited to increase throughput in order to meet the increasing demands for real-time operation. Examples of increasing demands placed upon computers in real-time environments include speech recognition, image processing, and computer vision applications.

The instrument system hardware is normally limited in its capacity (or bandwidth). Capacity is further eroded by the *overhead* or additional time burdens and delays placed upon it by the software. One of the first parameters to be determined is the ultimate capacity of the computer. This is normally measured by the number of instructions that can be completed by the processor in a given time interval. One measure of this is defined by the number of MIPS that can be completed. The unit MIPS is defined by

Number of MIPS = (instructions per cycle)(cycles per second) $\times 10^{-6}$.

The first term (instructions per cycle) depends on the word size of the computer; the larger the word size, the greater a computer's ability to complete a task (e.g. get two numbers from memory, add them together, and store the answer in a third location) in one 'tick' of the computer's clock. The second term (cycles per second) is the (clock) speed at which the computer is running. This depends on the type of technology that was used in the manufacture of the computer (e.g. CMOS chips). A computer manufacturer normally runs the system's processor at the maximum clock speed that is compatible with the underlying technology; this helps to maximize performance (MIPS). (The rapid increase in the performance of the modern computer—the number of MIPS it can achieve—has forced some changes in the way that performance is measured. What has become important is the computer's ability to complete tasks—its throughput. Therefore, the time taken to complete a 'standard' task, a *benchmark* program, has become one of the specifications that is often noted in the literature. The computer's performance is a very sensitive function of the nature of the benchmark. At present, there are no absolute rules that can be used to determine whether or not one architecture is better than another.)

In view of the definition of the MIPS shown above, it is possible to conclude that a rough measure of processing power is the width of the address and data paths multiplied by the clock frequency at which the processor runs. Because of this, computer system buses have grown in width from 8-bit to 16-bit and to 32-bit. There are, however, other ways to maximize performance:

- **Change the number of instructions in a program**. If the program cannot achieve the requisite real-time operation, one may examine the algorithm; sometimes a simple change can greatly affect the throughput, sometimes a completely new approach is required (see section 5.4).
- **Replace software with hardware**. Advances in software often end up as specialized hardware that speeds overall operation. For example, originally computers had no hardware provision for such arithmetic operations as multiplication and division; such calculations were carried out with software routines. Once efficient multiplication and division software was refined, a specialized circuit to perform the necessary calculations became possible (the *co-processor*). Further advances now permit this circuit to be included within the CPU itself, further increasing performance (MIPS).

Yet another example of how hardware has improved the performance of computer-based instrument systems comes from an algorithm that performs a *fast Fourier transform* (FFT) [2]. This has greatly enhanced the computer's ability to process data *(digital signal processing* or DSP) in such applications as image processing, speech processing, and (digital) signal filtering. Single-chip DSPs emerged in the early to mid-1980s, originally as 16-bit devices programmed in assembly code (see chapter 4) and capable of operating only on integer data. Currently, 32-bit, floating-point (real-number) chips are available and can perform 25 to 50 million floating-point operations per second. This makes them suitable for intensive real-time data processing. Larger memory sizes allow programming applications in HLL, and direct memory access channels (see section 5.4.3) greatly facilitate input/output operations, one of the bottlenecks in computer-based instrument systems.

DSP chips differ somewhat from conventional microprocessors, although they bear some resemblance to the RISC architectures discussed in chapter 3. They are designed with multiple internal buses to take advantage of pipelining. As with RISC devices, DSP chips depend on register files (groups of registers). The number of registers is generally smaller than for RISC chips, to allow faster (context) switching in response to interrupts (see section 5.4.2). DSP hardware supports single-cycle operation of typical DSP algorithms such as convolution, correlation, and filter design. High-speed multiply-and-accumulate hardware enables many DSP chips to perform single-cycle floating-point multiplications. As an example, current DSP chips can perform a 1024-point FFT in about 1 or 2 ms. A typical family of DSPs is the Texas Instruments TMS320Cxx series with the following characteristics (320C30):

Number of bits: 32.
Type of operation: floating-point calculations, 32- and 40-bit formats.
Speed: 50 ns (10^{-9} s) cycle time at 40 MHz clock speed.
Size of accumulator: 40 bits.
On-chip memory: two, 4 kbyte RAMs; one, 16 kbyte ROM.
Other: a newer model (320C40) is compatible with parallel processing and includes six, 8-bit DMA channels each capable of 20 Mbyte s^{-1} data rates.

(A number of chips with competing characteristics are available: DSP96002 (Motorola), DSP32C and DSP3210 (AT&T), ADSP-21020 (Analog Devices), mPD77230 and mPD77240 (NEC).)

A further way of maximizing performance is now described:

- **Execute more than one instruction at a time**. An instrument processing task may be partitioned into a number of smaller problems. These may be executed on a number of 'less powerful' (and less costly) processors. This increases the first term in the definition of the number of MIPS; it increases the number of instructions that can be executed in a single cycle. (A programming challenge that ensues from parallel architectures is the potential for reduced efficiency; some of the processors may be greatly underutilized.)

Consider an example in which real-time constraints preclude complete processing of the experimental data. An alternative becomes one in which the data are recorded and subsequently processed at slower speed. Throughput may be enhanced by employing two processors as shown in figure 5.1. A functional picture is provided in figure 5.1(a) which includes two processors labelled 'PROC A' and 'PROC B', together with a DUAL-PORT MEMORY module and a system bus. Dual-port memory includes control features that permit access to a RAM location from two distinct sources via the system bus. The control logic resolves timing conflicts arising from concurrent memory requests. PROC A is dedicated

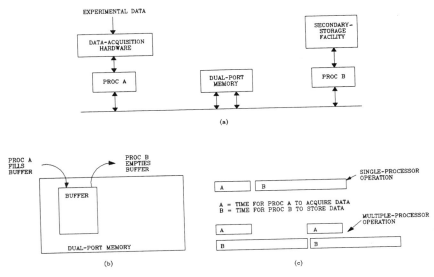

Figure 5.1 A multiple-processor architecture for enhancing instrument capacity. (a) Hardware organization. (b) A schematic diagram of the software. (c) Throughput enhancement.

to control and communication of the data-acquisition hardware which, in turn, receives the experimental data. PROC B provides control and communication with the SECONDARY-STORAGE FACILITY. Data may be stored here for off-line processing at a later time. Figure 5.1(b) shows how the processors interact with the DUAL-PORT MEMORY. Part of the memory space is set aside as a buffer which is filled by PROC A, with experimental data, and emptied by PROC B. (A circular-queue memory organization may be used as an alternative to the buffer.) Often, PROC B will not be able to empty the buffer as fast as PROC A stores the experimental data. However, normally there will be a gain in throughput as shown in figure 5.1(c). In a serial configuration, with a single proc ssor, data are acquired from the experiment ('A') and are subsequently stored in the secondary-storage system ('B'). (Notice that B is considered to require more time than A which is often the case.) With a multiple-processor architecture, data acquisition and storage may proceed in parallel at least part of the time, thus increasing overall system throughput.

5.2 ORGANIZING DATA (DATA STRUCTURES)

One task to be completed when developing a computer-based instrument is to give consideration to the organization of the experimental data as it relates to the system capacity, and the ultimate efficiency of the structure of the data in light of the application. The organization of data is directly related to the ways

ORGANIZING DATA (DATA STRUCTURES)

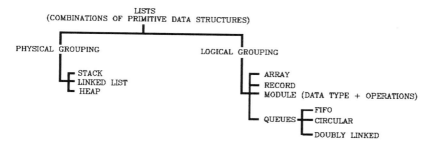

Figure 5.2 Data organization.

in which memory (both main memory and secondary memory) can support or retain such information. Figure 5.2 shows how data may be configured within a computer. Data organization may be viewed from a logical or a physical point of view. Logical organization refers to the functional representation—how we think of the data—while physical organization refers to the way in which such information is stored within the computer's memory [3].

All computers (when considered with an assembler, debugger, editor, compiler or other program) recognize some form of primitive data structures; these include real numbers, integers, and characters. Most computers have a variety of machine language instructions to manipulate these entities in a direct manner (e.g. convert an alpha-numeric (ASCII) character into an integer). There exist one or more memory locations within the computer for each of these primitive types; the exact number of locations and format differ across installations. These primitives may be combined into larger entities as shown in the figure. (More abstract arrangements such as those found in a data-base can use these structures as building blocks but that level of conception is not included in the figure.)

A list is an ordered set consisting of either a fixed or variable number of elements to which, with rare exceptions, additions and deletions can be made; in any case the values of the list elements can be changed. A list that displays the relationship of physical adjacency is called a *linear* list. Each element is a linear list assumed to contain one or more *fields*. A field can considered to be a primitive (string, integer, pointer (address)). Important operations that can be performed on linear lists include:

- Find the size or number of elements in a list.
- Search for a particular element containing a field having a certain value.
- Copy a list.
- Sort the elements of a list into ascending or descending order, depending on values of one or more fields within each element.
- Combine two or more lists to form a new one.
- Divide (split) a list into several sublists.

A pointer (sometimes called a link) is an address or reference to a data

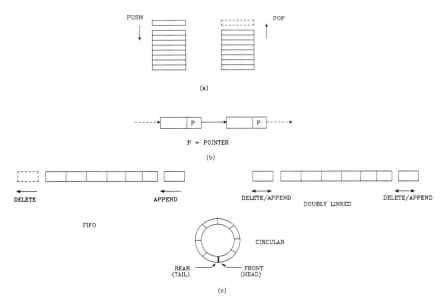

Figure 5.3 A schematic representation of data structures. (a) A stack. (b) A linked list. (c) Queues: FIFOs, circular, doubly linked.

structure. Pointers are fixed-size integers that permit the referencing of any data structure, regardless of its complexity, to be made in a uniform or consistent manner. There are two ways that can be used to obtain the address of an element in a data structure:

- Computed address method—calculate the address of the elements in the list. For efficiency reasons, it is desirable to organize data in the computer's memory so that a particular element of these data can be referred to by computing its address rather than performing a search.
- Link or pointer address method—store the address somewhere in the computer's memory. To access an element of a particular structure (pointer method) load the pointer value.

Some data structures use a combination of computed and link addresses.

Figure 5.3 depicts a schematic representation of all data structures described below.

- **Physical grouping**
- **Stack(s)**. This is a linear list of variable size. Recall that the most general form of linear list permits the insertion and deletion of an element at any position in the list. If we restrict occurrence of insertions and deletions to one end of a linear list, the subclass is called a stack. Often, the machine includes a *push* directive (assembly language instruction) to insert an element—place it on the stack—and a *pop* command to delete an element. The resulting arrangement

ORGANIZING DATA (DATA STRUCTURES) 151

is referred to as a last-in–first-out (LIFO) organization which describes the fact that elements can only be removed from the stack in the opposite order from that in which they were inserted into the stack. Such a structure is complementary with a vector whose size should be large enough to handle all insertions that can be made to the stack. Many HLLs allow explicit control, through special instructions, of when and how much storage is to be allocated for a certain stack—*controlled* storage. Additional instances (or occurrences) of a stack can be created by special statements. The programmer does not need to keep track of a separate stack index (element location) because only the most recently created element is available. Stack applications include sorting (not discussed) and problem solution, which uses *recursion*.

Example. Binary search

Suppose a series of datum points in a list is to be searched for a particular value. (The list could be the result of an experiment that represents results as a function of time and it is desired to find the point in time at which the trace exceeds a given value—the point at which a significant event has been detected.) The following recursive algorithm could be used for a search, and a stack structure is a convenient data structure:

1. If the current search interval is a single element whose value is not equal to the key—the value being sought—then the search has failed, else obtain the position of the midpoint element in the current search interval.
2. If the key value and midpoint element are equal, the search is successful.
3. If the key value is less than the midpoint element, repeat the search using the upper half of the current interval. (The fact that the search procedure initiates (or calls) itself in this repeated manner makes the algorithm recursive.)
4. If the condition is step 3 is not satisfied, repeat the binary search (recursive call) of the lower half of the current interval.

- **Linked lists**. Cases where storage requirements are unpredictable—there may be many items in the list or relatively few items in the list—and cases where stored data must be manipulated extensively are best handled by *linked lists*. While the addresses of items in a stack are arranged in a sequential manner, a linked list uses pointers or links to refer to elements of a linear list. The address of the successor of a particular element is stored as part of the item itself. The elements of the linear list need not be physically adjacent in memory and the resultant arrangement is called a linked list allocation. A pointer (which might be called FIRST) contains the address of the location of the first node (item) in the list. The last node in the list does not have a successor; its pointer field does not contain an actual address; instead it contains a flag that signals a NULL or empty address. If a list contains no elements, it is empty, and FIRST is assigned the value NULL to indicate this condition.

A common operation that presents problems for sequential lists is insertion and/or deletion of a datum. Lots of data must be moved in order to insert an item into the middle of the list. By contrast, insertion/deletion operations in linked lists are relatively easy; they both involve simply changes in pointer values:

- *Insertion*. Change the value of the pointer to that of the new datum at the point of insertion; assign to the pointer of the new datum (to be inserted) the value originally retained at the insertion point.
- *Deletion*. Replace the pointer of the datum preceding the deleted item with the value of the pointer of the deleted item.

Linked list management is important; the (software) manager should maintain a list of available nodes—list links. A node is the tuple consisting of (`pointer, datum_value`), and the list of nodes is updated as nodes are added or deleted. It may be maintained as a linked list itself—a list of free nodes. The advantage of memory management of this kind is that at any particular time, the only memory space that is actually used is what is really required. Management is also possible with a stack arrangement which keeps track of a stack of available nodes.

- **Heap**. A stored file of data with no additional access structure (e.g. array index) may be referred to as a *heap*. New records are inserted into a heap wherever there is room. For small files this structure may be very efficient [4]. Two storage algorithms may be used to find a place for the information, *best fit* or *first fit*. The names fairly reflect the nature of the approach. When using best fit, the secondary storage is searched to find the closest match between the amount of free space and the quantity of data to be stored.

For first fit the data are stored in the first available free space that will accommodate the data to be stored. Best fit may help to keep the secondary-storage system from becoming *fragmented* into small but widely separated regions of memory allocation. On average however, the first-fit method is faster. In this latter method the first available region that is large enough to accept the data will be allocated. Best fit and first fit are not exclusively associated with heap structures; they may be used with many other memory organizations.

- **Logical grouping**
- **Array**. A vector is one of the simplest data structures that makes use of computed addressing. Several contiguous memory locations are sequentially allocated to the vector. Each element occupies one word of memory and its location is given by:

$$\text{location}(A_i) = L_0 + c(i - 1)$$

where L_0 is the address of the first element in the vector and c is the number of words per element. A_i is the position of the ith element of the vector. Building upon this it is possible to create arrays consisting of multiple vectors.

These higher-order arrays may be stored as rows and columns with arbitrary lower and upper bounds on the starting location (index).

Example. Filtering noisy experimental data [5].

One of the simplest reduction techniques for smoothing noise in experimental data is achieved by reducing the number of data points; in doing so some of the noise is eliminated or reduced. The original data are retained in a one-dimensional array. Each set of n consecutive points is replaced by a single point that is the sum or average of its neighbouring points. A set of 1200 points might be reduced (or 'bunched') to produce 400 points by averaging points 1, 2, and 3 (to produce the first new point), then by averaging points 4, 5, and 6 to obtain the next point, etc. If the noise in the data is randomly distributed, then the noise in the data will be reduced (approximately) by the square root of the number of points being bunched. The raw data and the reduced data may all be retained in an array, particularly if this operation is performed off-line, after the experiment has been completed. Because of the ease with which the data are addressed, this technique may be used while the data are being collected (real time) in which case only the reduced data are retained in an array. The programmer is not required actually to compute the data address. In HLLs, a reference to an array element $A(i)$ will automatically access the proper datum point; the compiler (translator) will replace i with the address as described above.

- **Record.** This logical structure consists of a list in which each element is a combination of primitive data types (e.g. a combination of integers, real numbers, character variables). Records may appear as individual objects (entities) or as part of a larger data structure (e.g. an array of records).

Example. A general problem encountered in many laboratory environments is the detection and measurement of peaks or local extremes in continuous data

Spectroscopic and chromatographic data often have such characteristics and experimental outcomes often include many peaks. Experimental results of this kind may be retained in a number of ways. One could, for example, retain the entire sequence of experimental outcomes including regions that contain peaks as well as those that do not; this may require considerable storage space. Alternatively, one could summarize the data by storing only information that is essential to describe the peak. Each description would form a record, with fields shown by the following instance:

```
Key field: either a number or a unique name given to the peak
location: the value of the independent variable corresponding
     to the location of the centre of the peak. It might be
     a wavelength value, voltage value or other suitable
     unit of measure.
width: a number representing the peak spread.
height: a measure of the dependent variable which
     corresponds to the peak reading.
comment: a field which might include unusual
     characteristics of the peak such as the presence
     of high noise or overlapping peaks.
```

Taken together, the fields define one record of a peak; other experimental outcomes and objects may be defined in this way. In fact, some HLLs permit the programmer to define a new kind of data object in this way. The record above could be defined as a *peak type* and variables with these characteristics defined within a HLL program. For example, the variable spectroscopic_peak could be defined within a program; it is then convenient to refer to 'spectroscopic_peak.width' to access (read or write) the width field of the entity.

- **Queues**. The linked list arrangement discussed above may be used to implement a queue. A *queue* is a subclass of linear lists in which deletions are made at the front (head) of the list while additions are appended to the end, or tail of the list.

Example. A series of automated tests are to be performed

Some of these are to be completed during a time when the system is unattended (at night). These tests can be programmed and listed (queued) in the order in which they are to be performed. Each new experiment is taken from the head of the list while new experiments are appended to the tail of the queue.

Queues are referred to as *first-in-first-out* or FIFOs because of the way in which they function. A queue of tasks awaiting execution on a computer (as in the example above) may be referred to as a *priority queue* when factors other than strict position in the queue are used to determine which task is chosen next. A queue can also be viewed as a vector with a sufficiently large number of elements to handle variable length applications. If F and R represent pointers to denote the front and rear position elements, then insertions and deletions can be described as follows:

```
queue_insert:
     check for overflow
     increment R
     insert new element
     if queue was empty, set F=1.
```

```
queue_delete:
        check for underflow
        delete front element
        If queue is empty (F=R)
        then
                set F=R=0
                end operation.
        else,
                increment F
```

The formulation of the queue cited above should only be used when it can be expected to empty periodically; otherwise it produces a very inefficient utilization of memory. To get around underflow, one could use a *circular queue*. This can be viewed as a queue where the last element references (points to) the first element. A *double-ended* queue permits insertions and deletions from both ends of the list.

- **Module.** Some HLLs permit the programmer to define a new type of data as well as the operations that can be performed on the data. For example, in FORTRAN90 it is possible to define a type of data that is not intrinsically part of the syntax and also to define a set of operations associated with the data type—a matrix, combined with matrix multiplication and inversion. The result is placed in a module where it is available for use by many programs.

5.3 TIME OR FREQUENCY BASIS OF MODELLING

One of the important purposes of an experiment is to derive a model of some naturally occurring phenomena. The goal is to define a set of mathematical operations on, or relationships between, a group of input data and a system of responses or output(s). Inputs may include some form of incident energy (light, mechanical stimulation, electrical) with one or more output signals possibly consisting of variations in energies of the type mentioned. The mathematical operations may be represented as a series of blocks that signify what calculations are to be performed as well as the order (or position) in which they are to be carried out. (As such it has the characteristics of a program and often a computer is used to simulate the model.) Once the operations—sometimes called transformations—have been resolved, it is possible to predict the output response to a particular input. Of interest are those models (or systems) in which the transformations include a combination of integration, differentiation, scalar multiplication, and summation. Models include: those whose inputs, outputs and component responses are continuous functions of the independent variable (e.g. time, or variations in input energy); and similar systems but with discrete characterizations—where information is known only at distinct (discrete) incremental values of the independent variable, most notably time. Other terms that are applicable to such systems include [6]:

Linearity. If **T** represents the transformations, u and v represent different inputs, a and b are constants and y is the output, then a system is considered to be linear if

$$y = \mathbf{T}(au + bv) = a\mathbf{T}(u) + b\mathbf{T}(v).$$

($\mathbf{T}(x)$ is interpreted as meaning: apply the transformations specified by **T** to the variable x.) It is useful to cite a counter-example to help explain linearity. Suppose that a system squares the input that it receives ($y = t^2$). If $a = 1$, $b = 1$, $u = 2$, and $v = 2$, then it is clear that the system is non-linear because it does not satisfy the formula above ($(2 + 2)^2 \neq 2^2 + 2^2$), and hence squaring is a non-linear operation.

Time invariance. Systems in which the response does not depend on the absolute time at which the input starts but only on the time that has elapsed since the start of the input are referred to as time invariant. Mathematically, this means that if $\mathbf{T}\{x(t)\} = y(t)$ then $\mathbf{T}\{x(t - a)\} = y(t - a)$.

Causality. A system is causal if its response to an input does not depend on values of the input at later times (future times); it is also known as *non-anticipatory* or physically *realizable*.

Linear, time-invariant, causal systems (LTI) may be studied in a direct manner using two different, but complementary, techniques. One is to consider their *time domain* responses and the second is to study their *frequency* responses. The underlying analytic principles are summarized as follows.

Time domain analysis. A standard time-varying test signal is applied to the system and its response is noted. Application of any other time-varying input can be predicted by performing a mathematical operation on the given input using the response to the standard input; this calculation is known as *convolution*. The standard test signal is called an *impulse* and the response of the system to this input is called the *impulse response*. If the impulse response of the system is denoted as $h(t)$, and if the given input is $x(t)$ then the system output $y(t)$ is calculated from the *convolution integral* as:

$$y(t) = \int_{-\infty}^{+\infty} x(\tau) h(t - \tau) \, d\tau.$$

Practically speaking, an impulse is considered to be a very short, very strong (amplitude) signal. An alternative notation for convolution is

$$y(t) = x(t) * h(t).$$

This integration may also be carried out pictorially (graphically).

Frequency domain analysis. The system is tested using (a series) of sinusoidal inputs with different frequencies and the responses are recorded. This response is called the *frequency response* of the system. The response of the system to an arbitrary input may then be predicted by the following analysis:

the given input is decomposed into a sum of sinusoids whose frequencies are the same as the test frequencies; the response of the system to each of these frequencies is found from the original test results; the individual responses are summed to produce the overall system response to the given input.

The time domain analysis and the frequency domain analysis are summarized in figure 5.4. What is important to note is the fact that the frequency response may be derived from the impulse response and vice versa. The frequency response and the impulse response are transforms of each other. In particular, the Laplace transformation of the impulse response produces the system function. The Laplace transformation is defined as

$$H(s) = \int_{-\infty}^{+\infty} h(t)e^{-st}\,dt$$

where $H(s)$ is the system function and $h(t)$ is the impulse response. This is summarized in figure 5.5 for a component whose function is to perform integration from 0 to t.

Example. A Physiological Model

When the left ventricle of the heart contracts, it expels blood into an elastic arterial system within the human body. A pressure wave is produced by this action and the peak pressure wave is called the systolic pressure. Eventually the ventricle relaxes and the pressure falls as the blood drains into the peripheral circulatory system. Measurement of the peak (systolic) and relaxed (diastolic) arterial blood pressure gives diagnostically useful information about the heart, the elasticity of the arterial system, and the state of the peripheral circulation.

A number of conditions may give rise to high systemic arterial pressure. Some of these include: myocardial infarction ('heart attack'); malignant hypertension; and congestive heart failure. Hypertension also follows intra-cardiac operations.

Pharmacological agents are often used to regulate blood pressure, particularly after surgery. Manual control of the administration of these agents has several disadvantages: it is difficult and time consuming; the quality of control suffers because human monitors may be distracted; errors are more likely, particularly after long periods of concentration [7]. Such circumstances lend themselves to automated control. A block diagram of such a system is shown in figure 5.6. A catheter is used to detect blood pressure; this information is converted into an electrical signal by a transducer (see chapter 6). The signal is smoothed and averaged in the signal conditioning circuits and provides a measure of the present value of the patient's blood pressure for the computer-based controller. A second signal received by the computer is the desired blood pressure, called variously the (pressure) *set point* or *reference* signal. The difference between the observed and target signals is computed and this is used to generate an output that increases or decreases the rate at which the hypotensive agent is administered. The algorithm to control the infusion rate must take into account a number of factors: the delays introduced by the patient's physiological responses; the

158 DATA-PROCESSING CONSIDERATIONS

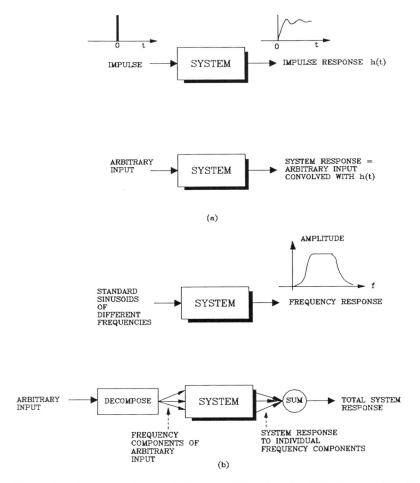

Figure 5.4 System analysis techniques. (a) Time domain. (b) Frequency domain.

patient's sensitivity to the agent; delays introduced by the distance between the drug and the infusion site.

If the automated system fails to adopt or take into account all of these factors then the time course of the pharmacological agent in the body and hence the blood pressure may result in undesirable patterns:

- The blood pressure may oscillate wildly producing irregular heart rhythms.
- The blood pressure may not stabilize fast enough.
- The blood pressure may greatly exceed the reference point and impair the ability of the heart to pump.

Because of these and other factors a number of control strategies have been studied [8]. Using computer simulation, the performance of the control strategy

TIME OR FREQUENCY BASIS OF MODELLING 159

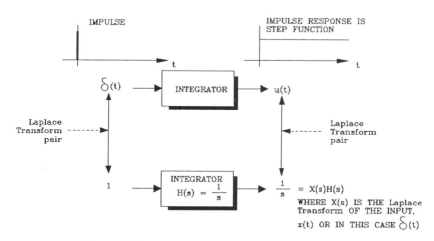

Figure 5.5 Analytic considerations for an integrator.

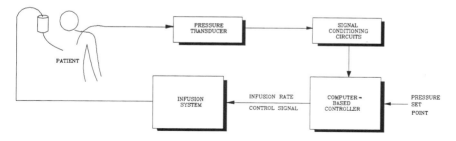

Figure 5.6 A control system for regulation of blood pressure using pharmacological agents.

may be studied if an accurate model of physiological responses is available. In theory, either of the methods described above (impulse response, frequency analysis) may be used to obtain a model of the physiological responses. If, for example, the impulse response method is used to study responses to hypotensive agents, a bolus injection of the agent is employed and blood pressure responses are measured. A problem with this method lies in its potential for systemic overload. Instead, frequency domain analysis may be employed because the injected quantities may be more carefully controlled; the system remains within its linear region. The input signal must reflect the frequency characteristics of an

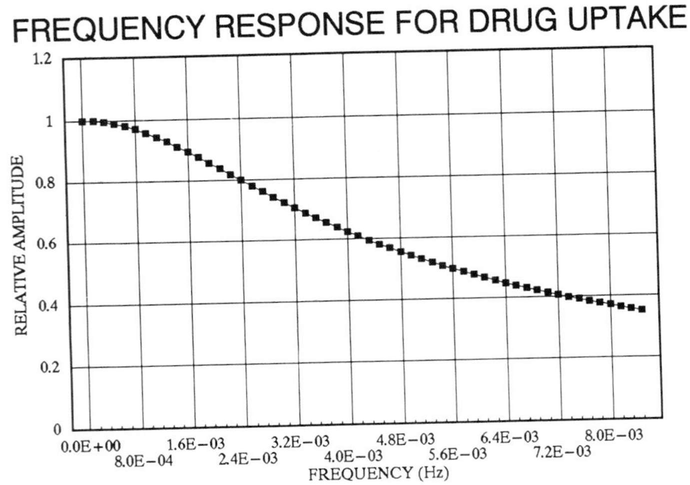

FREQUENCY RESPONSE FOR DRUG UPTAKE

Figure 5.7 Frequency response for drug uptake.

impulse function. Recall that the transform of the impulse function is 1 which implies that it has a (frequency) spectrum that is constant over all frequencies. A technique for obtaining the response is to use white noise—the spectrum of white noise is constant over some suitably wide frequency range—and to calculate the cross-correlation between the random input and the system (blood pressure) response. As an example, one model (system function) for drug uptake is described by

$$G_d(s) = \frac{Ke^{-T_i s}(1 + \alpha e^{-T_c s})}{1 + \tau s}$$

with the following meanings for the parameters:

K = patient sensitivity to the drug (= 1 in what follows)
α = recirculating fraction of the drug (= 0.4 in what follows)
τ = lag time constant for uptake, distribution, metabolism (= 50 s below)
T_i = transport time of drug to infusion site (30 s in what follows)
T_c = recirculation time (= 45 s in what follows).

The magnitude of this spectrum is plotted in figure 5.7; terms such as $e^{-T_i s}$ have a magnitude of 1—they contribute a phase shift, which is not shown. Even though a plot of the phase shift is not shown, it must be considered for a complete description of the system because a linear phase shift as a function of frequency corresponds to a pure systemic delay. (A term like $e^{-T_i s}$ has a has a frequency response magnitude of 1 but contributes a phase shift of $T_i \omega$, which is a linear function of ω and corresponds to a pure delay in time.)

Taking the inverse transformation of this function yields the impulse response which is given by

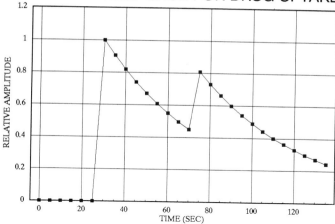

Figure 5.8 Impulse response for drug uptake.

$$h(t) = \frac{K}{\tau}\left(e^{-(t-T_i)/\tau} + \alpha e^{-(t-[T_i+T_c])/\tau}\right)$$

and is sketched in figure 5.8. These results are derived from the following Laplace transform theorems:

$$x(t-t_0) \Leftrightarrow X(s)e^{-st_0}$$

$$\frac{t^n e^{-\alpha t} u(t)}{n!} \Leftrightarrow \frac{1}{(s+\alpha)^{n+1}}.$$

$u(t)$ is a unit step function, and $x(t-t_0)$ is a function $x(t)$ that has been delayed in time by t_0. The unit step function is 0 for $t < 0$ and 1 for $t \geqslant 0$—this ensures that the function starts at t_0. Notice the following about the impulse response: the drug first appears in the blood after a delay of 30 s which is due to the transport delay introduced by the distance to the infusion site; a second drug bolus appears at 75 s because of the recirculation affect.

Many laboratory applications use frequency analysis to study data and include physiological data such as electroencephalographic (EEG) experiments, and electromyographs (EMG) records. EMG records may also be examined using the time domain. The spectral signature of various mechanical systems can be used to detect deficiencies; a chipped gear tooth gives rise to a characteristic frequency that does not appear in a smoothly operating gear train. Harmonic distortions are indications of the non-linear nature of a system; if a sinusoid of known frequency is an input to a system and frequencies other than the test signal are detected at the output then the system exhibits non-linear characteristics. Time records are useful when testing the response of various vehicle suspension systems to simulated (or actual) road conditions. The impulse response may be used to detect faults in concrete structures. It also has applications in oil exploration and other geological studies (e.g. of earthquakes).

Table 5.1 Characteristics of data-processing schemes.

System	Relative speed	Level of control
Polled	Slow	Highest
Interrupt driven	Medium	Medium
DMA	Fast	Lowest

5.4 SOFTWARE ARCHITECTURES FOR INPUT/OUTPUT

Issues regarding the data organization (structure) of the experiment are often related to other parameters of the system. In particular, the utilization of the computer (CPU) as well as the speed at which data must be processed—the so-called *system throughput*. There are three generally recognized methods for organizing processing and communications between the various parts of a computer-based instrument system: programmed or *polled*; *interrupt driven*; *direct memory access* (DMA) [9]. These schemes are distinguished according to the characteristics shown in table 5.1.

The 'relative speed' refers to the throughput or capacity of the system while the 'level of control' indicates how the CPU functions. In general, a high level of control indicates that the CPU follows a strict sequence of instructions whose timing is under rigorous constraint. In circumstances where control is medium or low, the CPU relinquishes its control to another element in the system whose operation then governs the time and rate at which data transfer occurs. Overall throughput varies from slow (polled operation) to fastest (DMA).

Hardware architectures that can be organized according to the methods cited above include those previously described in chapter 3. A simplified version of such systems is shown in figure 5.9. (As noted in chapter 3, systems that include more than one processor can increase throughput; only a single processor is included here for simplicity.) An important addition to the block diagram is the explicit inclusion of (at least) two registers in each peripheral that is under computer control. These registers are logical elements of a communication/control environment. The *status register* maintains the condition of each peripheral and may include such indicators as: power on; ready/busy; acknowledge; peripheral error, among others. The *data register* is involved in data-transfer procedures—it contains the information actually being transferred. The information may be of either of two types: data and control. The data type includes experimental or calculated results while the control type determines the commands to be executed (e.g. take a measurement, start the transfer of the data). (See the discussion on IEEE 488 for additional details.)

A simple program for acquisition of data may be described according to the following pseudocode:

SOFTWARE ARCHITECTURES FOR INPUT/OUTPUT

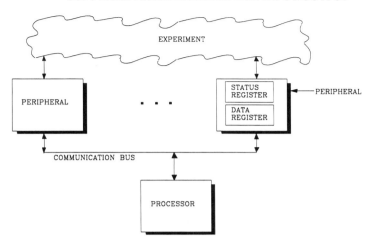

Figure 5.9 An instrument system consisting of a processor (computer), one or more peripherals, and a communications bus.

```
Input module:
    {accept user inputs}
    {initialize experimental parameters}
Control module:
    {Collect data}
    {Process data}
Output module:
    {Report results}
```

A program of this kind may be logically decomposed into three modules: the input module; the control module; and the output module. Briefly stated, these modules have the following purposes:

- *Input module*. Permits a user to enter appropriate experimental parameters such as the sampling interval, the duration of the experiment, various scaling settings and a variety of other pertinent information.
- *Control module*. This is the portion of the program in which the user's intentions are carried out. Normally, a sequence of repeated procedures is executed to acquire and store the information. Additional processing of data may take place during acquisition or after accumulation depending on timing considerations.
- *Output module*. The information is summarized, and results presented to the user either in graphical or tabulated form. Presentation of results may take place on the monitor, a hard (printed) copy provided, or both.

As noted above, the principal consideration is the method by which data 'are collected' or more broadly how they may be processed. (The term data processing is used in its broadest sense; it includes measurement, manipulation, and transmission.) In each of the methods to be described below, the operations will be discussed from two points of view: what occurs in the peripheral; and what happens in the computer or processor.

5.4.1 Polled (programmed) operation

For *polled operation*, the processor never relinquishes control of data processing. The pseudocode shown below describes a simple polling procedure for the purpose of acquiring data from a group of peripherals that would be executed by the processor. It further refines the '{Collect data}' operation cited above specifically for a polled architecture:

```
Repeat
    If {time_to_poll} then
        {initialize first_peripheral_to_be_polled}
        repeat
            {query next peripheral}
            If {data_available} then
                {transfer data}
            endif
        until {no_more_peripherals}
    endif
until {data_collection_terminates}
```

The logic of this sequence instructs the computer to wait until the time at which a sample is to be taken; proceed to query each peripheral for the presence of data; if data are available, transmit (store) these data in the computer's memory where they may wait for additional processing once the polling is complete.

An alternate form might include a further {process data} procedure following the {transfer data}. The code shown above represents a 'single sample' from each peripheral. If the sample is to reflect what happened 'exactly' at the sampling time then it is imperative to poll and transfer data as fast as possible as there is a time delay between the combined query and transfer operation for each successive peripheral. If, however, the user is confident that the data 'will not change' for the entire duration of the sample, it may be possible to process data as they are received from each peripheral. The result is more efficient use of the processor.

Among the various computers, there are two types of instruction that can be brought to bear to acquire data from a peripheral; these are the so-called *memory-mapped* and *input/output* or *I/O commands*.

Memory-mapped architectures. A specific portion of the computer's memory space is set aside to receive data from the peripherals. To initiate a transfer, an instruction of the form

```
Mov    dest,src
```

where 'src' is the address of a port (within the peripheral) that contains the data and 'dest' is the address within the memory where the data are to be stored is given. (See 'assembly language programming' in chapter 4.) Memory-mapped techniques have a speed advantage over I/O commands but 'pay a price' for having to set aside specific, 'pre-defined' memory address locations, which may restrict the ability to develop programs that can be moved (ported) from one computer to another.

Input/output architectures. Some processors include a group of commands known as I/O instructions. A typical command might be

```
out     456
```

and the following sequence of events occurs in response to this command:

- The contents of the computer's accumulator—whatever they may be—are transmitted to the data portion of the computer's communication bus.
- The number '456' is placed on the address portion of the computer's communication bus.
- One of the control lines of the communication bus is enabled to indicate that the computer wishes to write information to the device whose 'address' is 456. Had the computer wanted to take in data, the instruction would have been

```
in      456
```

and the control line would reflect the fact that this is a read operation as opposed to a write operation.

From the point of view of a peripheral, the following pseudocode reflects the logic that must be supported either with hardware, or with a dedicated program (firmware) supported by a controller in modern systems (a controller is a computer whose instruction set includes a number of commands designed to enhance the ability of a programmer to control the operation of a peripheral):

```
repeat
    if {address_identified} then
        if {operation_is_write} then
            {take_in_data_from_bus}
        else
            {put_data_on_bus}
        endif
    endif
until {forever}
```

The peripheral continually monitors the address lines on the communication bus. If it recognizes its own address, it will either accept the information on the bus's data lines—a write operation—or put its own data onto the data lines of the communication bus—a read operation. Depending on the address being sent, the peripheral may report the information in its *status register* or its *data register*. The computer will normally query the peripheral regarding its status prior to accepting data from the data register in order to confirm that the data are 'valid'.

5.4.2 Interrupt-driven systems

A representative memory-mapped data-acquisition application may require 8 to 10 μs to query and transfer information from each peripheral, even for the newer (32-bit) processors operating at clock speeds of 50 MHz. When other data processing (e.g. data smoothing) is added to this time, throughput may be reduced to 10 000 characters per second (80 000 bits per second). Some improvement may be possible if the system operates in interrupt-driven mode. In this method of operation the computer may proceed to complete its tasks without the need to query a peripheral. When a peripheral needs to be serviced or attended to, it signals the processor to this effect—in other words it 'interrupts' the computer. The computer suspends its current task, attends to the peripheral(s) and, when this has been done, returns to the suspended operation exactly where it left off. This description may be summarized as follows:

```
interrupt:
        {complete_instruction_being_executed}
        {save_contents_of_critical_CPU_registers}
        {disable_less_important_interrupts}
{transfer_control_to_interrupt_service_routine}
        if {interrupt_service_routine_not_done} then
            {continue}
        else {return_from_interrupt}
        endif;
```

An interrupt signal may occur any time during a computer's instruction cycle; the CPU must complete the instruction being executed before taking any action. Next, the computer must save the contents of certain critical registers; these registers, taken together, permit the computer to return to the point at which it was interrupted without degradation of the program it was executing prior to the interruption. The computer may then proceed to disable (future) interrupts, which are designated as having lower priority than the one just initiated. (Interrupts having higher priority may still interrupt the computer and when this happens the procedure noted above is repeated—this is an example of nested interrupts and computers are equipped to handle a large number of such instances.) After disabling lower-priority interrupts, control of the operation of the instrument is transferred to another section of code commonly referred to as the interrupt service routine or ISR. The ISR carries out all necessary steps in order to respond

SOFTWARE ARCHITECTURES FOR INPUT/OUTPUT 167

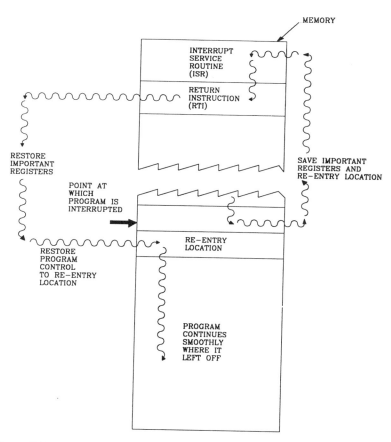

Figure 5.10 A software schematic diagram for the interrupt-driven architecture.

to the interrupt. For example, a clock may be used to generate an interrupt signal periodically. The 'ticks' may be set to occur at times at which the data are to be sampled. The ISR would then basically consist in querying appropriate peripherals for the data—this is the equivalent of a polling procedure but carried out in an interrupt-driven format. Alternatively, each peripheral may generate separate interrupt signals only when data are available, in which case the ISR includes relatively compact code for interrogating a single peripheral. This increases efficiency because such code can be executed rapidly. The ISR may also have to store additional information regarding the program previously in progress to provide for a 'smooth' return. The interrupt-driven software architecture is represented schematically in figure 5.10. The interrupt logic within the CPU automatically provides for saving the important registers and the re-entry point. When the ISR is (logically) complete a *return from interrupt* (RTI) instruction is

168 DATA-PROCESSING CONSIDERATIONS

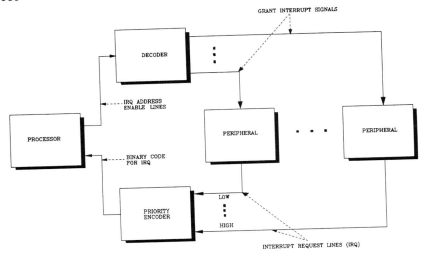

Figure 5.11 Simplified hardware architecture of interrupt-driven instrument systems.

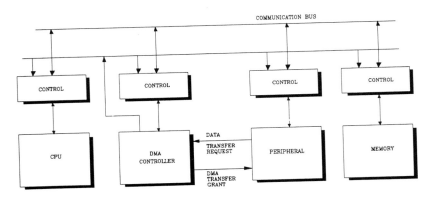

Figure 5.12 A block diagram of the DMA architecture.

executed; this instruction automatically restores the critical registers as well as transferring control back to the point at which the program was interrupted as shown in the figure.

The simplified hardware organization of a typical interrupt-driven system is shown in figure 5.11. Each peripheral is equipped with an *interrupt request line* (IRQ), and these are combined in a *priority encoder* which generates a binary code corresponding to any of the IRQ lines that may become enabled by a peripheral. The lines on the priority encoder are arranged according to priority; the user assigns priorities to the peripherals in this manner. The binary code produced by the priority encoder is connected to the processor which normally includes hardware (and logic) to support an interrupt structure. Whenever an

IRQ code is received by the processor it is compared—hardware comparison—to any prior interrupt that may have been received. If the new interrupt request has higher priority, it is *acknowledged* by generating a corresponding binary code on a series of IRQ *address enable lines*. This binary code is subsequently decoded—converted from a binary code into an enabling signal on one of the grant interrupt signal lines—and returned to the original peripheral where data may be transferred to the computer's communications bus (not shown in the figure). (In some systems, the peripheral returns the starting address of the ISR in response to the grant interrupt signal; such a scheme is referred to as a *vectored* architecture.)

Some users (programmers) avoid interrupt-driven systems for two reasons. One is a question of style or 'psychology' wherein the interrupt represents the 'unexpected' event, which can sometimes disrupt a program in unanticipated ways. The second reason is (perhaps) more practical; in a system that includes a number of peripherals all competing for the computer's resources on an interrupt basis, it is certainly possible that one of the peripherals may end up 'hogging' the processor at the expense of one or more of the other peripherals which may never receive service. In spite of such considerations, the interrupt architecture can increase throughput even though some software *overhead*—time required to save and restore register values—is needed every time an interrupt is encountered. Interrupt-driven architectures provide a 'medium' level of control; the processor does not relinquish control. Control is retained in two ways: some processors have additional control resources that may disable all interrupts (or all but the most important, such as one that may occur due to a power failure) while some important, 'non-interruptable' code segment is being executed; control is not actually relinquished, rather it is transferred to another portion of the memory—even though it is a 'disruption' of the normal program sequence. (In vectored systems program control is determined by a peripheral and in this case the computer does indeed give up its unequivocal control of the program—consider the 'nasty' case in which a peripheral mistakenly transfers control to an inappropriate address.)

5.4.3 Direct memory address (DMA)

This is the fastest architecture, which also results in almost complete loss of control of processor resources by the CPU. The scheme relies on additional circuitry in the form of an integrated circuit known as a *DMA controller* as shown in figure 5.12. Two new operative elements have been added to a block diagram that has been previously used as a functional description of a computer-based instrument. The first is the DMA controller and the second is a switch ('CONTROL') which is under direct control of the DMA circuit. The computer's communication bus can be accessed by either the CPU or a peripheral; bus control is jointly shared by the CPU and the DMA controller. DMA operation proceeds according to the following:

```
DMA Control:
    {Peripheral makes data transfer request}
    {CPU loads data origin (starting address) into
        DMA Controller}
    {CPU loads number of words (or bytes) to be
        transferred into DMA}
    {CPU relinquishes control of bus to DMA}
    {DMA places target data address on bus address lines}
    {DMA grants transfer request to peripheral
        (read/write)}
    if {transfer_not_complete} then
        {continue}
    else
        {issue interrupt to CPU}
    endif
end DMA Control;
```

A peripheral requests permission from the DMA controller to transfer (read or write) data. The target address and the number of words (or other data division such as bytes) to be transferred are supplied to the DMA controller which includes registers for storing such information. The CPU transfers control of the communications bus to the DMA controller. This element, in turn, places the target address on the address lines of the communications bus and signals the peripheral that data can be transferred. The control lines of the communication bus are now under the supervision of the DMA controller which oversees the complete transfer of the requisite information. Once the transfer is complete, the DMA controller hands back control of the system to the CPU.

As a practical matter the DMA controller is more complicated than the functional view just described. For example, a DMA controller may include interface facilities for several peripherals. In such instances it may include several register pairs for retaining the target address and amount of data to be transferred, one pair for each peripheral. (These registers may be loaded prior to any transfer and entirely under CPU control.) If more than one peripheral makes a data transfer request the DMA must be able to resolve such queries. The potential difficulties associated with such architecture are similar to those mentioned with respect to CPU interrupts.

There are two modes of operation for the transfer of data: the *burst mode*; and the *cycle steal* mode. As the name suggests, the burst mode accomplishes data transfer in the 'shortest' time and under the complete supervision of the DMA controller. The bus is 'seized' for the duration of the transfer and is only limited (in speed) by the time delays associated with reading or writing imposed by the slowest component—either the memory or the peripheral. For the cycle steal(ing) mode, the peripheral and the CPU essentially share the communications bus so that the computer can complete additional tasks while data are being transferred. One machine cycle now includes a portion of time set aside so that the DMA/peripheral combination can access a memory facility and the CPU can also access the same (or other) system resource.

Ever-increasing demands have been placed on system resources, particularly capacity, as instrument applications evolve. A principal limitation of many computer systems is the bandwidth limitations of the communications bus. Both interrupt-driven, and DMA architectures have emerged as methods for achieving maximum utilization of the communication bus capacity. It has already been noted that newer instrument architectures seek to avoid the limitations of the communications bus altogether. (For example, the transputer architecture, consisting of a number of processors that can communicate directly with one another, can increase throughput. In effect, multiple communications buses have replaced the single 'bottleneck' architecture of the more traditional systems.) In this regard, some arrangements have come to include a separate processor between the main computer and one or more peripherals. Its main function is to supervise transfer of data to/from the peripheral. Names such as *I/O channel*, *front-end processor* (FEP), and *peripheral processing units* (PPU) have been applied to these schemes. The techniques described above (polling, interrupt, DMA) may be employed within the I/O channel. System capacity is enhanced because the (main) processor need not be aware of the parametric minutiae associated with each peripheral—it need only send a single, 'more abstract' command to the peripheral. As previously noted, additional capacity may be achieved through a system of preprocessing of incoming or outgoing data.

REFERENCES

[1] Stone H S 1990 *High-performance Computer Architecture* 2nd edn (New York, NY: Addison-Wesley)
[2] Weiss R 1991 *32-bit Floating Point DSP Processors* (New York, NY: EDN)
[3] Tremblay J-P and Bunt R B 1981 *An Introduction to Computer Science: An Algorithmic Approach, Short Edition* (New York, NY: McGraw-Hill)
[4] Date C J 1990 *An Introduction to Database Systems* 5th edn, vol 1 (New York, NY: Addison-Wesley)
[5] Gates S C and Becker J 1989 *Laboratory Automation using the IBM PC* (Englewood Cliffs, NJ: Prentice-Hall)
[6] Irwin J D 1993 *Basic Engineering Circuit Analysis* 4th edn (New York, NY: MacMillan)
[7] Sheppard L C and Jannett T C 1988 Automatic control of blood pressure *Encyclopedia of Medical Devices and Instrumentation* ed J G Webster (New York, NY: Wiley-Interscience)
[8] Sheppard L C 1980 Computer control of the infusion of vasoactive drugs *Ann. Biomed. Eng.* **8** 431–44
[9] Clements A 1993 *Principles of Computer Hardware* (Kent: PWS)

6

Data Acquisition and Instrument Control Resources

Modern computer-based integrated instrument systems include some means for converting the physical and/or chemical property to be measured into a form that lends itself to further processing by the computer; normally the form is some binary-coded representation of the measurement. In addition the computer may be required to control various aspects of the experimental and/or manufacturing environment. This chapter includes discussions of how such conversions are accomplished and what is needed to control devices and instruments external to the computer.

Figure 6.1 is a functional diagram depicting those elements that are associated with data conversion and encoding. Each component has a well defined purpose although not all may be needed in every application. The most general functions include:

Transduction (frequently called sensor action). The generic name for a device that senses either the absolute value of, or change in, a physical quantity (e.g. temperature, pressure, flow rate, pH, intensity of light, sound, radio waves) and converts such information into useful signals for an information gathering/processing system.

Excitation. Some transducers require a precise source of energy (e.g. the reference signal) to operate properly; such sensors work by altering the reference in accordance with the information contained within the experiment or process (e.g. strain gauges, linear variable differential transformers).

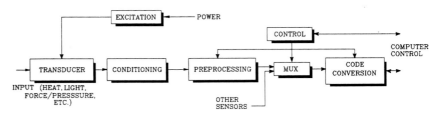

Figure 6.1 The functional diagram of the information transduction system.

Conditioning. The transducer signal may contain some contamination (noise); alternatively the user may only be interested in certain restricted aspects of the experimental information (e.g. only those information variations that lie within a certain band of frequencies). Conditioning components that extract appropriate information or eliminate noise may be included in the instrumentation system.

Preprocessing. In some circumstances it may be more efficient to perform some information processing using elements that are external to the computer itself. This is particularly true if such processing requires inordinate time for a program to complete. For example, computer calculations could be reduced by using an integration circuit to convert gas flow into a volume measurement; the integration circuit is considered to be a preprocessing component. In some cases the preprocessor may be a single-chip computer, particularly if such an architecture greatly reduces the preprocessing circuitry.

Code conversion. This converts information into a form that is suitable for a computer. Often the information that the converter receives (on a single input line) is continuous or 'analogue' in nature and it produces corresponding discrete or 'digital' data either on a single line (serial output) or on several lines (parallel output). Such devices are commonly known as analogue-to-digital converters, or A/D(s).

Multiplexers ('Muxes'). The A/D component may be expensive; consequently it may be shared by a number of transducers. Functionally the Multiplexer or Mux is considered to be an ideal switch, which receives inputs from many transducer sources and transmits the appropriate signal to the converter. The *control* elements of the system direct the Mux's operations (e.g. which signal to transmit and when to send it). In addition, the control elements initiate conversion and direct preprocessing operations (e.g. setting the gain of the preprocessor if that is appropriate).

6.1 TRANSDUCERS

A transducer normally converts a physical or chemical property that is to be measured into an appropriate electrical signal. The result of the transduction process is an information signal that varies in amplitude or frequency in accordance with the experimental data; this is depicted in figure 6.2. The frequency domain is characterized by signals whose amplitude excursions are fixed and whose repetition rate varies with the independent variable and conveys the underlying information. Transduction is accomplished when some change in the energy of the phenomena to be measured affects the characteristics of the transducer itself. While new and improved transduction techniques are continually being reported, there are a number of mature methods; these are represented in table 6.1 which includes the physical (energy) basis of the transduction method as well as some specific implementations.

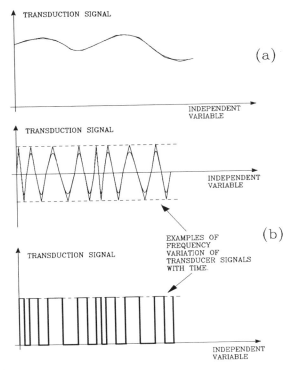

Figure 6.2 Transduction is characterized by (a) amplitude or (b) frequency variation of information with systematic variation of the independent variable.

Table 6.1 Transduction methods and devices.

Basis	Transducers
Heat	Thermistor, thermocouple, platinum (Pt) resistor, solid state
Light	Phototubes, phototransistors, photoresistors, solid state (charge coupled devices–CCD)
Force, pressure	Strain gauge, load cell, piezoelectric
Ionic activity	Electrochemical sensors (electrodes)
Position	Shaft encoders, potentiometers, variable transformers

6.1.1 Temperature transducers

Temperature is one of the most commonly measured physical quantities and is an important indicator of how much heat energy is being absorbed or dissipated. It can be measured by the mechanical changes that it produces (e.g. the height of a column of mercury), but these are not be compatible with computer-based processing. Temperature changes may alter other transducer characteristics that

Table 6.2 Summary of temperature transducer characteristics.

Type	Accuracy/sensitivity[a]	Range (°C)	Size	Speed
Thermistor	0.25–1 °C; high sensitivity	−100 to 50 50 to 150 150 to 300	Variety of shapes; very small	Microseconds for small sizes
Wire (platinum, iridium)	Most accurate and linear; 0.1 °C; moderate sensitivity	−160 to 2000	Fairly large; typically 2 × 25 mm	Slow
Thermocouples	1% to 3%; low sensitivity	−270 to 2500	Small	Fast, using plated devices
Semiconductor	0.5 °C at 25 °C	−55 to 150	Large	Slow

[a] Accuracy may be improved using temperature calibration.

are more compatible with automation such as: resistance; voltages (produced by the transducer); and colour or transparency of the measurement material. We will limit discussion to the most common types of temperature transducers: thermistors (resistance); platinum wire (resistance); thermocouples (potentials); and semiconductor integrated circuits (potentials). The characteristics of such devices are summarized in table 6.2.

Thermistors. These are actually semiconductors made by the sintering of metal oxides (ceramic materials). Their resistance varies with temperature according to the following formula:

$$R(T) = R(25) e^{\beta(1/T - 1/298)}$$

where:

$R(T)$ is the thermistor resistance at a temperature of T (in degrees Celsius);
$R(25)$ is the thermistor resistance at 25 °C (supplied by the manufacturer);
$b = E_g/2k_B$ with
    k_B = Boltzmann's constant (8.67×10^{-5} eV K^{-1});
    E_g = size of the band gap of the semiconductor material.
(β is supplied by the manufacturer, typically 2800 to 5500 K).
(The derivation of the formula can be found in [1].)

From this formula it can be seen that once the thermistor's resistance is known its temperature can be determined:

$$T = \frac{\beta(298)}{298 \ln[R(T)/R(25)] + \beta}.$$

Computing T may be accomplished by the computer once $R(T)$ is obtained by measurement.

Measurement of the thermistor's resistance—indeed any of the transducers whose resistance varies with temperature—follows directly from Ohm's law:

$$R = \frac{\text{voltage across transducer}}{\text{test current through transducer}}.$$

By passing a known current through the transducer—the denominator in the equation above—and measuring the resultant voltage across the transducer, the resistance R may be determined. This value is subsequently used by the computer to calculate the temperature using the equation shown above. One of the severe limitations encountered when using the thermistor is the so-called *self-heating* effect. Simply stated, the test current passing through the device generates heat which raises the thermistor's temperature, and, *unless it is successfully carried away by convection*, ultimately yields a false temperature reading. It may also end up (self-) destroying the transducer because the resistance of the thermistor falls; this *may* cause additional current to flow and hence increased heating.

A further limitation of its use follows from the *non-linear* nature of its response. A goal of any transducer is to produce a response of the form

$$\text{Transducer output} = A \times \text{Transducer input} + B$$

where A and B are known constants.

Among other virtues, this form permits a computer to calculate the underlying physical quantity in an efficient manner. This is accomplished either using a 'look-up table'—a listing of corresponding inputs and outputs—or by substituting each measurement (reading) into the equation and computing a result under program control. The equation above is considered to represent a *linear* ('straight-line') relationship between output and input. It is apparent from the prior discussion that the calculation for T does not fall into this category; the relationship between thermistor resistance and temperature is non-linear. However, when the experimental temperature is limited to a relatively small excursion, the thermistor response may be (acceptably) linearized by placing a second resistor in series with the transducer whose value is given by

$$R_m \frac{\beta - 2T_m}{\beta + 2T_m}$$

where

T_m is the midpoint of the expected temperature range; and

R_m is the resistance of the thermistor at T_m.

Thermocouples (TC). Such transducers produce a voltage that is linearly proportional to the temperature that it measures and is based on an old, well known observation, called the Seebeck effect. This is summarized in figure 6.3. *A current will flow in a loop of wire made of dissimilar metals if the two ends are at different temperatures.* Alternatively, if the loop is interrupted, a voltage difference will appear according to the following formula (to a first-order approximation):

$$\Delta V = \alpha(T_2 - T_1)$$

where

ΔV is the induced voltage; and

α is the constant of proportionality (52 μV K^{-1} for an iron/constantan thermocouple).

To ensure accurate readings, T_1 (the reference temperature) must be known. The reference temperature—known as the *cold junction*—may be room temperature but it must be known at the time at which the temperature of the hot junction is taken. (See figure 6.3(b).) In addition, it is possible for the reference temperature to drift or change over the course of time. To minimize such effects the output leads usually terminate on an isothermal block. Mounted on this block is another type of temperature sensor (e.g. a thermistor) which is used to measure the actual temperature of the cold junction, and an appropriate correction factor is applied—see figure 6.3(c). A variety of dissimilar wire configurations are possible and the United States National Institute of Standards and Technology (NIST) publishes standard TC tables.

Several additional features of TCs should be noted:

- They have very low sensitivity and their output signal requires amplification to be compatible with the needs of computer instrumentation.
- The wire itself has a very low resistance and as a consequence noise generated by the transducer will be low.
- It has an extremely wide range of operation and can be made small resulting in rapid responses to temperature changes.

In order to increase sensitivity, a number of TC junctions may be connected together (in serial fashion) creating what is known as a *thermopile*.

Platinum wire temperature transducers. These transducers have virtually linear resistance variation with temperature, which is based on the atomic stability of some metals, notably platinum. The wire is normally wound, and made mechanically secure, on a ceramic core. Care must be taken to see that there are no 'resistive deformations' as a result of any extraneous strain. A variety of resistance values and temperature coefficients are possible; standard values of 100 Ω (at 0 °C) and corresponding temperature coefficients (0.39 Ω K^{-1}) are useful in the biological temperature range [2] . Thin-film platinum devices have been used recently; these are intended to overcome a disadvantage of the traditional units, namely the relatively slow response to sudden temperature changes. Like thermistors, platinum devices suffer from potential self-heating effects when a test current is passed through the transducer. In addition, the wires leading to the transducer itself may add 'resistive'

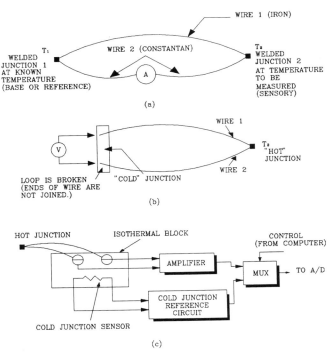

Figure 6.3 The Seebeck effect. (a) A single loop of dissimilar metals. (b) Modification using a single junction which produces a voltage output. (c) A block diagram of a measurement system.

errors to the measurement, and temperature changes will produce changes in these resistances. To obtain the greatest accuracy some platinum temperature transducers are equipped with four wires which can be used to measure and subtract the 'resistance error' from the measurement system.

Semiconductor integrated circuit temperature transducers. The *Zener diode* is a semiconductor device that produces a constant voltage drop when a constant current passes through the device; a range of constant currents will generate the same voltage drop to a close approximation. (For this reason such devices have a variety of electronic applications in addition to the one discussed here.) The voltage drop does, however, vary with operating temperature in a well defined manner and can therefore be used to measure temperature. The devices are fabricated in a manner to ensure repeatability. The voltage drop typically varies by 10×10^{-3} V K^{-1} (10 mV K^{-1}). These devices are physically large and therefore have considerable thermal inertia; they take a long time to respond to sudden changes in temperature (typically three minutes in free air). Even with such limitations they are useful for many temperature measuring applications because they are relatively accurate.

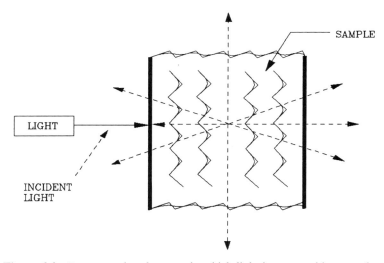

Figure 6.4 Demonstrating the ways in which light interacts with a sample.

6.1.2 Optical transducers

The interaction between light and various substances provides ways to measure a sample's physical properties. Figure 6.4 provides a basis for defining such interaction. Effects of the sample under test on the incident light are employed in a variety of commercial instruments designed to measure a sample's properties; they include the following:

- *absorption*: a part of the incident light energy is transferred to the sample;
- *transmission (transmittance)*: the extent to which light passes through the sample (formally, radiant power transmitted by the sample divided by the total radiant power incident upon the sample);
- *reflection*: the sample returns (reflects) part of the incident energy;
- *refraction*: the sample changes the direction of the incident light (particularly as it passes from one medium to another);
- *diffraction*: the confluence (algebraic addition of wave amplitudes) of incident light as it passes around an object or through openings (slits) in an object—a special case of the interference of two waves;
- *scattering*: the change in direction of light as it collides with particles in the sample;
- *turbidity*: cloudiness of a sample due to the presence of solid matter;
- *fluorescence and phosphorescence*: the substance absorbs some of the incident light energy and re-emits energy after a time delay (see figure 6.5).

Light may be constructively viewed in two ways according to the circumstances of the experiment. In some cases light is viewed as a wave that propagates at a velocity of 3×10^8 cm s^{-1} with a wavelength between 400×10^{-8}

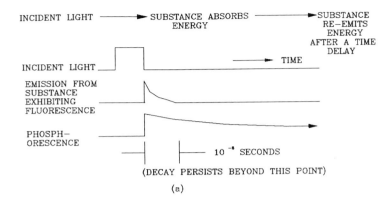

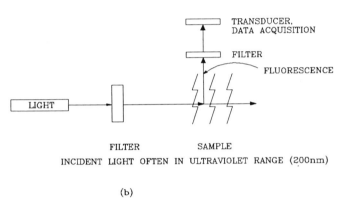

Figure 6.5 Differences between fluorescence and phosphorescence. (a) Time response. (b) A functional diagram of a fluorimeter.

and 700×10^{-8} m. In other cases light is considered to consist of discrete quanta of energy called *photons*.

Once the incident light has acted upon the sample it must be detected. A variety of transducers may be used: phototubes (photodiodes and photomultipliers) as well as semiconductor devices (phototransducers).

Phototubes (photodiodes and photomultipliers). These are used to detect photons and operate according to the principle of *photoemission*. Photons enter such devices at one end of an evacuated glass envelope and strike a metallic element called a *photocathode*. In doing so they release electrons which subsequently come under the influence of an electric field and drift towards a positively charged element—the *anode*—also within the evacuated glass envelope. A current is generated as a result and its time course mirrors the light received from the measurement. Phototubes with only one cathode and one anode are referred to as *photodiodes*: their sensitivity is normally quite low, although it is satisfactory for some applications. To overcome this limitation the tube may contain a number of additional 'cathodes' (called dynodes). Electrons emitted from one dynode are attracted to the next and when they strike the next one release additional electrons; ratios such as 1.8 can be found among commercial specifications and there may be ten dynodes within the envelope. This multiplier effect is carried along from one dynode to the next with the final element arranged to collect the resulting 'avalanche' producing a considerably more sensitive response; such tubes are called *photomultipliers*.

Photodetectors (semiconductor junction detectors). Potential differences are generated across junctions of dissimilar materials; this includes metal-to-metal, liquid barrier, and semiconductor junctions among others. Alteration of the charge equilibrium by energy generated within the experiment or process under control leads to a resulting current, which is a measure of the entity under test. When two dissimilar semiconductors are fused together a *junction* is formed with an associated potential difference across the boundary. This forms the basis of a variety of optical detectors: photodiodes, photovoltaic cells, and phototransistors (when there are two junctions within the device). (Other devices can be created such as light-emitting diodes but these will not be discussed here [3].)

Some materials are electrical insulators in their pure form (e.g. silicon) but can be modified—systematically *doped* with impurities—to become 'somewhat' conductive (a *semiconductor*). Generally such crystals include either an excess of electrons (*n-type* material) or a paucity of electrons (*p-type* material). A *p–n junction* is formed when semiconductor crystals of opposite doping are fused together. Under equilibrium conditions a potential barrier develops as a direct result of a natural process called *diffusion*. (The equilibrium condition is shown in figure 6.6(a).) If a photon impacts on the junction its energy is absorbed and negative (electron) and positive (*hole*) charges are created (figure 6.6(b)). There are two ways to detect this event, i.e. using a:

- *photovoltaic detector*: the potential at the output leads—as measured by a voltmeter—will change in response to absorbed light; or a
- *photoconductive detector*: place a current measuring circuit (*ammeter*) between the output leads; a current will flow (into the 'short circuit' presented by the ammeter) in response to the absorbed light.

The p–n junction just described only permits current to flow in one direction and the device is called a *diode*. The optical phenomena will generate currents in only one direction and in commercial form such sensors are called *photodiodes*. A *transistor* is created when two junctions are formed within the device (either pnp or npn). Transistors are inherently capable of amplification of any current that flows in the central (or *base*) region. Base currents can be generated electronically or light can be used to produce hole–electron pairs just as it does in the case of the photodiode. The resultant sensor is the phototransistor.

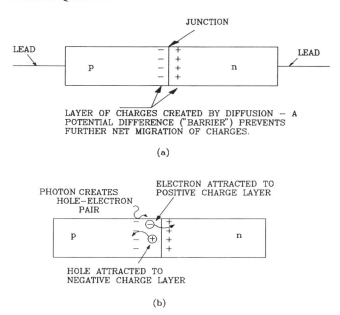

Figure 6.6 Semiconductor junction detectors. (a) A p–n junction in equilibrium. (b) Photons create hole–electron pairs that can be detected.

Junction transistors formed by the fusing of appropriately doped crystals have three leads with the current flowing through two terminals controlled by the current in the third (base) lead. Another type of three-terminal transistor, known as a *field-effect transistor* or FET, uses the voltage at the third terminal (the gate) to control the current flowing between the other two terminals. The voltage on the gate creates a double layer of charge within the body of the FET. The net effect is to create a capacitor, which can store charge for intervals of time. An array of such capacitors can be formed and the charges transferred from one 'capacitor' to another; such devices are referred to as *charge-coupled devices* (CCDs) [4]. CCDs have a variety of applications including signal processing (delay, filtering, multiplexing), memory, and logic arrays. Of particular interest are their possibilities for imaging; an array of such FET capacitors are arranged to form a pattern of ('stored') charge proportional to the light intensity that each element receives (the 'image'). If the array consists of a single row of CCDs then the optical image can be converted into electronic form by mechanically passing the row of sensors across the object under test. Alternatively a sensor array composed of $n \times n$ (e.g. 2048×2048) image-sensing elements can be scanned electronically by passing the stored charge in a systematic manner (e.g. one row at a time) to the A/D converter where a (binary-coded) numeric equivalent can be generated for further processing by the computer-based instrument system.

Optical fibre sensors. The idea of using fibre optic devices for information transduction is relatively recent. Fibre optics have been used in communication systems for some time; it was only when it became evident that the 'distortions' observed in such applications could be 'calibrated' that they became useful transducers. They have a number of potential advantages over some of the other techniques described in this chapter:

- They may be better suited to hostile environments.
- They are insensitive to electrical interference.
- They can be used in hard-to-reach locations.
- They have low cost, and are often disposable (for medical applications).

A simplified design for a pressure transducer based on fibre optic light modulation is shown in figure 6.7(a) [5, 6]. Incident illumination is reflected by a pressure-sensitive diaphragm. Depending on the wavelength of the incident light as well as the (pressure-dependent) size of the optical cavity, the amount of reflected light will vary. Figure 6.7(b) shows how an optical fibre is coupled to the transducer to provide a measure of the pressure. (In order to achieve useable linearity and range, a ratio of reflected light due to different wavelengths can be measured using a dichroic mirror as part of the system.)

A temperature measurement can be achieved if the diaphragm is made of silicon. The modified transducer is shown in figure 6.7(c). Each surface of the silicon diaphragm is partially reflecting. Thus the silicon functions as an optical cavity. The index of refraction of the silicon varies as temperature changes; this variation is substantial at certain characteristic frequencies of incident light. Furthermore, the same computer-based instrumentation can be used to interrogate the pressure and temperature sensors.

6.1.3 Force and/or pressure transducers

The application of force to, or pressure on, an object normally results in some kind of deformation. Forces that act across a (unit) area in a solid material that resists the separation, compacting or sliding that tends to be induced by external forces are considered to be mechanical *stress*; the resultant deformation is *strain*. These definitions are given by

$$\text{Stress} = \frac{\text{Applied Force}}{\text{Area}} = \frac{F}{A}$$

$$\text{Strain} = \frac{\text{Elongation}}{\text{Undistorted length}} = \frac{\Delta l}{l}.$$

In addition, it is to be noted that within certain limits [7] stress is proportional to the resultant strain. Combining these relationships yields

$$\Delta l = \frac{Fl}{AE}$$

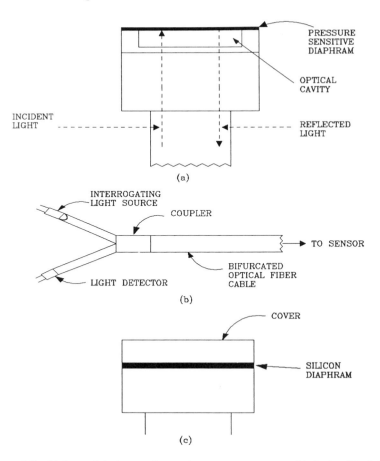

Figure 6.7 Light modulation can be used to measure pressure. (a) A simplified diagram of the transducer. (b) A schematic diagram of a measurement system. (c) The transducer adapted to measure temperature.

where E is a constant of proportionality. Notice that the elongation is proportional to the force that produced it. Moreover, having a measure of force permits a user to measure such quantities as velocity and/or position of an object, these being determined from physical definitions:

$$\text{Force} = \text{Mass} \times \text{Acceleration} = Ma$$

$$\text{Velocity} = \int a \, dt$$

$$\text{Distance} = \int v \, dt.$$

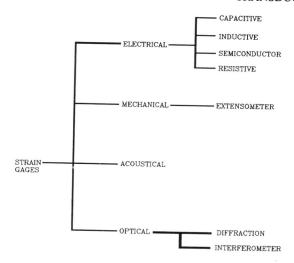

Figure 6.8 Types of strain gauges.

Strain gauges. These are the principal devices for the measurement of stress and strain. ('Forces' on an object may result from indirect causes such as temperature changes; consequently strain gauge applications include detection and measurement of: displacement (position); pressure; acceleration; applied force; and temperature.) A breakdown of strain gauge types is shown in figure 6.8. The particular choice of gauge depends on a number of parameters: accuracy and stability (as with temperature changes, for example) of the gauge; maximum deformation expected; location of the gauge. Tables 6.3 and 6.4 summarize some factors that must be considered when choosing an appropriate gauge. Although table 6.4 is not exhaustive it is representative of a variety of pressure transducers. In addition, there are other parameters that the user should note when specifying such sensors: excitation; temperature range and effects; sensitivity to shock and vibration; and price.

Resistive strain gauges. Deformation of an electrical (metallic) conductor in response to a force changes its length and cross sectional area; in other words it exhibits strain. The resistance of the conductor is proportional to its length and inversely proportional to its cross sectional area; a direct measure of strain (and hence applied force) is thus possible (see figure 6.9). Using the definition of resistance and the fact that the volume of the conductor must remain constant whether stretched or undistorted, it is possible to show that

$$\frac{\Delta R}{R} = 2\left(\frac{\Delta l}{l}\right).$$

That is, the change in resistance is approximately twice the strain. (The constant or proportionality, 2, is called the *gauge factor*.)

Table 6.3 Strain gauge types.

Type	Characteristics
Mechanical	Large size; relatively inaccurate; useful for large distortions; easy reading
Optical	Light rays substituted for mechanical levers; useful for some dynamic measurements; some systems insensitive to temperature variation (diffraction systems); reading-at-a-distance (laser/interferometer); good dynamic range
Acoustical	replacement for mechanical systems, with Vernier capability; can have large dynamic range; insensitive to temperature effects

Notice from the figure that the resistivity of the conductor, ρ, may also change when the conductor is distorted. For metallic conductors this effect is very small—ρ is constant. For **semiconductor strain gauges** (e.g. germanium) the resistivity is a function of the (systematic) concentration of the charged impurity carriers. For such materials the change in electrical resistance under deformation is much greater than that for metallic conductors. The gauge factor for such transducers is typically 20 to 90 times greater than for the metallic transducers. Semiconductor strain gauges have several other potential advantages: small size; high fatigue life (10^7 cycles with no damage); and low hysteresis (memory). However, semiconductor transducers are far more sensitive to temperature variations than metallic strain gauges although manufacturers supply the semiconductor versions with built-in temperature compensation.

Typically, strain gauge sensors are arranged in a *bridge circuit* as shown in figure 6.10. A temperature-compensating resistor R_c is added as shown. The bridge circuit is a common electrical arrangement used for measurement applications. A bridge circuit is often used when accurate measurements are required [8].

Piezoelectric sensors. The old phonograph needle is an excellent example of this type of sensor. The grooves in the record exert force on the needle which generates an output signal (voltage) as a direct consequence. The needle is a quartz crystal and for such material the following observation is apposite: *a crystal will deform if an electric field is placed across it and, conversely, a crystal will produce a voltage if it is deformed as a consequence of applied force*. In addition to quartz there are other materials that exhibit this *piezoelectric effect* including barium titanate ceramic and lead zirconate titanate ceramic. As a force sensor it is very linear; the (change in) voltage generated is proportional to the change in force. However, when a steady force is applied the generated energy (voltage) will eventually be dissipated in the material's internal resistance. Therefore such sensors are only suitable for measurements of transient ('AC')

Table 6.4 Pressure/strain gauge parameters.

Sensor	Output	Accuracy (%)	Range (PSI)[a]	Frequency response[b]	Stability per year (%)	Life (cycles)[c]
Capacitor	High level	0.05 to 0.5	0.01 to 5000	DC to > 100 Hz	0.05 to 1.0	> 10^7
Differential transformer	High level	0.05	30 to 10 000	> 100 Hz	0.25	> 10^6
Piezoelectric	Medium needs amplification	1	0.1 to 10 000	1 Hz > 100 kHz	1	Indefinite
Strain bonded or thin film	Low (3mV/(V of excitation))	0.25	0.5 to 10 000	DC to > 5 kHz	0.5	10^6
Semiconductor (bonded)	Medium (3–20 mV/(V of excitation))	0.25	5 to 10 000	DC to > 5 kHz	0.5	10^4

[a] PSI = pounds per square inch.
[b] DC = direct current ('0 Hz').
[c] Life is limited by the shift in calibration (between 0.05% and 0.5%).

Table 6.5 Characteristics of electrical strain gauges.

Type	Characteristic
Inductive/capacitive	Wide dynamic range; with modifications can be used to measure velocity, acceleration, pressure or flow rate
Semiconductor	Very sensitive (can measure very small strains); extreme temperature sensitivity (without compensation); only moderate linearity; less susceptible to failure under cyclical strain than resistive type
Resistive	Sensitive to temperature variation; low sensitivity (requiring amplification); versatile

phenomena.

Mechanical sensors. The first strain gauges were mechanical; however, a number of disadvantages (poor accuracy, large size) limit their usefulness. All versions operate as follows: two points of such devices (*knife edges*) are secured to the test object; a series of levers magnify the displacement, which permits a user actually to read the displacement, which can be converted into strain and stress. An electromechanical version (*extensometer*) exists and is used for industrial applications as well as materials testing. Reading strain directly from a dial can be convenient and when size is not important such gauges can be used.

Capacitive sensors. Capacitance is a measure of ability of a pair of conductors to retain electrical charge when a given amount of potential exists between them. Formally it is given by:

$$C = \frac{Q}{V}$$

where Q is charge in coulombs and V is the potential in volts. The units of capacitance (C) are farads (F) (or subdivisions such as microfarads (10^{-6} F)). The shape and dimensions of the conductors are important determinants of the capacitance. A number of possible shapes and their corresponding capacitances are shown in figure 6.11. These shapes are not exhaustive but it is clear that any measurement that alters an appropriate dimension will change the capacitance. Measurement of the sensor's capacitance proceeds in a way that parallels the measurement of resistance with one very important difference: alternating current and voltages must be considered in place of the direct currents used to test for resistance. The equations that govern the measurement are

$$I_{capacitor} = \frac{(\text{AC voltage across capacitor})}{X_{capacitor}}$$

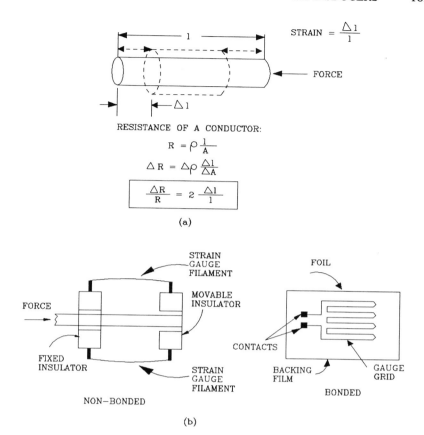

Figure 6.9 The basis of resistive strain gauges. (a) Resistive parameters. (b) Strain gauge configurations.

where $X_{\text{capacitor}}$ is given by:

$$X_{\text{capacitor}} = \frac{1}{2\pi f C}$$

where f is the frequency of the alternating current. By using the formulas given in figure 6.11 the relationship between test current and dimensions can be determined. For example, for the parallel-plate capacitive sensor the result is

$$E_c = \frac{I_c d}{2\pi (0.225) A k}.$$

In other words, for a known test current of I_c the measured voltage (E_c) is directly proportional to the distance between the plates (d). This distance

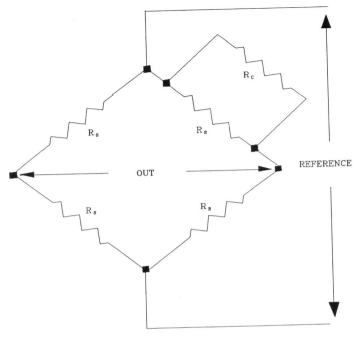

R_s ARE STRAIN GAGE RESISTORS ARRANGED IN TENSION/COMPRESSION

Figure 6.10 A bridge circuit for strain gauge measurements with temperature compensation.

can be changed as a result of applied forces. This leads to a variety of sensor applications: a strain gauge; a microphone (where the applied force is the air pressure produced by sound); displacement (physical movement of the plates); and the presence of an object between the two plates. As with the resistor strain gauge, the capacitance of the sensor can be measured using a bridge circuit.

Inductive sensors are discussed in section 6.1.5 (on position encoding) and their potential for stress measurement (strain gauge applications) is explained.

6.1.4 Electrochemical sensors

Electrochemical sensors respond selectively to (chemical) changes in the sample being tested [9]. Such sensors consist of two electrodes, one called the reference electrode and the second called the ion-selective electrode (ISE). Potentiometric transducers generate a potential difference between the electrodes in response to chemical activity. Conversely, in some cases a potential may be applied to the electrodes, producing a current that is related to the concentration of the analyte. (This arrangement produces a reduction–oxidation (redox) reaction.)

Features of ISEs include:

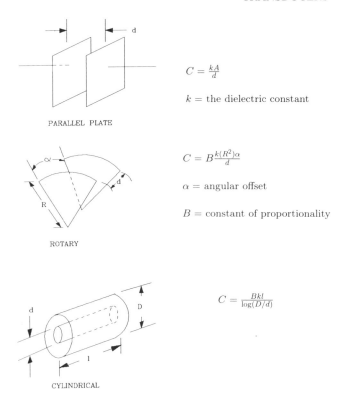

Figure 6.11 Capacitive strain gauges.

- They do not destroy or contaminate the sample.
- They do not require sample dilution or additional preparations.

The ISE operates like a storage battery with an important difference being the resultant potential, which must vary as the amount of analyte in the sample. Figure 6.12 is a schematic picture of an ISE. Operation of such electrochemical sensors is based on a fundamental observation of nature: *a potential difference is generated at the junction of dissimilar solutions (as well as metals)*. As shown in the diagram there are a number of 'junctions'. If all junction potentials—with the exception of the membrane/sample junction—remain constant, the resultant potential difference as measured by the voltmeter is a function of the (free) concentration of the analyte in the sample (c_x). The Nernst equation governs this measurement:

$$E = E_1 + \left(\frac{RT}{nF}\right) \ln c_x$$

where:

E = the measured result;

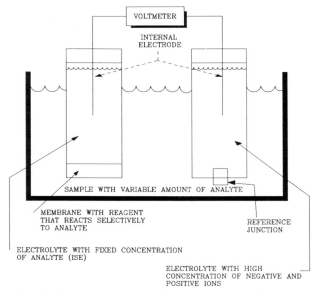

Figure 6.12 A schematic representation of an electrochemical sensor.

E_1 = the sum of all potential differences except membrane/sample—a constant;
R = the gas constant;
T = temperature in degrees Kelvin;
n = the charge on the chemical species of interest in the sample;
F = the Faraday number.

E is a linear function of $\ln c_x$; the slope of this line is a measure of the sensitivity of the device. Theoretically it is equal to 59.12 mV per decade change in activity for a monovalent positively charged ion (cation) at 25 °C. Types of ISE membranes include: glass (originally used in pH meters) with the composition determining ionic selectivity; wires coated with appropriate metal salt films; pressed silver sulphide pellets; reagent-impregnated polymers; organic liquids. The gas content of the sample can be measured using a scheme in which the ISE system is separated from the sample by a gas-permeable layer. Gases diffuse through the layer and generate a change in electrolyte composition, which is measured with the potentiometric sensor. The reference electrode must generate a stable, repeatable, junction potential that is not influenced by the sample. It must not interact chemically with the sample; silver/silver chloride wire is sometimes used as the reference electrode.

With amperometric sensors the selectivity is a function of the applied potential and/or use of a physical membrane between sensor and sample.

6.1.5 Position encoding

The mechanical position of an instrument part can, at times, be important. For example, the setting of a rotatable grating in a diffraction grating monochromator will determine the wavelength of the output light; this must be known accurately. While electronic conversion may be possible, it is sometimes simpler to use an electromechanical device. A number of such transducers have found application in modern instrumentation systems.

Shaft encoder. A series of brushes ride on a disc that contains a pattern of conducting and non-conducting regions (see figure 2.4). In operation, the brushes and disc move relative to each other—a fixed disc and moving brushes or the reverse. The position of the brushes relative to the pattern determines a uniquely coded representation of the position. One brush, called the *common*, may be used to determine which of the other brushes are, or are not, in contact with it. This produces a unique code which, for example, might be a straight binary function of the distance of the brushes from some (arbitrary) starting point. If the ends of the disc are joined together, forming a circle, the result is a rotating device which can be employed in a variety of ways.

Opaque and transparent segments may be used in place of the conducting and non-conducting elements. A light source is placed on one side of the disc and optical detectors placed on the other side. The outputs of the optical detectors produce a binary sequence similarly to the case just described. Such a device is called an *optical shaft encoder*. It has some advantages (e.g. less disc wear) and some disadvantages (e.g. higher power consumption) when compared to the brush system.

One of the requirements of shaft encoders described above is the need to have absolutely perfect alignment of the reading elements (brushes or lights/detectors). Figure 2.4(a) shows the potential for excessive error when alignment is imperfect. In the case shown, a very small misalignment can produce a coded result that is greatly in error. For the binary code shown it can be seen that, depending on which brushes are misaligned, any one of the potential codes may be generated. This limitation may be overcome by using a different code. The *Gray code* depicted in figure 2.4 is one example of a *unit-distance code*. For such codes only one bit changes between code patterns. Thus, brush misalignment can only result in an error of one incremental position of the shaft encoder; this is normally acceptable. (Misalignment of fewer than two discrete coded positions is not difficult to achieve in commercially available encoders.) While the error is eliminated it is to be noted that an additional burden is now placed on the computer. Any computation on the data (e.g. arithmetic calculations, data display, or simply tabulation of results in decimal format) requires an additional conversion from Gray code to normal binary.

The ability of an encoder to resolve the physical dimension depends on the number of bits in the code. Specifically, for n brushes (not counting the common brush), the resolution is 1 part in 2^n. (Eight brushes divides the 'ruler' into 256 parts.)

Potentiometer. A shaft—either rotatable or sliding—moves a wiper (brush) across resistive material. The resistance between the wiper and (either) end of the transducer thus varies in accordance with the position of the wiper. (The variation can be quite linear or in some cases it can be made to vary as the log of the position, providing not only transduction but built-in preprocessing which is useful in some cases (e.g. in some optical measurements the log of the transmittance is the parameter of interest).) Such transducers are popular because of their general economy, availability, and simplicity. Their disadvantages include: limited life; inherent high noise; and poor frequency response. (Because of the size and consequent inertia of such devices, they cannot be moved at particularly high speeds.) The life of such devices can be greatly increased if certain conductive plastics are used for the resistive material; the cost of such devices is high when compared to wire- or carbon-based potentiometers.

The resistance of the potentiometer—and hence the position of the shaft—must be converted into a signal that is suitable for further data processing. This may be accomplished as previously outlined: pass a known test current through the transducer and measure the resultant voltage generated.

Variable transformers. While these devices are described as position sensors it will become apparent from the discussion that they are also suitable as force transducers. Figure 6.13 includes schematic drawings of the variable transformer transducer which can be fabricated as rotating devices as well as linear devices. The underlying principle of operation comes from the fact that: *a changing current gives rise to a changing magnetic field and a changing magnetic field gives rise to a changing electric field*. The excitation (input voltage) produces a current, which then generates a magnetic field primarily confined to the windings within the housing. This changing magnetic field generates ('induces') voltages in each of the secondary windings. These windings are arranged such that their voltages subtract from each other. The magnitude of the induced voltage is greatly dependent on the proximity of the transducer's movable core to a particular secondary coil. The result is a linear relationship between the position of the core and the output signal. The magnitude of the output increases for both positive and negative displacements; however, the time correspondence ('phase') between primary (reference) and output changes for positive and negative displacements.

Application as a position transducer follows directly from the inherent operation of the variable transformer sensor. Its operation as a force transducer is less obvious—although a force must be applied to move the core. Figure 6.14 shows how such a device can be employed to measure forces (e.g. weight). (When used in this way the transducer may be referred to as a *load cell*.) The system included in the figure is referred to as a *servomechanism*. The objective of a servomechanism is to accept an input signal (called the reference or reference input) and generate an exact replica (called the *controlled* signal), which is capable of considerably more work than the reference signal. In this

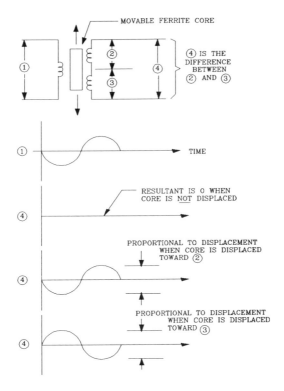

Figure 6.13 Variable differential transformers—principles of operation.

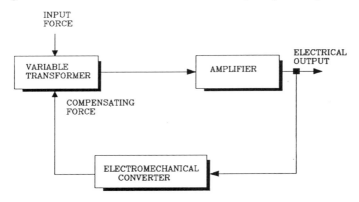

Figure 6.14 A load cell for measuring force using a variable transformer ('load cell').

example the applied force (possibly in the form of the weight of the object to be measured) is the 'reference' and the voltage needed to balance this input (eventually) exactly is the 'controlled' output. The (servomechanism) system works as follows: the core of the variable transformer can be (mechanically) actuated from both sides; if it is not in a neutral position, a net signal is generated (as described above) which is transmitted to the amplifier; this *error* signal is (greatly) amplified and produces an output that drives the opposite end of the variable transformer core; the direction of the drive reduces the error signal. The *feedback* process continues until the 'restoring' drive just equals the force (or weight) to be measured. The signal (voltage at the output) is a direct measure of the balancing force; hence the unknown input to be measured.

Other position-sensing systems. It may not always be convenient to locate the position transducer in very close proximity to the source signal. For example, to measure the position of a human subject (arms, legs, etc) it may not be possible to use a telemetry instrumentation architecture without interfering with the underlying behaviour—the transmitter may be too large and bulky (see section 6.3). However, a small light source that does not require a lot of power (e.g. a light-emitting diode (LED)) may be affixed to the moving part and the light emitted may be detected by a remotely mounted TV system. In another example, a grid of LEDs and oppositely sited phototransistors may be used to locate the coordinates of the object within the field of view. When the light beam is interrupted by the presence of the object, two phototransistors signal this event thus identifying the X- and Y-coordinates of the location. An acoustic equivalent of such systems makes use of ultrasonic sources and detectors in place of the optical elements.

A number of other sensing applications are regularly reported in the literature. Examples include:

- A position-sensing scheme where the location is sensed by voltage changes that occur when the object straddles adjacent regions of the observation field [10].
- A sensing scheme that determines whether or not a test point has been touched using changes in capacitance (between the terminal and ground) [11].

6.2 SIGNAL CONDITIONING

Signal-conditioning components form the interface between the sensor/transducer and the A/D converter or preprocessing circuitry if this is present in the system. The conditioning elements 'shape' the signal for further processing. Examples include: amplifying the signal to improve resolution; eliminating contamination by limiting the frequency content of the information; producing an 'event marker' from raw sensor information; converting sensor data to 'engineering units'; and extracting a portion of the sensor signal for further processing. Such transformations are achieved through the use of operational amplifier ('op amp') circuits (either linear or non-linear) [12].

An op amp is an electronic circuit that, in combination with other components such as resistors, capacitors, diodes and transistors, can generate linear as well as non-linear signals in response to the input signal. A short list of the functions that can be formed in data-acquisition, automation, and instrument control applications includes:

- Among the linear examples:
    amplification (gain (with/without phase inversion) for voltages and/or currents), either single ended or differential;
    impedance matching or signal isolation;
    conversion of current input to voltage output (as well as voltage input to current output);
    integration or differentiation of the input;
    filters.
- Among the non-linear examples:
    oscillators (sinusoidal, square-wave);
    multipliers, dividers;
    rectifiers (half-wave, full-wave);
    peak detectors;
    sample and hold circuits;
    comparators.

Figure 6.15 includes the generally accepted symbol for the op amp. Electronics within the op amp includes an amplifier with very high gain ($\sim 10^5$ or 10^6) whose output is the amplified difference between signals at the input terminals. When passive (e.g. resistors) and/or active (e.g. transistors) components are added to the basic circuit the resultant feedback arrangement generates functional operations on the input(s) to the combination. Figure 6.15 includes some simple representative op amp circuits; it is by no means an exhaustive list. Moreover these are 'ideal' circuits in that they do not include compensation for the limitations (or errors) inherent in practical op amps. Some of the limitations in op amp specifications that must be considered when choosing a suitable integrated circuit—together with their nominal ('ideal') values in parentheses—are: input offset voltage (0), input offset current (0), input bias current (0), input resistance (∞), common mode rejection ratio (ideally the CMRR is ∞ dB). (Practical deviations from the ideal values will introduce inaccuracies into the functional calculations that are performed.)

As an example of how such signal-conditioning circuits might be used, consider the following. An automobile manufacturer uses a computer-based automated quality control system to test the transmission. A simplified block diagram of this is shown in figure 6.16 together with the representative signal-conditioning circuits. A transmission is selected (according to quality assurance rules) and placed under test. This system is tested for a number of hours and sampled to measure a number of parameters. Two significant parameters are the temperature (at various places within the transmission) and vibration. (Only single-measurement sites are shown.)

Recall that a thermistor temperature transducer requires a constant current

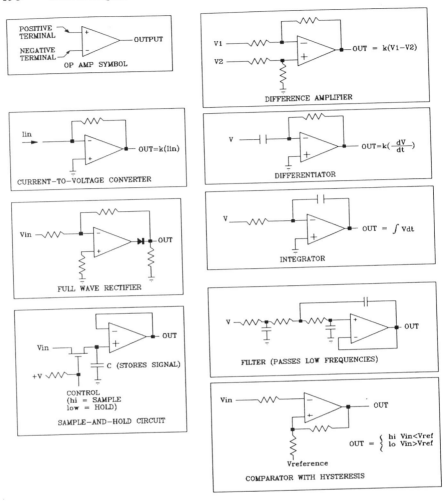

Figure 6.15   Sampling of operational amplifier (op amp) circuits.

to prevent thermal runaway. The constant-current ('Howland') circuit shown in the figure supplies a constant current to the thermistor. The constant-value test current supplied to the thermistor generates a voltage ($V = IR$), which is dependent on the resistance value of the sensor; this, in turn depends on the temperature of the transmission under test. The potential across the thermistor is further amplified so that the resultant voltage is directly related to the underlying temperature. (An additional resistor that produces approximate linearity may be included but is not shown.)

The piezoelectric crystal will generate a voltage in response to mechanical stress. Vibrations within the transmission will cause mechanical deformations of

SIGNAL CONDITIONING 199

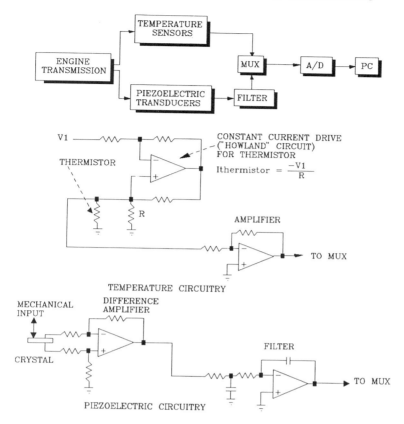

Figure 6.16 A simplified block diagram of an automated quality control system.

the crystal which generates a corresponding voltage. This potential is amplified using a difference amplifier circuit. The output may then pass to a filter, which limits the highest frequencies that can be passed to the Mux (and ultimately the computer-based acquisition system). For example, the information from the transmission may be sampled at a 5 kHz rate. This limits the highest rate of information to 2.5 kHz—see chapter 2. Thus, any frequencies beyond 2.5 kHz are considered to be contaminants. (This, of course, might be an unwarranted assumption but it is inherent in the way in which this system is configured.) One artefact that may be introduced occurs because of subtle signal paths introduced by the 5 kHz sampling—via the power supplies, for example. The low-pass filter shown in the figure will eliminate all frequencies beyond 2.5 kHz if appropriate values for the resistances and capacitors are selected. (The term 'low-pass' refers to the fact that this circuit will only transmit or 'pass' signals whose frequencies are below the filter's 'cut-off frequency'. Other filters include 'high-pass' and 'bandpass' with corresponding connotations.)

6.3 TELEMETRY

Physical connection between the computer and the entity it is intended to measure or control is not always possible or desirable. For example, precipitation measurements from the top of a mountain are not directly feasible. On other occasions the data to be gathered may be in a hostile environment (e.g. where there is high radioactivity or an extremely corrosive atmosphere). Or how is one to track bird migration and navigation by computer? In these, and other cases, telemetry is a useful or necessary alternative.

Figure 6.17 shows two possible telemetry architectures but many others are possible. Figure 6.17(a) includes a sensor whose output provides the information to a transmitter. At some remote distance this information is received where it is converted into signals that are similar to the original source data. From there it is converted into a form that is suitable for computer processing. (This is performed by the encoder which is discussed in section 6.4.) A more elaborate system is contained in figure 6.17(b). This involves several sensors whose outputs are multiplexed and then encoded into a discrete (digital) form. The transmitter sends these signals to the remote data-processing facility. In addition, the telemetering station can receive information from the computer (via a transmitter at the computer resource), and convert the data into a series of control signals which are used locally.

Power for the telemetering stations is not shown but this is an important consideration when specifying station requirements. Power may be supplied from any of a number of possible sources: storage battery; gasoline-powered generator; atomic energy, solar power; wind power; and wave power. If the remote telemetering station is accessible for periodic servicing, a storage battery might be suitable, but if the facility cannot be readily reached, solar energy might be a better alternative.

The transmitter contains the essential communication elements. A communication system—even wire communication—can be described as having three principal elements:

- *Modulation*: this is the information to be transmitted.
- *Carrier*: the energy on which the modulation is attached.
- *Medium*: the means by which the carrier and its resultant modulation reach the destination.

There are two widely recognized communication protocols for information transmission. One is to alter the amplitude of the carrier in accordance with the data, which is known as *amplitude modulation* or simply AM. The second is to vary the carrier's repetition rate, which results in *frequency modulation* or FM. There are a variety of communication arrangements and these differ mainly in the way in which the information is encoded prior to the modulation process. (Communication systems may include 'suppressed carrier' or 'single-sideband' configurations, but these specialized concepts will not be described.) Figure 6.18

TELEMETRY   201

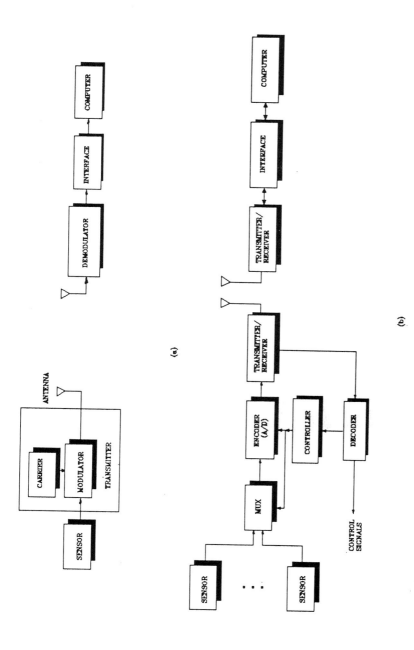

Figure 6.17 Telemetry configurations for data acquisition and process control. (a) A simple system. (b) High-data-capacity application.

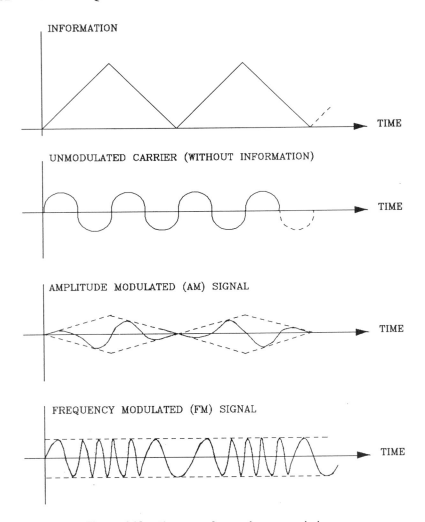

Figure 6.18 Elements of AM and FM transmission.

depicts the essential principles of AM and FM.

6.4 DATA CONVERSION

Digital computers operate on digital quantities—normally binary numbers—while experiments and processes generate analogue information. Thus, components must be included in any instrumentation system to convert from one form (analogue) to the other (digital). There are a variety of such devices including those employing an electronic counter as well as mechanical devices (shaft encoders). (Shaft encoders were described in section 6.1.5, on position encoding.) Additionally, there are circumstances in which the computer supplies a binary number that must be converted to a voltage or current that is required by an instrument.

6.4.1 Analogue-to-digital (A/D) conversion

Basically, this conversion process is achieved by performing the opposite calculation! That is, an electronic counter generates a sequence of binary numbers; these are used to generate an analogue signal, which is then compared to either the signal to be converted or to some reference (like zero). The binary number at which the counter stops when the comparison is satisfied is the digital equivalent of the unknown signal. This general principle gives rise to a number of different specific methods: binary counting scheme, successive approximation, flash converter, dual slope integration. Each of these is now briefly described [13].

Binary counting scheme. Figure 6.19 is a simplified model employing this method. The switches represent mechanical equivalents of the electronic counter. There are four possible combinations for the two switches each having two positions. Figure 6.19(b) shows the signal (voltage) transmitted to the comparator circuit for each of the possible switch settings. (To verify this, redraw the resistor network for each setting and replace the switch with a battery of either V V or 0 V (ground)—then calculate the voltage at the junction of the resistors.) According to the table the switches sequence as follows: 00, 01, 10, 11; this is the binary equivalent of 0, 1, 2, and 3. If one imagines for the moment that the input is restricted to a value between 0 and $+V$, then the switch settings (in combination with the comparator) can be used to compute its value. This is accomplished according to the algorithm shown in figure 6.20(a). According to this procedure the switches will assume a binary setting that is equivalent to the unknown signal when the computation is complete. (In this very simple model the unknown can only be resolved into one part out of four; however, the scheme can be extended to much higher resolving power. Add more switches and resistors (whose values double for each successive resistor added). The resultant resolution will be 1 part in 2^n where n is the number of switches.) One limitation with this method is the time that it takes to reach an answer. If, for example, the input had a value of $+V$ then it would require four switch-setting steps to obtain the result. A modified algorithm—called the successive-approximation method—requires n steps (where n is the number of switches) to arrive at the result; in the current example that would be two switch adjustments instead of the four illustrated above.

Successive-approximation method. Using the same system of resistors, switches and comparator circuits as the binary counting scheme, refer to the algorithm in figure 6.20(b). Starting with the most significant one, each switch in turn is set to 1. If the resistor network output exceeds the unknown the switch

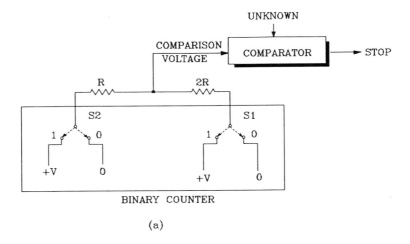

Figure 6.19 A simplified model of analogue-to-digital conversion. (a) The block/circuit diagram. (b) Comparison voltage versus switch position for a binary counting scheme.

is reset to 0. If this procedure is followed for the example shown above (input $= +V$) then the switch sequence will be 10 followed by 11. Thus it would have required two iterations to arrive at the (closest) answer as contrasted with four iterations for the straight binary counting scheme.

The resistor network used to illustrate the A/D methods just discussed is limited in several important ways: for higher-resolution conversions the resistor values quickly reach impractical values (recall the need to double the resistor value for each increase in resolution); the resistors require tight tolerances—

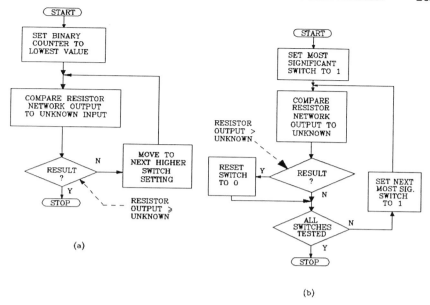

Figure 6.20 A/D procedures. (a) A binary counting scheme. (b) The successive-approximation method.

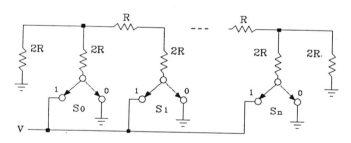

Figure 6.21 An R–$2R$ ladder for A/D conversions.

high accuracy—in order to maintain the accuracy of the scheme; high resistance values tend to introduce considerable noise. In place of the network shown an 'R–$2R$ ladder' network is normally used. An R–$2R$ ladder is shown in figure 6.21 and the output voltage is given by

$$V_{\text{out}} = \sum_i \frac{V_i}{2^{n-i}}$$

where the index i ranges from 0 to n (the number of counting elements) and V_n is either $+V$ or 0 depending on the individual switch setting, which may follow a straight binary sequence or a successive-approximation sequence.

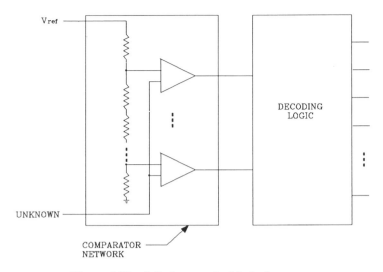

Figure 6.22 A flash converter block diagram.

Flash converter. Instead of using successive approximations to resolve the unknown signal, the comparisons can be made simultaneously. The obvious advantage is the dramatic improvement in the time required to complete the conversion. However, such converters are expensive and often have somewhat lower resolution than the other methods. Figure 6.22 is a block diagram of such converters. The reference signal is divided into n distinct voltages where n is the number of discrete binary numbers that the A/D can reproduce (e.g. 256). The unknown is compared against these distinct voltages using n comparator circuits. The comparator outputs will be determined by the relationship between the unknown and each of the distinct voltages derived from the reference. These comparators generate binary results ('yes', the unknown is greater than the derived reference, or 'no', the unknown is not greater than the derived reference) which are passed to a decoder network. This network transforms the comparator results into one of n distinct binary numbers, the binary number being equivalent to the unknown signal. Flash converters may be able to complete a conversion within 100×10^{-9} s (100 ns) corresponding to data rates of 100 MHz. It is to be noted that computers can rarely process data at this rate and thus such converters are reserved for specialized applications. (A series of very rapid conversions are made and stored in a local memory (buffer); they are subsequently processed at a slower rate by the main computer.)

Dual slope conversion. This is somewhat different from the methods just described and has better resistance to noise on the input; figure 6.23 depicts the method. The unknown signal is applied to an electronic integrator for a predetermined period of time. The voltage out of the integrator is thus proportional to the unknown voltage V_x at the end of this period. The integrator

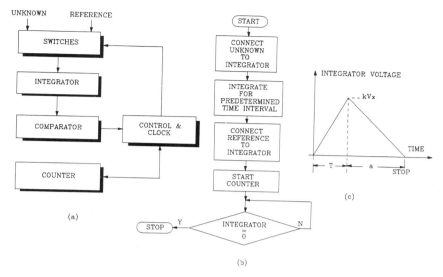

Figure 6.23 Dual slope A/D conversion. (a) A block diagram. (b) A flow diagram of the algorithm. (c) The signal description

then accepts a reference input and continues the integration (with negative slope); at the same time a counter is started. When the integrator reaches zero the counter is stopped and the number in the counter is proportional to a (in figure 6.23(c)) and to the unknown signal V_x.

6.4.1.1 Practical A/D converters. Because the A/D technology is well developed, a variety of such devices are available in integrated circuit form. A typical block diagram of such devices is shown in figure 6.24. Large-scale integration of electronic circuits has led to the development of multipurpose devices. (As shown in the figure, A/D converters in very-large-scale integrated (VLSI) form include a multiplexer, which was formally a separate component in figure 6.1.) The VLSI circuit contains: address lines (normally derived from the computer) which determine which one of the analogue signals is to be converted; the analogue inputs themselves; a control line, which initiates the conversion process; a status line, which informs the computer when the conversion has been completed; a clock (for sequencing the counter); a voltage reference; the data output lines, which contain the binary conversion; an output enable line. This last line is used to connect the A/D to the computer's data lines under computer control. When the output is disabled other devices may be safely connected to the data lines without interference.

While a number of A/D architectures are possible [14] several parameters, which are common to most systems, must be defined and/or understood when specifying an A/D converter. These are now described.

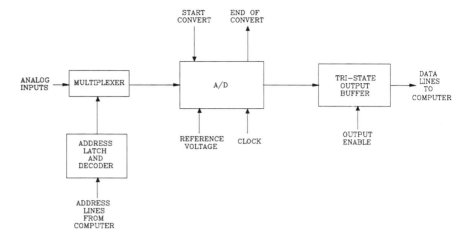

Figure 6.24 A block diagram of the integrated circuit version of an A/D converter.

Code. The digital code at the output of the converter. Some of the popular codes in addition to normal binary are:
- *Offset binary*—the input analogue signal is offset by half of the full scale plus one half of a least significant bit.
- *Sign magnitude*—the most significant bit of the binary number is a sign bit (e.g. 0 corresponds to a positive number and 1 corresponds to a negative number). The rest of the binary number represents the magnitude of the conversion.
- *BCD code*—although not an efficient coding scheme, it is somewhat more convenient for displaying results in decimal number applications (e.g. stand-alone instruments). The result appears as a sequence of 4-bit numbers with each group corresponding to one decimal digit (e.g. 0010 1001 is equivalent to 29 decimal).
- *Twos complement*—convenient for arithmetic operations. There are several ways to find the twos complement of a binary number. One simple algorithm is to start at the LSB (of a normal binary number), and copy over each bit up to and including the first 1. From that point on (to the MSB) complement (invert) each bit of the original number.

Errors. A number of errors within VLSI converters will contribute to inaccuracies in the results. The user should be aware of the errors as shown in figure 6.25; the effects are greatly exaggerated. An important specification requires that when the input is increased the output increases as well. This specification is characterized by the term *monotonicity*.

Limitations on the input(s). The A/D converter manufacturer should include such characteristics as:

- Number of channels—single-ended or differential.

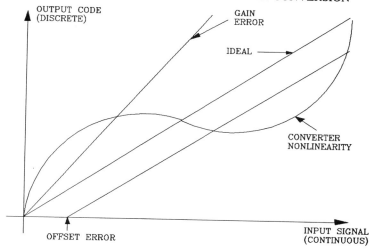

Figure 6.25 Error effects in an A/D converter.

- Permissible ranges—often several (under computer control).
- Maximum values.
- Input resistance and input capacitance.
- Bandwidth—the range of frequencies that will not contribute materially to inadequate response (normally specified as the frequency at which the system amplification has fallen by 3 dB (decibels).
- Amount of isolation between channels—the extent to which signals on one channel are 'carried over' (cross-talk) to an adjacent one.
- Size of the binary output—the number of bits of resolution, which indicates how 'finely' the signal can be measured.
- Conversion rate—how much time is required to complete a conversion.
- Accuracy—often specified in terms of the confidence of the result in bits.
- Temperature considerations—how much the gain and offset drift with temperature.
- Environmental—such characteristics as operating and storage temperature, and humidity, as well as the weight.

As an example, shown in table 6.6 are specifications of two different A/D converters, both of which are compatible with PCs; one is intended for extremely high-speed applications while the second is a low-cost, general-purpose device.

Commercial A/D converters such as the high-performance unit outlined in table 6.6 are often supplied with programming aids (in the form of a 'driver') which permit the user to set all the necessary parameters readily. One such driver provides a display on the computer's monitor similar to the one shown in figure 6.26. The user may conveniently employ the computer's mouse to select the parameters to be set; the keyboard may also be used.

A trigger signal initiates the conversion process. Such signals may be supplied

Table 6.6 Specifications for A/D converters.

	Parameter	High performance	Low cost
Input specification	No of channels	4, single ended[a]	8, single ended
	Range(s)	$0 \rightarrow +5$; $0 \rightarrow +10$ ± 2.5; ± 5; ± 10	± 5
	Maximum value	± 25	± 35
	Resistance	100 K	No specification
	Capacitance	10×10^{-12} F	No specification
	Bandwidth	6 MHz	—[c]
	Isolation	62 dB	No specification
A/D specifications	Resolution	12 bits	8 bits
	Conversion time	1 μs	30 μs
	Accuracy	2 bits	0.2% of reading ± 1 bit
	Linearity	No specification	± 1 bit
Miscellaneous specifications	Temperature drift offset	15 PPM K^{-1}	20 μ V K^{-1}
	Temperature drift gain	25 PPM K^{-1}	50 PPM K^{-1}
	Temperature drift linearity	1.5 PPM K^{-1}	No specification

[a] Sequence is under computer control.
[b] Input leakage current specified as 100×10^{-9} A, maximum.
[c] Requires 10 μs to allow A/D conversion to track signal.

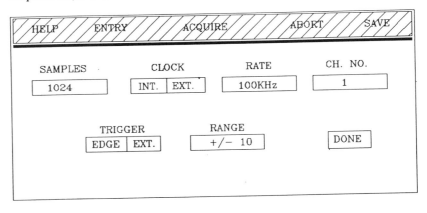

Figure 6.26 A sketch of display panel for an A/D converter.

from the computer or they may come from another instrument, perhaps a signal generator, which produces the stimulus to the experiment or preparation. The

A/D unit may trigger either on the rising or falling edge of the trigger signal, or the A/D unit may include a feature that permits the user to set the level of input signal on which to start the conversion. Some A/D units come with more than one triggering mode, which can be set (programmed) from the computer.

Modern instrumentation systems intended for laboratory or automation need to process information at very high rates. As noted above, flash converters may have data rates exceeding a megabyte per second (10^6 byte s^{-1}). This imposes formidable information capacity requirements on the computer. In some cases relatively high data rates may be achieved via *direct memory access* (DMA; see chapter 5). The resources for this operation may be included with the commercial A/D converter [14]. The CPU sends the starting (memory) address and number of words to be transferred to the DMA device. The A/D converter will fill the memory as rapidly as possible with the designated number of samples (words). Data rates of 100×10^3 byte (100 kbyte) appear as practical upper limits but new computer developments can be expected to increase this number. While the computer is executing a DMA operation the CPU (computer) is not available for other operations—it is entirely devoted to writing to (or reading from) its memory.

When DMA is not available, practical, or cannot handle the requisite information capacity, an alternative architecture may prove to be feasible. Some A/D units are available with additional memory built into the converter system itself, which serves potentially useful purposes. In these cases, the samples may be stored in the converter's memory; the computer may subsequently process this information at it own rate. (Of course it should be noted that the local A/D unit memory cannot be refilled until the computer has processed prior information. This leads to applications requiring short bursts of high capacity interleaved with periods of very low capacity.)

Because the conversion time is finite, however small, the user must be assured that the analogue input will remain constant (to within the limits of resolution) in order to avoid serious conversion error. Commercial A/D converters may be equipped with a special circuit to ensure that the signal supplied to the converter remains constant during the conversion process. This is called a *sample-and-hold* circuit.

6.4.2 Digital-to-analogue (D/A) conversion

It has been pointed out that the process of A/D conversion may have included an operation that performed the opposite conversion, namely *digital-to-analogue* or D/A. The conversion used an R–$2R$ resistive divider which is useful in its own right. For example, a robotic arm may require an analogue positioning or control signal. In a modern integrated instrumentation system the source of this control is likely to be the computer, which generates a digital output. Thus the need arises for conversion from digital to analogue form. D/A converters require a finite time to convert from the digital input to the equivalent analogue output

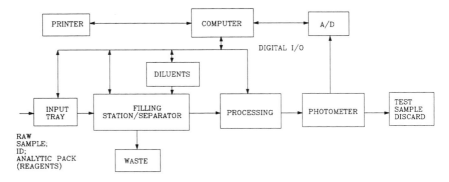

Figure 6.27 A block diagram of a serum analyser.

signal. Commercially supplied D/A converters will therefore specify a parameter called the *settling time*. From the start of the convert command until the end of the settling time, the user cannot depend on the accuracy of the analogue signal. This time may take a few microseconds and depends on how much the output is expected to change.

6.5 INSTRUMENT CONTROL: DIGITAL I/O

Laboratory instrument systems intended for use in automated environments will need to sense and control a variety of switches as well as electromechanical components. A block diagram of a representative application is shown in figure 6.27 [15] (additional examples can be found in chapter 1). Analysis of blood serum is one example of an automated assay system and includes functional components that can be found in other applications (e.g. analysis of urine). The components function as follows:

- *Input tray*. Accepts the sample to be analysed, performs sample identification (possibly on the sample in the form of a product code), includes a pack of reagents for the desired tests to be performed.
- *Filling station*. The sample ID is read and transmitted to the computer as a digital code; a precision pipette is energized from the computer (digital on/off signal) and draws a small amount of serum sample; this is forced into the reagent pack; the valve on the appropriate diluent is opened (digital on/off signal) resulting in a serum wash operation.
- *Processing*. The test sample is warmed; the reagent pack is pierced (digital on/off control); tapping mixes the reagent and the (washed) serum; the drive chain is stopped (digital on/off control) for a period of time to permit reactions to occur.
- *Photometer*. The test sample moves into photometer; the test sample is forced into a cuvette (digital on/off control); a filter wheel is rotated to select

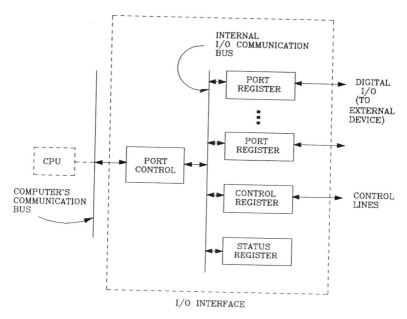

Figure 6.28 A block diagram of a digital I/O port interface for laboratory and process control applications.

the correct colour filter (digital on/off stepping of wheel); the absorption measurement is completed (data to the computer via A/D conversion); the test sample is discarded.

It is clear that such applications require circuits capable of accepting or supplying discrete (digital, or 'on/off') information under computer control in addition to the A/D (and D/A) resources previously described. In modern instrument systems such communication is provided by one or more digital I/O interface units [16]. Figure 6.28 is a block diagram showing the major functional units typically found in a digital I/O interface. Included are:

- *Port control*. This supports communication between the I/O interface and other elements of the computer such as the CPU or computer memory.
- *Port register*. This unit supplies (accepts) digital, on/off information to (from) the external device being controlled (e.g. the valve on one of the diluent reservoirs). There may be one or more of such ports in the I/O interface and each one may contain several lines (bits). Some computers may have three port registers with 8-bit lines in each for a total of 24 potential control lines. It is not unusual for some computers intended for laboratory or process control applications to have 72 (or more) I/O lines available.
- *Control register*. This provides two service functions. In one instance it may control an individual port register. Here it would control which port bits are

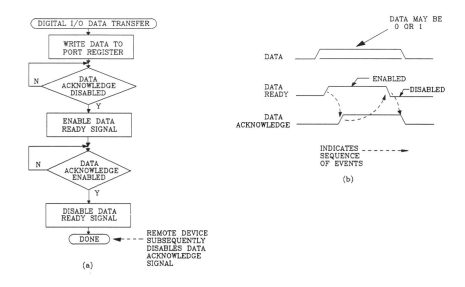

Figure 6.29 A typical handshaking and synchronization protocol for a digital I/O data transfer. (a) A flow diagram. (b) A timing diagram.

to receive data (input) and which are to supply data (output). At any point in a program the computer can issue a command to reverse the direction of any of the port bits. In the second case the control register is used to synchronize the transfer of data to/from the remote instrument. This may be necessary when the data capacity of the external device is not compatible with the I/O interface. (A more detailed description of the synchronization ('handshaking') is discussed below.)

- *Status register.* If one of the ports receives a byte from the remote device, one of the bits of this register is used to signal that fact. In addition, this register may contain a bit that assures the computer that the information transfer occurred without error (e.g. a parity error did not occur). An error may occur if the power to the external device is turned off and the I/O is trying to communicate. The I/O will continue to attempt communication to the external device 'forever'—it will 'hang-up'. Programs can take this into account by waiting a finite time and taking corrective action if 'time-out' occurs—that is, the completion bit in the status register has not been set in the requisite time period.

The I/O interface depicted in figure 6.28 can handle simple and complex input/output operations and control of external devices. For example, suppose

Table 6.7 Bytes written to the register port.

Function	Byte to register port
No reservoir enabled	1 1 1 1 1 1 1 1
Reservoir A enabled	0 1 1 1 1 1 1 1
Reservoir B enabled	1 0 1 1 1 1 1 1
Reservoir C enabled	1 1 0 1 1 1 1 1
Reservoir D enabled	1 1 1 0 1 1 1 1
Reservoir E enabled	1 1 1 1 0 1 1 1
Reservoir F enabled	1 1 1 1 1 0 1 1
Reservoir G enabled	1 1 1 1 1 1 0 1
Reservoir H enabled	1 1 1 1 1 1 1 0

there are eight diluent reservoirs to be controlled in the application cited above. To control these chemicals one could use a single port of an I/O interface. The computer would have first to set all eight bits of this port to output mode. Then, to control the reservoirs the computer would write a word (byte) to the port register according to the following schedule in table 6.7.

Each valve would normally be controlled by an electromechanical device (e.g. a commercially available solenoid valve) and it is assumed that only one diluent at a time is used. (In the schedule in table 6.7 each bit controls one diluent valve; to energize more than one at a time it is only necessary to write a combined byte such as 00111111 to energize reservoirs A and B together.)

Some external devices may be more complex in nature. For example, a spectrophotometer instrument might include a light source, (measurement) wavelength selector, carousel sample rack (with samples in cuvettes), and light detector. A variety of control commands might be required to obtain proper measurements (e.g. select wavelength, carousel sequence, etc). In this case it is desirable (or necessary) to be assured that commands and data are transferred properly. The handshaking signals provided by the I/O interface may be used for this purpose. In addition to the data lines (from one of the port registers), two additional control lines pass between the units to support reliable and flexible synchronization between systems. One is the *data ready* signal and the second is the *data acknowledge* signal; the flow diagram and timing diagrams shown in figure 6.29 show how the control register of the I/O interface functions in these circumstances (from the point of view of the computer).

REFERENCES

[1] Normann R B 1988 *Principles of Bioinstrumentation* (New York, NY: Wiley)
[2] *Temperature Measurement Handbook and Encyclopedia* 1986 (Stamford: Omega Engineering)

[3] Wilson J and Hawkes J 1983 *Optoelectronics: An Introduction* (Englewood Cliffs, NJ: Prentice-Hall International)
[4] Streetman B G 1990 *Solid State Electronic Devices* 3rd edn (Englewood Cliffs, NJ: Prentice-Hall)
[5] Wolthuis R A, Mitchell G L, Saaski E, Hartl J C and Afromowitz M A 1991 Development of medical pressure and temperature sensors employing optical spectrum modulation *IEEE Trans. Biomed. Eng.* **BE-38** 10
[6] Mims F M 1991 Optical fibre sensors *Computercraft* July
[7] Shigley J E and Mischke C R 1989 *Mechanical Engineering Design* 5th edn (New York, NY: McGraw-Hill)
[8] Wolf S and Smith R F M 1990 *Student Reference Manual for Electronic Instrumentation Laboratories* (Englewood Cliffs, NJ: Prentice-Hall)
[9] Fogt E J 1988 Electrochemical sensors *Encyclopedia of Medical Devices and Instrumentation* ed J G Webster (New York, NY: Wiley–Interscience)
[10] Giulian D and Silverman G 1975 Solid state animal detection system: its application to open field activity and freezing behavior *Physiol. Behavior* **14** 109–12
[11] Strohmayer A, Silverman G and Grinker J 1980 A device for the continuous recording of solid food ingestion *Physiol. Behavior* **24** 789–91
[12] Miner G F and Comer D J 1992 *Physical Data Acquisition for Digital Processing* (Englewood Cliffs, NJ: Prentice-Hall)
[13] Meiksin Z H and Thackray P C 1984 *Electronic Design with Off-the-shelf Integrated Circuits* 2nd edn (Englewood Cliffs, NJ: Prentice-Hall)
[14] Keithley Metrabyte 1991 *Data Acquisition and Control* vol 24
[15] Eggert A A 1988 Automated analytical methods *Encyclopedia of Medical Devices and Instrumentation* vol 1, ed J G Webster (New York, NY: Wiley–Interscience)
[16] Mano M M 1988 *Computer Engineering Hardware Design* (New York, NY: Prentice-Hall)

Part 2

Applied Instrumentation Automation

The primary goal of part 2 is to apply a variety of computer-aided tools in order to analyse instrumentation circuits, and to the acquisition, processing, and presentation of laboratory data. Emphasis is placed on spectral and statistical analysis, as well as data recovery in a noisy environment.

Circuit analysis and digital signal processing tools (SPICE and DSPlay) are contrasted with the more general mathematics equation-solver and spreadsheet packages (MathCAD, MATLAB^{TM}†, and Lotus 1-2-3). The graphical user interface (GUI) is illustrated via a software development system for creating data-acquisition and instrument control applications (LabWindows). The GUI concept is then reinforced by exploring the Windows environment of the PC. The tools are applied to two examples from the area of optics, using measurements actually taken in the laboratory. Part 2 is concluded by an exploration of the application of newer tools such as artificial intelligence and neural nets in the context of the laboratory environment.

The first five chapters of part 2 (chapters 7 to 11) each deal with a different software tool. Although chapter 8 primarily deals with MathCAD, an introduction to MATLAB is included in the last section. The capabilities of each package are reviewed in the introductory section of each chapter. Where appropriate, the mechanics of getting started with a particular tool is described with the aid of simple examples. The goal is to motivate the novice to become familiar with the tool, without trying to replicate a more comprehensive user manual. Other early examples are included to highlight a particular feature of the tool. In the case of LabWindows (chapter 11), use of the tool requires knowledge of either QuickBASIC or C. Hence, no attempt is made to introduce 'simple' programming exercises.

Beyond the introductory exercises, subsequent examples place stress on instrumentation circuit and data-processing applications. Most of these examples are used to illustrate several of the tools, providing reinforcement to the reader and illustrating trade-offs among the tools. A summary of application examples and the tools used for each is presented below:

---

† MATLAB is a registered trademark of The MathWorks, Inc., 24 Prime Park Way, Natick, MA 01760-1500, USA. Phone: +508-653-1415. Fax: +508-653-2997. E-mail: into@mathworks.com.

- Instrumentation circuits
    Differential amplifiers (SPICE, MathCAD, MATLAB, Lotus 1-2-3)
    D/A converters (SPICE, DSPlay)
    Logarithmic amplifiers (SPICE)
- Data acquisition and presentation
    Acquisition hardware and driver programs (LabWindows)
    Graphing (all tools applied to various examples)
- Data processing
    Spectral analysis
        Fourier series (SPICE, MathCAD, MATLAB, DSPlay, Lotus 1-2-3)
        Discrete convolution, DFT, FFT (MathCAD, MATLAB, Lotus 1-2-3)
- Statistical analysis
    Signal detection in noise (MathCAD, MATLAB, DSPlay, Lotus 123, LabWindows)
    Simulated temperature readings (MathCAD, MATLAB, LabWindows)
    Simulated EMG muscle contraction (LabWindows)
    Simulated instrument behaviour shaping (LabWindows)
    Linear regression (MathCAD, MATLAB, Lotus 1-2-3)

Chapter 12 introduces the increasingly popular PC Windows environment, emphasizing data exchange between application programs. Whereas DOS only permits data exchange through file transfer, Windows also offers the convenience of exchanging through the common area called the clipboard. Additionally, dynamic data exchange (DDE) is a feature of Windows, permitting real-time transfer. As an example, DDE can be used in the laboratory environment to enable data acquisition to be performed, along with immediate processing and presentation.

We should emphasize that the software tools chosen are not intended to promote particular products, but rather reflect the authors' experiences in both educational and applied environments. Notice, however, that products such as MathCAD, MATLAB, and LabWindows are widely used in PC-based data acquisition and control (based on recent annual reader surveys by *Personal Engineering* magazine). The Windows environment is showing growing popularity as well.

The goal of chapter 13, on fully integrated applications, is to present the results of two optical laboratory experiments conducted with the tools presented in earlier chapters. In one experiment data on light intensity from a laser source are acquired and graphed (i.e. presented graphically) as a function of source position, to verify the expected theoretical diffraction pattern. The data are analysed and presented with the aid of a spreadsheet. A separate tool, Lotus Measure, automates the acquisition of the data and directs it to the Lotus 1-2-3 spreadsheet. The second experiment utilizes LabWindows to acquire and process data from a laser source coupled to an optical fibre. The fibre bandwidth is computed with the aid of a DFT routine written in a high-level language supported by LabWindows.

Chapter 14 presents an overview of two emerging tools which can be expected to play important roles in laboratory automation: artificial intelligence, and neural nets. These tools are first explained and the descriptions are subsequently followed with practical applications that have been reported. Once the technology has matured, automata of this type can be expected to learn, understand, interpret and arrive at conclusions regarding the data in a manner that would be considered to be intelligent if a person were doing it. In one of the examples from the laboratory, the automata (the expert system) can recognize the 'signature' spectrum of materials within a sample analysed by a spectrometer. In a second application a neural net(work) is trained to diagnose and recommend treatment for hypertension.

7
Design Aids—SPICE

7.1 INTRODUCTION

SPICE (an acronym for *simulation program with integrated circuit analysis*), was developed at the University of California (Berkeley) in the early 1970s. Since that time, SPICE has been widely used in industry as well as at universities, and various versions of the software have evolved. PSpice® from MicroSim Corporation is the version that runs on the IBM PC and compatibles [1, 3, 4]. An extensive library of analogue components is available with current versions; additionally, a digital device simulation option is available with PSpice for machines with sufficient memory.

The analogue component library below enables SPICE to analyse a variety of electronic circuits:

- Sources
    Independent voltage and current sources
    Dependent voltage and current sources
- Passive devices
    Resistors, capacitors, inductors, mutual inductors
- Passive and active semiconductor devices
    Diodes
    Transistors—bipolar junction, junction field-effect (JFET), MOS field-effect (MOSFET)
- Other devices
    Transmission lines, transformers

A circuit is simulated by first creating a SPICE input file consisting of a title line, circuit description statements, control lines, and optional comments. Control lines permit SPICE to perform various types of voltage or current analysis and either print or plot the results. For example, a DC analysis permits a source to be varied over a specified range and presents the results at each source value. In contrast, an AC analysis is performed over a specified frequency range, and a transient analysis over a desired time interval. A Fourier analysis can be performed on the transient waveform, computing the amplitude and

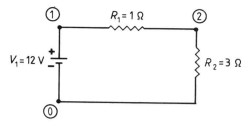

Figure 7.1 A simple DC circuit.

phase of nine harmonics of a specified fundamental frequency, in addition to the DC component. SPICE can also perform DC transfer function or small-signal sensitivity analysis, as well as distortion and noise calculations in conjunction with AC analysis.

Independent sources can be DC or one of five time-varying functions. A single pulse or train of pulses can be specified along with rise and fall times, resulting in pulse, square-wave, triangle, or sawtooth inputs. A sinusoid can be specified with optional exponential damping, as well as a single pulse with exponential rise and decay. Additionally, a single-frequency FM waveform is available, and a piecewise-linear waveform can be constructed to approximate nearly any signal desired.

7.2 ELEMENTARY CIRCUIT EXAMPLES

To illustrate the use of SPICE consider the simple DC circuit shown in figure 7.1.

A battery is connected to two resistors in series. SPICE requires that circuit elements be described with the use of nodes—the circled numbers in the diagram. An arbitrary reference node, designated as node 0, is always required. A SPICE input file to simulate this circuit is given below. Notice that SPICE requires upper case in general.

```
SIMPLE DC CIRCUIT
.WIDTH OUT=80
.OPTIONS NOPAGE
V1 1 0 DC 12
R1 1 2 1
R2 2 0 3
.END
```

The first line is a mandatory title line ignored by the processor. Optional comment lines can be inserted anywhere by preceding the comment with an asterisk (*). The lines that start with a period (.) are control lines. The .WIDTH and .OPTIONS commands are not required but are generally recommended. They format the output to an 80-column width and suppress extra pages of printout.

222   DESIGN AIDS—SPICE

The next three lines obviously describe the circuit, and the .END command signifies the end of the input file.

The general format for a statement describing an independent constant-voltage source is

$$\text{Vxxx..} \quad +\text{node} \quad -\text{node} \quad \text{DC} \quad \text{value}$$

The source name appearing first may be one to eight characters in length and must start with the letter V. The positive and negative node numbers follow along with a DC specifier. If the specifier is omitted, the source is assumed to be DC. The last entry is the source value in volts. Optional prefixes can be used—for example, 5K for 5000. Additionally for documentation purposes SPICE ignores other non-prefix characters that follow the value. Thus, in the example shown the source value could have been specified as 12V or 12VOLTS just as well. Prefixes and units will be used in subsequent examples for clarity.

A resistor description has a similar format given by

$$\text{Rxxx..} \quad \text{node} \quad \text{node} \quad \text{value}$$

Notice that the name must start with R and that the node numbers can be specified in either order since the resistor's polarity depends on the current direction. The value is in ohms in the absence of a prefix. Incidentally, capacitors and inductors have a similar format using Cxxx.. and Lxxx.. as component names respectively. Initial conditions can also be specified after the element value (capacitor voltage or inductor current).

When SPICE is run, the input file is processed and an output file is generated. All the examples shown in this chapter were run with an educational evaluation version of PSpice. The output file generated for the simple DC circuit is shown below, with some editing performed to condense it.

```
*** 04/30/91 *** Evaluation PSpice (January 1990)
*** 10:28:09 ***
 SIMPLE DC CIRCUIT
 ****      CIRCUIT DESCRIPTION
 ********************************************************
 .WIDTH OUT=80
 .OPTIONS NOPAGE
 V1 1 0 DC 12
 R1 1 2 1
 R2 2 0 3
 .END
```

ELEMENTARY CIRCUIT EXAMPLES   223

```
**** SMALL SIGNAL BIAS SOLUTION   TEMPERATURE = 27.000 DEG C
NODE   VOLTAGE    NODE   VOLTAGE    NODE   VOLTAGE    NODE   VOLTAGE
(   1)   12.0000  (   2)    9.0000
     VOLTAGE SOURCE CURRENTS
     NAME         CURRENT
     V1          -3.000E+00
     TOTAL POWER DISSIPATION   3.60E+01  WATTS
         JOB CONCLUDED
         TOTAL JOB TIME          2.31
```

We see that the output file includes the input file statements as well as an operating point analysis of the simple circuit. For the remaining examples in this chapter we will only present the relevant results from the output file, rather than the entire printout. For this first example, the operating point values are displayed even in the absence of any control lines requesting a particular type of analysis. The two node voltages are calculated with respect to the reference node (12 and 9 V), as are the overall source current (3 A) and total power dissipated by the resistors (36 W). If the input file contains any errors these are clearly identified in the output file, and the analysis is generally aborted.

7.2.1 DC analysis

Suppose we want to analyse the circuit of figure 7.1 for battery voltages ranging from 4 to 12 V in steps of 1 V, for example. The modified input file shown below accomplishes this.

```
SIMPLE DC CIRCUIT WITH VARYING BATTERY VOLTAGE
.WIDTH OUT=80
.OPTIONS NOPAGE
V1 1 0
R1 1 2 1
R2 2 0 3
.DC V1 4 12 1
.PRINT DC V(1) V(2) V(1,2)
.END
```

The source description line now only identifies the two nodes, and a separate control line starting with .DC sweeps the source through the required values. The .PRINT command displays a table of values for the source voltage and the voltages across the two resistors. Obviously the two resistor voltages should sum to the source value. Notice that a node voltage is referenced by (i.e. referred to as) V(+node -node). If the negative node is the reference node it may be omitted.

The results from the output file are shown below.

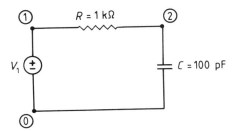

Figure 7.2 An *RC* circuit.

```
**** DC TRANSFER CURVES    TEMPERATURE = 27.000 DEG C
    V1           V(1)           V(2)          V(1,2)
 4.000E+00     4.000E+00      3.000E+00      1.000E+00
 5.000E+00     5.000E+00      3.750E+00      1.250E+00
 6.000E+00     6.000E+00      4.500E+00      1.500E+00
 7.000E+00     7.000E+00      5.250E+00      1.750E+00
 8.000E+00     8.000E+00      6.000E+00      2.000E+00
 9.000E+00     9.000E+00      6.750E+00      2.250E+00
 1.000E+01     1.000E+01      7.500E+00      2.500E+00
 1.100E+01     1.100E+01      8.250E+00      2.750E+00
 1.200E+01     1.200E+01      9.000E+00      3.000E+00
```

A disadvantage of SPICE is that it lacks an interactive capability that would permit the changing of component values at the keyboard while the program is running. The .DC command facilitates varying a voltage source value, but varying a resistor value is more difficult. PSpice allows the modelling of a resistor and varying its value as well as other parameters. Modelling is required for circuits with semiconductor devices, and will be discussed in section 7.4.

7.2.2 Transient and AC analysis

Figure 7.2 shows a simple *RC* circuit that will be used to illustrate both transient and AC response. The SPICE input file below applies a pulse input to the circuit and requests a plot of the transient response of the capacitor's voltage.

```
TRANSIENT ANALYSIS OF AN RC CIRCUIT
.WIDTH OUT=80
.OPTIONS NOPAGE
V1 1 0 PULSE (0 1 0 0 0 0.1US 1US)
R 1 2 1K
C 2 0 100PF
.PLOT TRAN V(2)
.TRAN 0.01US 0.3US
.END
```

ELEMENTARY CIRCUIT EXAMPLES   225

The voltage source now produces a repetitive pulse input to the circuit. The parameters following PULSE allow us to specify the initial and pulsed values in volts, delay, rise, and fall times, pulse width, and pulse period respectively. All times are, by default, in seconds. For this example, the specified pulse rises instantly from 0 to 1 V at zero time and returns to zero after 0.1 μs. The period is specified as 1 μs, although the circuit is only to be analysed during the first period. The control lines after the circuit description request a plot of capacitor voltage at node 2 with respect to time. The .TRAN command specifies a time increment of 0.01 μs and a maximum time of 0.3 μs. The plot automatically starts at zero time.

The time plot written to the output file is shown below.

```
****       TRANSIENT ANALYSIS                TEMPERATURE =    27.000 DEG C
   TIME       V(2)
   (*)----------   0.0000E+00   2.0000E-01   4.0000E-01   6.0000E-01
                - - - - - - - - - - - - - - - - - - - - - - - - - -
 0.000E+00    0.000E+00 *            .            .            .
 1.000E-08    4.826E-02 .  *         .            .            .
 2.000E-08    1.385E-01 .       *    .            .            .
 3.000E-08    2.205E-01 .           .*            .            .
 4.000E-08    2.949E-01 .           .        *    .            .
 5.000E-08    3.618E-01 .           .            *.            .
 6.000E-08    4.226E-01 .           .            . *           .
 7.000E-08    4.777E-01 .           .            .      *      .
 8.000E-08    5.273E-01 .           .            .           * .
 9.000E-08    5.723E-01 .           .            .             *.
 1.000E-07    6.131E-01 .           .            .             . *
 1.100E-07    6.500E-01 .           .            .             .     *
 1.200E-07    6.344E-01 .           .            .             .   *
 1.300E-07    5.742E-01 .           .            .          *  .
 1.400E-07    5.195E-01 .           .            .     *       .
 1.500E-07    4.701E-01 .           .            .  *          .
 1.600E-07    4.253E-01 .           .            .*            .
 1.700E-07    3.848E-01 .           .         *. .             .
 1.800E-07    3.482E-01 .           .       *    .             .
 1.900E-07    3.150E-01 .           .    *       .             .
 2.000E-07    2.850E-01 .           .  *         .             .
 2.100E-07    2.580E-01 .           *            .             .
 2.200E-07    2.334E-01 .         * .            .             .
 2.300E-07    2.111E-01 .        .* .            .             .
 2.400E-07    1.911E-01 .      *.    .            .            .
 2.500E-07    1.729E-01 .     * .    .            .            .
 2.600E-07    1.564E-01 .   *   .    .            .            .
 2.700E-07    1.415E-01 .  *    .    .            .            .
 2.800E-07    1.280E-01 . *     .    .            .            .
 2.900E-07    1.158E-01 . *     .    .            .            .
 3.000E-07    1.048E-01 .*      .    .            .            .
```

For this circuit the capacitor begins charging when the pulse is applied. Analysis shows that for the duration of the pulse the capacitor voltage is given by

$$V(2) = 1 - e^{-t/RC}.$$

Since the RC time constant and the pulse width are both equal to 0.1 μs the voltage rises to

$$1 - e^{-1} = 0.632\,12$$

before the pulse is removed. The capacitor then discharges through the resistor and its voltage decays toward zero before the pulse is reapplied in the next period. The exponential rise to a peak value of approximately 0.63 V is verified by the plot from the output file. If you have access to SPICE try this example changing the capacitor first to a smaller value (say 10 pF) and then to a larger value (1000 pF). The smaller time constant results in shorter rise and fall times, and the capacitor voltage nearly duplicates the input pulse. In contrast a large time constant produces a slow exponential rise to a smaller fraction of the pulsed voltage followed by a very slow decay. The rise appears linear in fact and the result is an approximate integration of the input waveform. A more accurate integrator will be illustrated shortly using an op amp circuit.

The RC circuit can also be viewed as a low-pass filter. To illustrate this, SPICE can provide a sinusoidal input to the circuit and then perform an AC analysis over a specified frequency range. The input file below demonstrates this capability.

```
AC ANALYSIS OF AN RC CIRCUIT
.WIDTH OUT=80
.OPTIONS NOPAGE
V1 1 0 AC 1V
R 1 2 1K
C 2 0 100PF
.PLOT AC V(2)
.AC DEC 5 10K 100MEG
.END
```

Now the source is specified to be sinusoidal (AC) with an amplitude of 1 V. The .PLOT and .AC control lines request that the magnitude of the capacitor voltage be plotted as a function of frequency. The plot is to be on a logarithmic (decade) scale with five points per decade, over a frequency range from 10 kHz to 100 MHz. If a linear frequency scale is desired, simply replace DEC by LIN. The values then identify the total number of points on the curve followed by the lowest and highest frequencies respectively.

The frequency plot written to the output file is shown below.

```
****     AC ANALYSIS                    TEMPERATURE =    27.000 DEG C
   FREQ        V(2)
(*)----------      1.0000E-02    1.0000E-01    1.0000E+00    1.0000E+01
      - - - - - - - - - - - - - - - - - - - - - - - - - - - - - - -
1.000E+04   1.000E+00  .              .              *              .
1.585E+04   1.000E+00  .              .              *              .
2.512E+04   9.999E-01  .              .              *              .
3.981E+04   9.997E-01  .              .              *              .
6.310E+04   9.992E-01  .              .              *              .
1.000E+05   9.980E-01  .              .              *              .
1.585E+05   9.951E-01  .              .              *              .
2.512E+05   9.878E-01  .              .              *              .
3.981E+05   9.701E-01  .              .              *              .
6.310E+05   9.296E-01  .              .              *              .
1.000E+06   8.467E-01  .              .             *               .
1.585E+06   7.086E-01  .              .          *.                 .
2.512E+06   5.352E-01  .              .       *      .              .
3.981E+06   3.712E-01  .              .    *         .              .
6.310E+06   2.446E-01  .              .  *           .              .
1.000E+07   1.572E-01  .              . *            .              .
1.585E+07   9.992E-02  .             *               .              .
2.512E+07   6.323E-02  .           * .               .              .
3.981E+07   3.995E-02  .         *   .               .              .
6.310E+07   2.522E-02  .       *     .               .              .
1.000E+08   1.591E-02  .    *        .               .              .
```

From steady-state AC analysis of this circuit, the magnitude–phase relationship of output to input is given by

$$V(2) = \frac{V_1}{1 + i\,2\pi f RC}.$$

Since the magnitude of the source voltage is unity and the RC time constant is 0.1 μs, the magnitude of the capacitor voltage is

$$|V(2)| = \frac{1}{\sqrt{1+k^2}}$$

where

$$k = \frac{2\pi f}{10^7}.$$

The magnitude of $V(2)$ is obviously one at a frequency of zero (the DC case) and decreases with increasing frequency. The filter's bandwidth is generally defined by the frequency where the magnitude is down to 0.707 of the peak value, or equivalently down by 3 dB. This corresponds to $k = 1$ and a bandwidth of

$$B = \frac{10^7}{2\pi} = 1.591\,55 \text{ MHz}.$$

We see from the plot and accompanying table that $V(2)$ has a magnitude of 0.7086 at 1.585 MHz, thus closely matching the expected bandwidth.

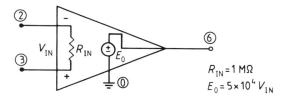

Figure 7.3 An idealized op amp model.

7.3 OPERATIONAL AMPLIFIER SUBCIRCUITS AND APPLICATIONS

An operational amplifier (op amp for short) is a high-gain, high-input-impedance device available as an integrated circuit. These characteristics make a variety of application areas appropriate—such as summing circuits, isolators, differentiating or integrating circuits, and differential instrumentation amplifiers. Additionally, op amp filters of all kinds can be implemented without the need for bulky inductor components. For the purpose of SPICE simulation we will implement an op amp as a subcircuit, using both an idealized model and the model of a real, commercially available component.

7.3.1 An idealized op amp

The idealized model shown in figure 7.3 is actually a small-signal model at low frequencies. Node 2 is the inverting input, node 3 the non- inverting input, and node 6 the output. A real op amp also requires two supply voltages (called V_{CC} and V_{EE}), not required by the idealized model. The resistance R_{IN} represents the input impedance and is arbitrarily chosen as 1 MΩ, a relatively high value. The source E_0 is a voltage-controlled voltage source (VCVS), one type of dependent source easily handled by SPICE. Whereas an independent source must have a name beginning with V, SPICE requires that a VCVS name start with E. This voltage is dependent on the input voltage V_{IN}, with the gain factor arbitrarily set to 50 000.

Creating a subcircuit in the SPICE input file has the advantage of separating the details of the device's implementation from the rest of the circuit, thereby improving readability. More importantly, it provides a single description of the device, which can be referenced more than once for circuits using several identical components. In this sense the subcircuit is analogous to a subroutine referenced by a main program written in a high-level or assembly language. For the idealized op amp model, the subcircuit can be described in the input file by the statements below:

OPERATIONAL AMPLIFIER SUBCIRCUITS AND APPLICATIONS

```
.SUBCKT OPIDEAL  2   3   6
* PINOUTS        -IN +IN OUT
RIN 2 3 1MEG
EO 6 0 3 2 50K
.ENDS OPIDEAL
```

The description always starts with a control line starting with .SUBCKT followed by an arbitrary subcircuit name. The nodes that follow are only those non-reference nodes external to the device. Although not the case for this example, a subcircuit may have additional internal nodes used in its description. Those node numbers are known only to the subcircuit itself and therefore can overlap with node numbers used for the main circuit. Internal subcircuit nodes are therefore analogous to local variables in a program, and main-circuit nodes to global variables. After describing the subcircuit a control line starting with .ENDS indicates the end of the subcircuit to the processor. Including the name on the line is optional. Notice that the VCVS is described by two pairs of nodes, the nodes for the source followed by the nodes corresponding to the dependent voltage, V_{IN} in this case. The gain factor then follows.

7.3.2 An inverter circuit

An op amp inverter circuit is shown in figure 7.4. Nodes 4 and 7 have been included for the supply voltages required by the model for the real op amp. The external source V_1 is applied arbitrarily to the inverted input. It can be shown that for the assumptions of very high-input impedance and op amp gain the output is approximately

$$V_{OUT} = -\frac{R_2}{R_1} V_1$$

independently of the load resistance R_L. The resistor R_3 improves the balance for a real op amp when its value is close to the parallel combination of R_1 and R_2.

The SPICE input file below simply applies a 2 V DC input to the inverter. Notice the op amp description line referencing the subcircuit (the line before

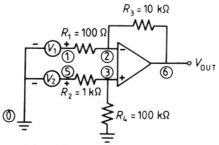

Figure 7.4 An op amp inverter.

230   DESIGN AIDS—SPICE

.END). The device is given a name which must start with X. Following the name are the subcircuit nodes in proper order and the matching subcircuit name.

```
OP-AMP INVERTER (USING IDEALIZED MODEL)
.WIDTH OUT=80
.OPTIONS NOPAGE
.SUBCKT OPIDEAL  2   3    6
* PINOUTS        -IN +IN OUT
RIN 2 3 1MEG
E0 6 0 3 2 50K
.ENDS OPIDEAL
V1 1 0 DC 2V
R1 1 2 1K
R2 2 6 10K
R3 3 0 0.909K
RL 6 0 1K
XOPAMP 2 3 6 OPIDEAL
.END
```

The calculated node voltages written to the output file are as follows:

```
NODE VOLTAGE NODE  VOLTAGE  NODE  VOLTAGE   NODE  VOLTAGE
( 1)  2.0000 ( 2) 400.3E-06 ( 3) 363.5E-09 ( 6) -19.9960
```

Since the ratio of R_2 to R_1 is 10, the 2 V input is multiplied by 10 and inverted. The actual output of -19.996 V is very close to the expected value of -20 V. Notice also that dividing the 20 V output by the op amp's internal gain of 50K gives us 400 μV, very close to PSpice's calculated voltage difference between nodes 2 and 3.

7.3.3   An inverter using a non-idealized op amp

A commercial UA 741 op amp has been modelled by an assortment of resistors and capacitors, along with an inductor. An op amp requires several transistor stages along with diodes to provide the required gain and biasing. The model used emulates the semiconductor components with internal voltage and current sources, including one VCVS. The input file for the inverter using a UA 741 op amp model is shown below.

```
OP-AMP INVERTER (USING UA741 MODEL)
.WIDTH OUT=80
.OPTIONS NOPAGE
.SUBCKT UA741   2   3   6   7   4
* PINOUTS       -IN +IN OUT VCC VEE
RP 4 7 10K
RXX 4 0 10MEG
IBP 3 0 80NA
```

OPERATIONAL AMPLIFIER SUBCIRCUITS AND APPLICATIONS 231

```
RIP 3 0 10MEG
CIP 3 0 1.4PF
IBN 2 0 100NA
RIN 2 0 10MEG
CIN 2 0 1.4PF
VOFST 2 10 1MV
RID 10 3 200K
EA 11 0 10 3 1
R1 11 12 5K
R2 12 13 50K
C1 12 0 13PF
GA 0 14 0 13 2700
C2 13 14 2.7PF
RO 14 0 75
L 14 6 30UHY
RL 14 6 1000
CL 6 0 3PF
.ENDS UA741
VCC 7 0 10V
VEE 4 0 -10V
V1 1 0 2V
R1 1 2 1K
R2 2 6 10K
R3 3 0 0.909K
RL 6 0 1K
XOPAMP 2 3 6 7 4 UA741
.END
```

The subcircuit includes two additional external nodes for the supply voltages V_{CC} and V_{EE}, which have been set to +10 V and −10 V, respectively, in the main program. The node voltages calculated by PSpice are shown below.

NODE	VOLTAGE	NODE	VOLTAGE	NODE	VOLTAGE	NODE	VOLTAGE
(1)	2.0000	(2)	.0010	(3)	-72.23E-06	(4)	-10.0000
(6)	-19.9880	(7)	10.0000	(XOPAMP.10)	34.62E-06		
(XOPAMP.11)	106.8E-06			(XOPAMP.12)	106.8E-06		
(XOPAMP.13)	106.8E-06			(XOPAMP.14)	-19.9880		

The output of −19.988 V is nearly as close to the expected value of −20 V as was the case using the idealized model. Since the differential input to the op amp between nodes 2 and 3 is approximately 1 mV, the op amp gain is then 20 V/1 mV or 20K, as compared to 50K for the idealized model. This probably accounts for the slight decrease in accuracy from the idealized model's output. Notice also that PSpice lists the main circuit's node voltages first followed by those internal to the subcircuit, a useful feature for debugging a problem with the simulation.

7.3.4 The integrating circuit

A differentiator can be implemented with the circuit in figure 7.4 by replacing R_1 by a capacitor of appropriate value. Alternatively, if the capacitor replaces R_2, an integrator results. The input file below simulates the integration of a square-wave input.

```
OP-AMP INTEGRATING CIRCUIT (USING IDEALIZED MODEL)
.WIDTH OUT=80
.OPTIONS NOPAGE
.SUBCKT OPIDEAL   2   3   6
* PINOUTS       -IN +IN OUT
RIN 2 3 1MEG
E0 6 0 3 2 50K
.ENDS OPIDEAL
V1 1 0 PULSE (-1 1 1MS 0 0 1MS 2MS)
R1 1 2 1K
C  2 6 1UF IC=0V
R3 3 0 1K
RL 6 0 1K
XOPAMP 2 3 6 OPIDEAL
.TRAN 0.1MS 2MS UIC
.PLOT TRAN V(6)
.END
```

Analysis shows that the inverted output should be very close to the integral of the input voltage V1 starting from time $t = 0$, divided by the time constant R_1 times C. The pulse alternates between -1 V and $+1$ V every 1 ms. The time constant is also chosen to be 1 ms to give us a triangular waveform peaking at unity. Notice that the .TRAN command includes the option UIC which tells it to use the initial condition specified for the capacitor. Since the default value is zero, the IC=0V specification could have been omitted.

OPERATIONAL AMPLIFIER SUBCIRCUITS AND APPLICATIONS 233

The time plot produced by PSpice is shown below.

```
TIME           V(6)
(*)----------  0.0000E+00   5.0000E-01   1.0000E+00   1.5000E+00
               - - - - - - - - - - - - - - - - - - - - - - - - -
0.000E+00      0.000E+00 *
1.000E-04      1.000E-01 .    *
2.000E-04      2.000E-01 .       *
3.000E-04      3.000E-01 .          *
4.000E-04      4.000E-01 .             *
5.000E-04      5.000E-01 .                *
6.000E-04      6.000E-01 .                   *
7.000E-04      7.000E-01 .                      *
8.000E-04      8.000E-01 .                         *
9.000E-04      9.000E-01 .                            *
1.000E-03      1.000E+00 .                               *
1.100E-03      9.998E-01 .                               *
1.200E-03      8.998E-01 .                            *
1.300E-03      7.998E-01 .                         *
1.400E-03      6.998E-01 .                      *
1.500E-03      5.998E-01 .                   *
1.600E-03      4.998E-01 .                *
1.700E-03      3.998E-01 .             *
1.800E-03      2.998E-01 .          *
1.900E-03      1.998E-01 .       *
2.000E-03      9.979E-02 .    *
```

Nearly identical results are obtained for the integrator when the idealized op amp model is replaced by the UA 741 model. An exception is that at $t = 0$ the output voltage is approximately -0.24 V rather than the 0 V value of the ideal model. The difference is apparently a result of the internal biasing of the UA 741.

7.3.5 The differential amplifier

For amplifiers used in bioinstrumentation applications it is desirable that the input impedance, the gain, and a feature called the common-mode-rejection ratio all be high. Sufficiently high impedance is necessary to avoid loading of transducer output signals, whereas high gain is needed to enhance small input signals such as electroencephalogram (EEG) potentials typically in the μV range. Common-mode rejection permits the cancelling out of the noise or other interference when a differential measurement technique is used. For example, a hospital patient's body picks up relatively large-amplitude 60 Hz interference with respect to ground. An electrocardiogram (ECG) can be taken by grounding the patient's leg and measuring the difference between the wrists, with the heart in between the measurement points. The 60 Hz interferences are nearly equal on the two wrists, so a circuit amplifying the difference signal cancels out most of the 'common-mode' interference.

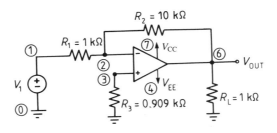

Figure 7.5 A one-op-amp differential amplifier.

A bioinstrumentation amplifier should therefore be able to amplify the difference between two input signals. Figure 7.5 shows a simple differential amplifier using only a single op amp. The load resistor has been omitted for simplicity, since the results are nearly independent of its value anyway. Analysis of this circuit shows that if the resistors are chosen such that

$$\frac{R_1}{R_2} = \frac{R_3}{R_4}$$

the output is given by

$$V_{OUT} = \frac{R_3}{R_1}(V_2 - V_1)$$

and is thereby proportional to the difference in inputs as desired.

The SPICE input file below implements the differential amplifier using the idealized op amp model.

```
OP-AMP DIFFERENCE CIRCUIT (USING IDEALIZED MODEL)
.WIDTH OUT=80
.OPTIONS NOPAGE
.SUBCKT OPIDEAL   2    3     6
* PINOUTS         -IN  +IN   OUT
RIN 2 3 1MEG
E0 6 0 3 2 50K
.ENDS OPIDEAL
V1 1 0 DC 2MV
V2 5 0
R1 1 2 100
R2 3 5 1K
R3 2 6 10K
R4 3 0 100K
XOPAMP 2 3 6 OPIDEAL
.DC V2  -5MV   5MV   1MV
.PRINT DC V(5,1) V(6) I(V1) I(V2)
.END
```

OPERATIONAL AMPLIFIER SUBCIRCUITS AND APPLICATIONS

Input V_1 is fixed at 2 mV and V_2 is DC swept from -5 mV to $+5$ mV. The resistors are chosen to satisfy the proportionality requirement shown above. Since the ratio R_3/R_1 is 100, the circuit amplifies the difference $V_2 - V_1$ by this factor. This difference exists between nodes 5 and 1 and is therefore referenced as V(5,1) in the SPICE file. The output table produced by the .PRINT control line is shown below.

```
****   DC TRANSFER CURVES   TEMPERATURE = 27.000 DEG C
V2           V(5,1)        V(6)         I(V1)        I(V2)
-5.000E-03  -7.000E-03   -6.986E-01   -6.937E-05    4.952E-08
-4.000E-03  -6.000E-03   -5.988E-01   -5.948E-05    3.962E-08
-3.000E-03  -5.000E-03   -4.990E-01   -4.960E-05    2.971E-08
-2.000E-03  -4.000E-03   -3.992E-01   -3.972E-05    1.981E-08
-1.000E-03  -3.000E-03   -2.994E-01   -2.984E-05    9.907E-09
 0.000E+00  -2.000E-03   -1.996E-01   -1.996E-05    3.952E-12
 1.000E-03  -1.000E-03   -9.980E-02   -1.008E-05   -9.899E-09
 2.000E-03   0.000E+00   -6.546E-17   -1.980E-07   -1.980E-08
 3.000E-03   1.000E-03    9.980E-02    9.683E-06   -2.971E-08
 4.000E-03   2.000E-03    1.996E-01    1.956E-05   -3.961E-08
 5.000E-03   3.000E-03    2.994E-01    2.945E-05   -4.951E-08
```

Including the source currents in the table enhances understanding of the circuit and highlights a problem with the input impedance. As an example, for the last line of the table the differential voltage $V_2 - V_1$ is 3 mV and the output of 299.4 mV is very close to the expected 300 mV. The current through the 2 mV source V_1 results in an additional 2.945 mV across R_1, and consequently a voltage of 4.945 mV at inverting input node 2. The smaller current, through both the 5 mV source V_2 and R_2, is leaving the positive side of the source. The drop of 0.0495 mV across R_2 gives us 4.951 mV at the other op amp input. The differential input to the op amp is therefore very close to zero as expected.

For comparison purposes, the idealized op amp was replaced by the UA 741 model and the circuit was run by SPICE. The results were not close to the expected gain of 100. However, when higher input voltages were used the gain was found to be very close to the theoretical value. The problem was traced to the voltage source VOFST in the UA 741 subcircuit description. This source sets the device's input offset voltage to 1 mV, a value comparable to the input voltages used. The offset can be reduced with appropriate external circuitry when a real UA741 is used. For the purpose of our examples, however, we will simply change the subcircuit's offset value to 1 μV. The following output table results.

**** DC TRANSFER CURVES   TEMPERATURE =   27.000 DEG C

V2	V(5,1)	V(6)	I(V1)	I(V2)
-5.000E-03	-7.000E-03	-7.065E-01	-7.025E-05	-2.919E-08
-4.000E-03	-6.000E-03	-6.066E-01	-6.035E-05	-3.919E-08
-3.000E-03	-5.000E-03	-5.066E-01	-5.046E-05	-4.919E-08
-2.000E-03	-4.000E-03	-4.067E-01	-4.056E-05	-5.919E-08
-1.000E-03	-3.000E-03	-3.067E-01	-3.067E-05	-6.919E-08
0.000E+00	-2.000E-03	-2.068E-01	-2.077E-05	-7.920E-08
1.000E-03	-1.000E-03	-1.069E-01	-1.088E-05	-8.920E-08
2.000E-03	0.000E+00	-6.913E-03	-9.816E-07	-9.920E-08
3.000E-03	1.000E-03	9.303E-02	8.913E-06	-1.092E-07
4.000E-03	2.000E-03	1.930E-01	1.881E-05	-1.192E-07
5.000E-03	3.000E-03	2.929E-01	2.870E-05	-1.292E-07

For the cases where the differential input is at least 2 mV the gain is within about 3% of the theoretical value. The offset is clearly a more significant factor for smaller input differences.

7.3.6 An optimally designed instrumentation amplifier

The single-op-amp differential amplifier unfortunately fails the requirement of very high input impedance. Analysis shows that the equivalent impedance between nodes 1 and 5, or between the sources, is approximately equal to $R_1 + R_2$. The table shows that for all cases the ratio of the differential voltage input $V_1 - V_2$ to the difference in source currents $I(V_1) - I(V_2)$ is a constant, approximately equal to the value of R_1 alone.

Figure 7.6 shows a differential amplifier which overcomes the input impedance problem, but at the expense of additional op amp and resistor components [2]. By inserting op amps between each voltage source and the relatively small resistances that follow, we achieve a very high differential impedance. The resistors R_a, R_b, and R_c permit flexibility in setting the differential gain, giving us an optimal design for our instrumentation amplifier.

Analysis shows that the balance condition resulting in a true differential output is the same as for the single-op-amp circuit, namely

$$\frac{R_1}{R_2} = \frac{R_3}{R_4}$$

for which the output is then given by

$$V_{OUT} = \frac{R_3(R_a + R_b + R_c)}{R_a R_1}(V_2 - V_1).$$

As compared to the single-op-amp differential amplifier the differential gain of this circuit is higher by the factor

$$\frac{R_a + R_b + R_c}{R_a}$$

OPERATIONAL AMPLIFIER SUBCIRCUITS AND APPLICATIONS

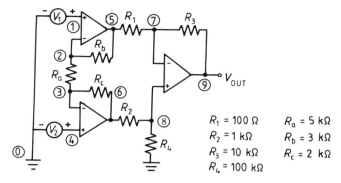

Figure 7.6 The optimally designed instrumentation amplifier.

and any of these resistors can be varied to set to the desired gain.
The SPICE file below implements the optimally designed amplifier.

```
OPTIMAL INSTRUMENTATION AMPLIFIER
*(USING IDEALIZED OP-AMP MODEL)
*REF.-Normann;"Principles of Bioinstrumentation"; Fig.2.16
.WIDTH OUT=80
.OPTIONS NOPAGE
.SUBCKT OPIDEAL    10   11   12
* PINOUTS         -IN  +IN  OUT
RIN 10 11 10MEG
EO 12 0 11 10 50K
.ENDS OPIDEAL
V1 1 0 DC 2MV
V2 4 0
RA 2 3 5K
RB 2 5 3K
RC 3 6 2K
R1 5 7 100
R2 6 8 1K
R3 7 9 10K
R4 8 0 100K
XOPAMP1    2   1   5    OPIDEAL
XOPAMP2    3   4   6    OPIDEAL
XOPAMP3    7   8   9    OPIDEAL
.DC V2 -5MV 5MV 1MV
.PRINT DC V(4,1) V(9) I(V1) I(V2)
.END
```

The source and resistor values of the single-op-amp circuit have been preserved, and the three additional resistors have been chosen to enhance further the differential gain of 100 by a factor of 2.

The DC transfer results from the output file are shown below.

```
****    DC TRANSFER CURVES      TEMPERATURE = 27.000 DEG C
   V2        V(4,1)       V(9)        I(V1)       I(V2)
-5.000E-03  -7.000E-03  -1.397E+00  -1.240E-14   1.560E-14
-4.000E-03  -6.000E-03  -1.198E+00  -1.120E-14   1.280E-14
-3.000E-03  -5.000E-03  -9.979E-01  -1.000E-14   1.000E-14
-2.000E-03  -4.000E-03  -7.984E-01  -8.800E-15   7.200E-15
-1.000E-03  -3.000E-03  -5.988E-01  -7.600E-15   4.400E-15
 0.000E+00  -2.000E-03  -3.992E-01  -6.400E-15   1.600E-15
 1.000E-03  -1.000E-03  -1.996E-01  -5.200E-15  -1.200E-15
 2.000E-03   0.000E+00   3.992E-10  -4.000E-15  -4.000E-15
 3.000E-03   1.000E-03   1.996E-01  -2.800E-15  -6.800E-15
 4.000E-03   2.000E-03   3.992E-01  -1.600E-15  -9.600E-15
 5.000E-03   3.000E-03   5.988E-01  -4.001E-16  -1.240E-14
```

As expected, the differential gain in each case is very close to 200. Also we see that the currents through the sources are several orders of magnitude smaller than for the single-op-amp circuit, thus satisfying the high-input-impedance criterion.

Again a comparison can be made to the same differential amplifier using the UA 741 op amp in place of the idealized model. The subcircuit's offset voltage (VOFST) was again reduced from 1 mV to 1 μV to improve the accuracy for inputs of the order of mV. The results from the output file are shown below.

```
****    DC TRANSFER CURVES      TEMPERATURE = 27.000 DEG C
   V2        V(4,1)       V(9)        I(V1)       I(V2)
-5.000E-03  -7.000E-03  -1.416E+00  -8.020E-08  -7.950E-08
-4.000E-03  -6.000E-03  -1.216E+00  -8.020E-08  -7.960E-08
-3.000E-03  -5.000E-03  -1.016E+00  -8.020E-08  -7.970E-08
-2.000E-03  -4.000E-03  -8.165E-01  -8.020E-08  -7.980E-08
-1.000E-03  -3.000E-03  -6.166E-01  -8.020E-08  -7.990E-08
 0.000E+00  -2.000E-03  -4.167E-01  -8.020E-08  -8.000E-08
 1.000E-03  -1.000E-03  -2.168E-01  -8.020E-08  -8.010E-08
 2.000E-03   0.000E+00  -1.693E-02  -8.020E-08  -8.020E-08
 3.000E-03   1.000E-03   1.830E-01  -8.020E-08  -8.030E-08
 4.000E-03   2.000E-03   3.829E-01  -8.020E-08  -8.040E-08
 5.000E-03   3.000E-03   5.828E-01  -8.020E-08  -8.050E-08
```

As for the case of the single-op-amp differential amplifier, the gain is within a few per cent of the theoretical value when using the real op amp model. The currents are considerably higher than for the idealized model, but I(V1) is still much smaller than for the single-op-amp circuit, whereas I(V2) is comparable. However measured, the input impedances are still of the order of several megohms.

OPERATIONAL AMPLIFIER SUBCIRCUITS AND APPLICATIONS

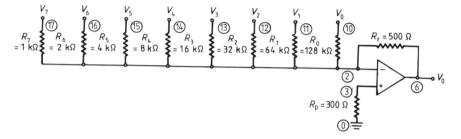

Figure 7.7 The D/A converter—the summing circuit.

7.3.7 The D/A converter—the summing circuit

A digital-to-analogue (D/A) converter provides us with another practical example of an op amp application. One such circuit is referred to as a summing circuit. Figure 7.7 implements an 8-bit D/A converter with the digital input represented by voltages V_7 to V_0. The inverted output of this circuit is equal to a weighted sum of all eight input voltages, where the weighting factor of each input is the ratio of the feedback resistor R_f to the input resistance in series with the source. That is,

$$V_{OUT} = -\left(\frac{R_f}{R_7}V_7 + \frac{R_f}{R_6}V_6 + \ldots + \frac{R_f}{R_0}V_0\right).$$

On choosing these ratios to be 1/2, 1/4, 1/8, ..., 1/128 the output represents the analogue equivalent to the digital input. We can then write the output as

$$V_{OUT} = -\frac{1}{2^8}(2^7 V_7 + 2^6 V_6 + 2^5 V_5 + \ldots + 2^1 V_1 + 2^0 V_0)$$

where each input voltage has a value corresponding to either binary 0 or 1.

The SPICE input file below implements the 8-bit D/A summing circuit for the binary digital input 10101011. Only the idealized op amp is used here. For the purpose of illustration an analogue range from 0 to 1 V is assumed, so the digital values simply correspond to 0 V for binary 0 and 1 V for binary 1.

```
8 BIT D/A CONVERTER - SUMMING CIRCUIT
.WIDTH OUT=80
.OPTIONS NOPAGE
.SUBCKT OPIDEAL   2    3     6
* PINOUTS        -IN  +IN   OUT
RIN 2 3 1MEG
EO 6 0 3 2 50K
.ENDS OPIDEAL
V7 17 0 DC 1V
V6 16 0 DC 0V
V5 15 0 DC 1V
V4 14 0 DC 0V
V3 13 0 DC 1V
V2 12 0 DC 0V
V1 11 0 DC 1V
V0 10 0 DC 1V
R7 17 2 1K
R6 16 2 2K
R5 15 2 4K
R4 14 2 8K
R3 13 2 16K
R2 12 2 32K
R1 11 2 64K
R0 10 2 128K
RF 2 6 500
RP 3 0 300
XOPAMP 2 3 6 OPIDEAL
.SENS V(6)
.END
```

The only type of analysis requested here is a sensitivity analysis, which permits us to see the effect of changes in the resistor values on the analogue output. As we have seen earlier the DC node voltages are automatically printed out as well. The relevant portion of the output file is shown below.

```
****  SMALL SIGNAL BIAS SOLUTION      TEMPERATURE =  27.000 DEG C
 NODE   VOLTAGE     NODE   VOLTAGE    NODE VOLTAGE   NODE  VOLTAGE
(  2) 13.36E-06   (  3)  4.008E-09   (  6)  -.6679   (10)   1.0000
 (11)    1.0000    (12)    0.0000    (13)  1.0000    (14)   0.0000
 (15)    1.0000    (16)    0.0000    (17)  1.0000
```

OPERATIONAL AMPLIFIER SUBCIRCUITS AND APPLICATIONS   241

```
****     DC SENSITIVITY ANALYSIS    TEMPERATURE =  27.000 DEG C
DC SENSITIVITIES OF OUTPUT V(6)
 ELEMENT          ELEMENT         ELEMENT         NORMALIZED
  NAME             VALUE         SENSITIVITY     SENSITIVITY
                                (VOLTS/UNIT)   (VOLTS/PERCENT)
  R7             1.000E+03        5.000E-04        5.000E-03
  R6             2.000E+03       -1.670E-09       -3.341E-08
  R5             4.000E+03        3.125E-05        1.250E-03
  R4             8.000E+03       -1.044E-10       -8.351E-09
  R3             1.600E+04        1.953E-06        3.125E-04
  R2             3.200E+04       -6.525E-12       -2.088E-09
  R1             6.400E+04        1.221E-07        7.812E-05
  R0             1.280E+05        3.052E-08        3.906E-05
  RF             5.000E+02       -1.336E-03       -6.679E-03
  RP             3.000E+02        2.666E-11        7.999E-11
XOPAMP.RIN       1.000E+06       -1.468E-14       -1.468E-10
  V7             1.000E+00       -5.000E-01       -5.000E-03
  V6             0.000E+00       -2.500E-01        0.000E+00
  V5             1.000E+00       -1.250E-01       -1.250E-03
  V4             0.000E+00       -6.250E-02        0.000E+00
  V3             1.000E+00       -3.125E-02       -3.125E-04
  V2             0.000E+00       -1.562E-02        0.000E+00
  V1             1.000E+00       -7.812E-03       -7.812E-05
  V0             1.000E+00       -3.906E-03       -3.906E-05
```

7.3.8 The D/A converter—the ladder circuit

An interesting comparison can be made to another D/A converter circuit, called an R–$2R$ ladder. This circuit, shown in figure 7.8, uses resistors of value R and $2R$ only. The value of R is chosen as 1 kΩ arbitrarily. Although more resistors are required than for the summing circuit, the range of their values is much smaller.

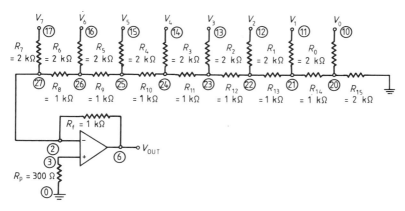

Figure 7.8 The D/A converter—the R–$2R$ ladder circuit.

A SPICE input file was created for the ladder circuit, with again only a sensitivity analysis being requested for the analogue output at node 6. The input file is omitted due to its length, but the DC and sensitivity analysis produces the following output:

```
****    SMALL SIGNAL BIAS SOLUTION    TEMPERATURE =  27.000 DEG C
 NODE   VOLTAGE     NODE   VOLTAGE    NODE   VOLTAGE    NODE   VOLTAGE
(  2)  13.36E-06   (  3)  4.008E-09  (  6)   -.6679    ( 10)   1.0000
( 11)   1.0000     ( 12)   0.0000    ( 13)   1.0000    ( 14)   0.0000
( 15)   1.0000     ( 16)   0.0000    ( 17)   1.0000    ( 20)    .5512
( 21)    .6024     ( 22)    .4548    ( 23)    .5347    ( 24)    .3818
( 25)    .4199     ( 26)    .1680


****    DC SENSITIVITY ANALYSIS      TEMPERATURE =  27.000 DEG C
DC SENSITIVITIES OF OUTPUT V(6)
  ELEMENT         ELEMENT         ELEMENT        NORMALIZED
   NAME            VALUE        SENSITIVITY     SENSITIVITY
                                (VOLTS/UNIT)   (VOLTS/PERCENT)
    R7           2.000E+03       2.500E-04       5.000E-03
    R6           2.000E+03      -2.100E-05      -4.199E-04
    R5           2.000E+03       3.625E-05       7.251E-04
    R4           2.000E+03      -1.193E-05      -2.386E-04
    R3           2.000E+03       7.271E-06       1.454E-04
    R2           2.000E+03      -3.553E-06      -7.107E-05
    R1           2.000E+03       1.553E-06       3.106E-05
    R0           2.000E+03       8.765E-07       1.753E-05
    R8           1.000E+03       8.398E-05       8.398E-04
    R9           1.000E+03       6.298E-05       6.298E-04
    R10          1.000E+03      -4.761E-06      -4.761E-05
    R11          1.000E+03       9.552E-06       9.552E-05
    R12          1.000E+03      -2.495E-06      -2.495E-05
    R13          1.000E+03       2.306E-06       2.306E-05
    R14          1.000E+03      -4.001E-07      -4.001E-06
    R15          2.000E+03      -1.077E-06      -2.153E-05
    RF           1.000E+03      -6.679E-04      -6.679E-03
    RP           3.000E+02       2.672E-11       8.015E-11
    XOPAMP.RIN   1.000E+06      -2.137E-14      -2.137E-10
    V7           1.000E+00      -5.000E-01      -5.000E-03
    V6           0.000E+00      -2.500E-01       0.000E+00
    V5           1.000E+00      -1.250E-01      -1.250E-03
    V4           0.000E+00      -6.250E-02       0.000E+00
    V3           1.000E+00      -3.125E-02      -3.125E-04
    V2           0.000E+00      -1.562E-02       0.000E+00
    V1           1.000E+00      -7.812E-03      -7.812E-05
    V0           1.000E+00      -3.906E-03      -3.906E-05
```

The DC node voltages show for both circuits that the differential input to the op amp, between nodes 2 and 3, is very small as expected. The analogue output at node 6 is calculated by SPICE to be -0.6679 to four significant figures.

Recalling that the binary digital input is set to 10101011, the expected analogue output is then

$$V_{OUT} = -\frac{1}{2^8}(2^7 + 2^5 + 2^1 + 2^0) = -\frac{1}{256}(128 + 32 + 8 + 2 + 1)$$
$$= -\frac{171}{256} = -0.66796875$$

The results of the sensitivity analysis are not entirely conclusive. The normalized sensitivity tabulates the variation of analogue voltage output for a one per cent change of each resistor individually. The overall effect of resistor tolerance is difficult to assess on a statistical basis from these data, but a worst-case voltage change can be obtained by simply adding all of the resistor sensitivities. This would give us the voltage change resulting from the highly unlikely event of each resistor varying by 1% either in a positive or negative direction.

From the tabulated data we can see that resistor R_7 causes a 5 mV swing in analogue output voltage for a 1% resistor change, for both D/A circuits. The other resistors have a smaller effect. Notice that for the summing circuit the resistors connected to the inputs set to 0 V have negligible effect as expected. Such is not the case for the ladder circuit, where the sensitivity analysis is more complicated. The results are also reassuring in that for both circuits the sensitivity to the non-inverting input resistor R_p is very small, as is the effect of changing the op amp's input impedance.

PSpice has the capability of performing a Monte Carlo (statistical) analysis on a circuit. This will be applied to the D/A converter circuits as an example in the next section, in conjunction with the modelling of a resistor.

7.4 DEVICE MODELLING

The examples in previous sections have shown that various sources and passive components can be described on a single line in a SPICE input file. Additionally a complex circuit can be set up and repeatedly referenced via a subcircuit. Semiconductor devices, such as diodes and transistors, can be simulated with SPICE through the use of a device model. The model permits us to specify a variety of parameters to enable the model to be closely matched to a real device. For example, diode parameters such as saturation current and reverse breakdown voltage can be specified. Similarly the maximum forward beta and the emitter ohmic resistance are two of approximately forty parameters that can be set for a bipolar transistor. In most applications relatively few of the device parameters are specified. SPICE will assign a default value for non-specified parameters, which will suffice for most applications. In most of our examples we will allow SPICE to set all the parameters to their default value.

SPICE has built-in models for three types of transistors as well as diodes. Included are bipolar junction transistors (BJTs), junction field-effect transistors (JFETs), and metal–oxide–semiconductor field-effect transistors (MOSFETs). PSpice has additional capability permitting the modelling of passive components (resistors, capacitors, inductors) as well as a gallium arsenide metal–semiconductor FET (GaAs MESFET), non-linear magnetics, and switches controlled by a voltage or current. Although modelling of a passive device is not required, it permits us to vary component values in a circuit. Statistical analyses can be performed on a circuit where resistor tolerances have been specified, as one example. Earlier versions of SPICE permit component value variation via an .ALTER statement, but this is relatively awkward to work with. A SPICE model is invoked with a .MODEL control line of the form:

.MODEL modelname typename (optional parameters and values).

The model name is arbitrary but the type name refers to a specific device. For example type names include D for diodes, NPN or PNP for bipolar transistors, and RES if resistors are to be modelled. As an example, for a circuit with two diodes the following statements might appear in the SPICE input file:

```
.MODEL SWITCH D
DIODE1 1 2 SWITCH
DIODE2 3 4 SWITCH
```

DIODE1, between nodes 1 and 2 (the first node being the positive or current entry node), and DIODE2, connected to 3 and 4, are the individual component names. Notice that a diode name must start with D, just as a resistor name must start with R or an independent voltage source with V. SWITCH is the model name that must be referenced by both diodes, and D the required diode type name. Notice that referencing a model is similar to referencing a subcircuit, except that the model is 'built into' SPICE. Default parameters are of course used here. Parameters could be added to the .MODEL statement in which case the same values would apply to both diodes. On the other hand, specifying different parameters for each diode would require separate .MODEL statements with two different model names.

7.4.1 The full-wave rectifier

We will illustrate a diode model application with the well known full-wave rectifier circuit shown in figure 7.9. For a sinusoidal input the path of current conduction goes through diodes D_1 and D_4 during the positive half-cycle, and through D_2 and D_3 when the input voltage is negative. The voltage across the resistor has the same polarity for both halves of the cycle.

DEVICE MODELLING   245

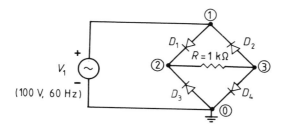

Figure 7.9 The full-wave rectifier.

The SPICE input file for the full-wave rectifier is shown below:

```
FULL WAVE RECTIFIER
.WIDTH OUT=80
.OPTIONS NOPAGE
.MODEL RECTIFY D
V1 1 0 SIN (0V 100V 60HZ)
D1 1 2 RECTIFY
D2 3 1 RECTIFY
D3 0 2 RECTIFY
D4 3 0 RECTIFY
R  2 3 1K
.TRAN 0.5MS 20MS
.PLOT TRAN V(2,3)
.END
```

The source is specified as a sinusoid with zero DC value, a peak of 100 V, and a frequency of 60 Hz (corresponding to a period of 16.67 ms). The transient output is plotted from 0 to 20 ms, over slightly more than one full cycle. The resulting output is shown below.

```
 ****     TRANSIENT ANALYSIS              TEMPERATURE =   27.000 DEG C
   TIME       V(2,3)
   (*)----------   -5.0000E+01   0.0000E+00   5.0000E+01   1.0000E+02
   - - - - - - - - - - - - - - - - - - - - - - - - - - - - -
 0.000E+00  1.602E-27 .            *            .            .            .
 5.000E-04  1.728E+01 .            .      *     .            .            .
 1.000E-03  3.524E+01 .            .            .    *       .            .
 1.500E-03  5.192E+01 .            .            .         *  .            .
 2.000E-03  6.677E+01 .            .            .            .  *         .
 2.500E-03  7.934E+01 .            .            .            .      *     .
 3.000E-03  8.876E+01 .            .            .            .         *  .
 3.500E-03  9.504E+01 .            .            .            .            *.
 4.000E-03  9.803E+01 .            .            .            .            *.
 4.500E-03  9.763E+01 .            .            .            .            *.
 5.000E-03  9.338E+01 .            .            .            .          * .
 5.500E-03  8.584E+01 .            .            .            .      *     .
 6.000E-03  7.534E+01 .            .            .            .  *         .
 6.500E-03  6.220E+01 .            .            .            *            .
 7.000E-03  4.658E+01 .            .            .        *.               .
 7.500E-03  2.933E+01 .            .            .  *         .            .
 8.000E-03  1.107E+01 .            .       *    .            .            .
 8.500E-03  5.381E+00 .            .  *         .            .            .
 9.000E-03  2.334E+01 .            .            *            .            .
 9.500E-03  4.096E+01 .            .            .       *.               .
 1.000E-02  5.713E+01 .            .            .            .  *         .
 1.050E-02  7.134E+01 .            .            .            .       *    .
 1.100E-02  8.273E+01 .            .            .            .          * .
 1.150E-02  9.117E+01 .            .            .            .            * .
 1.200E-02  9.646E+01 .            .            .            .            *.
 1.250E-02  9.842E+01 .            .            .            .            *
 1.300E-02  9.649E+01 .            .            .            .            *.
 1.350E-02  9.117E+01 .            .            .            .           * .
 1.400E-02  8.270E+01 .            .            .            .         *   .
 1.450E-02  7.134E+01 .            .            .            .      *      .
 1.500E-02  5.715E+01 .            .            .            .  *          .
 1.550E-02  4.096E+01 .            .            .       *   .             .
 1.600E-02  2.334E+01 .            .            *           .             .
 1.650E-02  4.886E+00 .            . *          .            .            .
 1.700E-02  1.107E+01 .            .   *        .            .            .
 1.750E-02  2.933E+01 .            .            .  *        .             .
 1.800E-02  4.656E+01 .            .            .        *. .             .
 1.850E-02  6.220E+01 .            .            .            .*           .
 1.900E-02  7.536E+01 .            .            .            .     *      .
 1.950E-02  8.584E+01 .            .            .            .         *  .
 2.000E-02  9.356E+01 .            .            .            .            *  .
```

7.4.2 The logarithmic amplifier

A useful circuit in instrumentation is one that converts a logarithmic input voltage to a linear output. Linearization is advantageous in general, but

DEVICE MODELLING   247

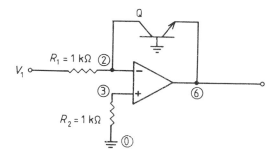

Figure 7.10 The logarithmic converter.

logarithmic conversion also eases the processing of input data that vary over a wide range of values. A simple logarithmic converter circuit is shown in figure 7.10 [5]. The circuit is actually a modification of the basic inverter of figure 7.4, with the feedback resistor replaced by a bipolar n-p-n transistor with grounded base. For the transistor connected as shown, the base-to-emitter voltage is proportional to the natural logarithm of the collector current. Since virtually all of the collector current is flowing through the input resistor R_1, it is easy to show that the output voltage at node 6 increases negatively in proportion to the log of the input voltage V_1.

A SPICE simulation file for the logarithmic converter is shown below.

```
LOGARITHMIC AMPLIFIER (USING IDEAL OP-AMP MODEL)
.WIDTH OUT=80
.OPTIONS NOPAGE
.SUBCKT OPIDEAL   2   3   6
* PINOUTS        -IN +IN OUT
RIN 2 3 1MEG
EO 6 0 3 2 50K
.ENDS OPIDEAL
.MODEL BIPOLAR NPN
V1 1 0
R1 1 2 1K
R2 3 0 1K
Q  2 0 6 BIPOLAR
XOPAMP 2 3 6 OPIDEAL
.DC DEC V1 0.01V 10KV 1
.PLOT DC V(6)
.END
```

The transistor model name is arbitrarily chosen as BIPOLAR, and the type name is NPN as required for this device. The device name on the circuit description line is Q (all bipolar transistor names must start with Q). The transistor is a three-terminal device and the nodes must be entered as collector, base, and emitter node in that order. To demonstrate the logarithmic conversion conveniently over a wide range of inputs, a DC sweep of V_1 over a logarithmic scale is used. The input is increased by a factor of 10 each time, and ranges from 10 mV to 10 kV. A linear DC sweep of the input over the wide range of inputs desired results in too many points to be calculated. Alternatively, the overall range can be broken up into several smaller ranges, but this requires the input file to be edited and rerun several times.

The output plot for the logarithmic converter is shown below.

```
 ****       DC TRANSFER CURVES          TEMPERATURE = 27.000 DEG C
  V(1)            V(6)
  (*)--------   -1.2000E+00   -1.0000E+00   -8.0000E-01   -6.0000E-01
             - - - - - - - - - - - - - - - - - - - - - - - - - - - -
 1.000E-02  -6.551E-01  .             .             .         *    .
 1.000E-01  -7.147E-01  .             .             .      *       .
 1.000E+00  -7.742E-01  .             .             .  *           .
 1.000E+01  -8.338E-01  .             .          *  .              .
 1.000E+02  -8.933E-01  .             .       *     .              .
 1.000E+03  -9.529E-01  .          .  *             .              .
 1.000E+04  -1.012E+00  .       *.                  .              .
```

Since the input voltage V_1 is plotted on a logarithmic scale, the straight line verifies that the output is proportional to the logarithm of the input as expected. From the tabulated values we see that each increase in input voltage by a factor of ten results in the output decreasing by approximately 60 mV. It was also found that when the idealized op amp model is replaced by the UA 741, nearly identical results are obtained.

7.4.3 Monte Carlo analysis of a D/A converter

As a final example in this section we will utilize PSpice's capability for modelling a resistor and performing a statistical or Monte Carlo analysis for random variations in resistance. Recall the two 8-bit D/A converter circuits presented in the previous section (see figures 7.7 and 7.8). The sensitivity analysis performed on these circuits showed the effect of individual resistor variation on the output voltage. With the aid of a Monte Carlo analysis the circuits can be processed a large number of times with random values assigned to as many resistors as desired. A realistic view of the effect of component manufacturing tolerances on the operation of a circuit results from this type of statistical analysis.

The modified SPICE input file for the summing circuit implementation of the 8-bit D/A converter is given below.

DEVICE MODELLING 249

```
8 BIT D/A CONVERTER - SUMMING CIRCUIT (USING
*RESISTOR MODEL)
.WIDTH OUT=80
.OPTIONS NOPAGE
.SUBCKT OPIDEAL  2   3    6
* PINOUTS         -IN +IN OUT
RIN 2 3 1MEG
EO 6 0 3 2 50K
.ENDS OPIDEAL
.MODEL RMOD RES (R=1 DEV=5%)
V7 17 0 DC 1V
V6 16 0 DC 0V
V5 15 0 DC 1V
V4 14 0 DC 0V
V3 13 0 DC 1V
V2 12 0 DC 0V
V1 11 0 DC 1V
V0 10 0
R7 17 2 RMOD 1K
R6 16 2 RMOD 2K
R5 15 2 RMOD 4K
R4 14 2 RMOD 8K
R3 13 2 RMOD 16K
R2 12 2 RMOD 32K
R1 11 2 RMOD 64K
R0 10 2 RMOD 128K
RF 2 6  RMOD 500
RP 3 0  RMOD 300
XOPAMP 2 3 6 OPIDEAL
.DC V0 1V 1V 1V
.MC 100 DC V(6) YMAX
.END
```

Notice that we have included a .MODEL statement and have added the model name RMOD to each resistor description line, before the component value. As indicated previously, the type name RES refers to a resistor model and two parameters are included in parentheses. The parameter R is a multiplying factor useful if you want to scale the values in your models. The deviation DEV applies the specified resistor tolerance independently to each resistor referencing the model. For this example we are not scaling any nominal resistor values, but are assuming that all resistors are subject to a ±5% tolerance.

The .MC statement performs multiple runs (100 in this case) of the selected analysis (DC). The first run is performed using nominal values of all components, to provide a basis for comparison for subsequent runs that utilize the specified tolerance. The keyword YMAX is required following the output variable on which

250 DESIGN AIDS—SPICE

the analysis is being performed, V(6) in this case. One other change from the original file is that the statement

VO 10 0 DC 1V

has been replaced by the two lines

VO 10 0
.DC VO 1V 1V 1V

The value of VO is still fixed at 1 V, but the change is necessary since the Monte Carlo analysis must be accompanied by either a DC, AC, or transient analysis control line.

After running the modified input file with PSpice the following output is produced.

```
**** SORTED DEVIATIONS OF V(6)  TEMPERATURE = 27.000 DEG C
                    MONTE CARLO SUMMARY
Mean Deviation = -155.0100E-06
Sigma          =     .0228
   RUN               MAX DEVIATION FROM NOMINAL
  Pass    25          .0522  (2.29 sigma)  lower   at VO =  1
                      ( 107.81% of Nominal)
  Pass     5          .0489  (2.15 sigma)  higher  at VO =  1
                      (  92.678% of Nominal)
  Pass    13          .0456  (2.00 sigma)  higher  at VO =  1
                      (  93.175% of Nominal)
                            :
                            :
                            :
  Pass    94         447.0300E-06  ( .02 sigma)  lower   at VO = 1
                      (100.07% of Nominal)
  Pass    50         396.1300E-06  ( .02 sigma)  higher  at VO = 1
                      (99.941% of Nominal)
```

The results of all 100 runs are printed out in a sorted order. We have only shown the first three and last two results. Recall that the nominal output voltage is 0.668 V or equivalently 668 mV for this example, corresponding to the conversion of binary number 10101011 to an analogue voltage scaled from 0 to nearly 1 V. In varying all resistors randomly, run 25 produced the maximum absolute deviation from this nominal value. The nominal output was lower than the resulting output by 0.0522 V or about 52 mV. The actual output for this run is therefore 740 mV or 107.81% of the nominal output.

Conversely, the lowest deviation occurred on run 50 resulting in an output deviation of only about 0.4 mV below the nominal value, or 99.94% of it. The mean deviation shown is very close to zero as would be expected, since the resistors are equally likely to be above and below their nominal values. The standard deviation (sigma value), however, ignores the sign of the deviation. Its value of 0.0228 V or 22.8 mV for a mean of 668 mV gives us a 'normalized' standard deviation of about 3.4%. This value seems reasonable since it is of the same order of magnitude as the resistor tolerance of 5%. Since the analysis is statistical the results will be more accurate for a larger number of runs, but at the expense of longer running time for PSpice.

The same Monte Carlo analysis was performed on the 8-bit D/A converter implemented with the R–$2R$ ladder (figure 7.8). For 100 runs the resulting standard deviation is 25.6 mV. Notice that this deviation is only slightly higher than that for the summing circuit even though there are about twice as many resistors in the ladder circuit exhibiting random variation. The analysis was also repeated for both D/A circuits replacing the 5% resistors by those with a 1% deviation. The resulting standard deviations observed for 100 runs are approximately 4.6 mV for the summing circuit and 5.1 mV for the ladder configuration. Notice that as the resistor tolerance has been decreased by a factor of 5, the standard deviations have decreased by about the same amount. Overall, the results indicate that the accuracy of both circuits is comparable, a conclusion somewhat less obvious from the sensitivity analysis shown in the previous section.

7.5 FOURIER ANALYSIS

SPICE can perform a Fourier analysis on a periodic waveform in conjunction with a transient analysis. A .FOUR control line automatically computes and writes to the output file the DC value as well as the first nine frequency components of the waveform.

A general periodic waveform $f_p(t)$ is shown in figure 7.11. For a period T the fundamental frequency f_1 in Hz is given by $1/T$, and the radian frequency is $\omega_1 = 2\pi f_1$. Any periodic waveform can be expressed as a Fourier series, consisting of a DC term plus an infinite sum of sinusoids at the fundamental frequency and multiples of it. The multiple-frequency terms are called harmonics. In many cases the periodic waveform can be approximated quite well by the DC and fundamental frequency component, along with a relatively small number of harmonic terms.

It can be shown that the Fourier series of the general periodic waveform can be written as

$$f_p(t) = a_0 + \sum_{n=1}^{\infty}[a_n \cos(n\omega_1 t) + b_n \sin(n\omega_1 t)]$$

where a_0 is the DC value given by

DESIGN AIDS—SPICE

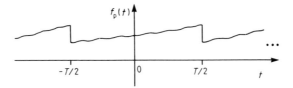

Figure 7.11 A general periodic waveform.

$$a_0 = \frac{1}{T} \int_{-T/2}^{T/2} f_p(t)\, dt$$

and the coefficients a_n and b_n are given by

$$a_n = \frac{2}{T} \int_{-T/2}^{T/2} f_p(t) \cos(n\omega_1 t)\, dt$$

$$b_n = \frac{2}{T} \int_{-T/2}^{T/2} f_p(t) \sin(n\omega_1 t)\, dt.$$

Often, the complex coefficient c_n is used instead, and this is given by

$$c_n = a_n - ib_n = \frac{2}{T} \int_{-T/2}^{T/2} f_p(t) e^{-in\omega_1 t}\, dt$$

where

$$f_p(t) = \frac{1}{2} \sum_{n=-\infty}^{\infty} c_n e^{in\omega_1 t}.$$

SPICE's Fourier analysis capability will be illustrated for the periodic square wave illustrated in figure 7.12. Within the period from $-T/2$ and $T/2$ the waveform is given by

$$f_p = A \qquad (0 < t < T/2)$$

and is zero for $t < 0$. For this waveform the coefficients are easily evaluated from the integrals above and become

$$a_0 = \frac{A}{2} \qquad a_n = 0 \qquad b_n = \frac{A}{n\pi}[1 - \cos(n\pi)].$$

Since the term $\cos(n\pi)$ for $n > 0$ alternates between -1 and 1 as n increases, the coefficients b_n become $2A/n\pi$ for odd n and zero for even n. The Fourier series is then

$$f_p(t) = \frac{A}{2} + 2\frac{A}{\pi}[\sin(\omega_1 t) + \tfrac{1}{3}\sin(3\omega_1 t) + \tfrac{1}{5}\sin(5\omega_1 t) + \ldots]$$

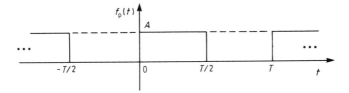

Figure 7.12 A periodic square wave.

where $\omega_1 = 2\pi/T$.

The SPICE input file below performs a Fourier analysis on the square wave of unit amplitude.

```
FOURIER ANALYSIS OF A PERIODIC SQUARE WAVE
.WIDTH OUT=80
.OPTIONS NOPAGE
VIN 1 0 PULSE (0 1V 0 0 0 1MS 2MS)
R 1 0 1K
.TRAN 0.01MS 2MS
.FOUR 0.5KHZ V(1)
.END
```

Since SPICE requires a closed circuit we have created a minimal one by placing an arbitrary 1 kΩ resistor R across a source. The source's square wave created by a PULSE input has unit amplitude ($A = 1$) and a 2 ms period, resulting in a fundamental frequency of 0.5 kHz. The transient analysis is performed over a full period using a total of 200 points. In general it is recommended that the number of points be at least 100 to get reasonably accurate Fourier coefficients. The .FOUR command is followed by the fundamental frequency and the voltage or current on which the analysis is performed.

The Fourier analysis written to the output file is shown below.

```
****    FOURIER ANALYSIS            TEMPERATURE = 27.000 DEG C
FOURIER COMPONENTS OF TRANSIENT RESPONSE V(1)
DC COMPONENT =    5.049873E-01
HARMONIC  FREQUENCY  FOURIER   NORMALIZED  PHASE     NORMALIZED
   NO       (HZ)    COMPONENT  COMPONENT   (DEG)     PHASE (DEG)
   1      5.000E+02  6.365E-01  1.000E+00  -1.802E+00  0.000E+00
   2      1.000E+03  9.972E-03  1.567E-02   8.594E+01  8.774E+01
   3      1.500E+03  2.119E-01  3.330E-01  -5.405E+00 -3.603E+00
   4      2.000E+03  9.966E-03  1.566E-02   8.188E+01  8.368E+01
   5      2.500E+03  1.269E-01  1.993E-01  -9.004E+00 -7.202E+00
   6      3.000E+03  9.955E-03  1.564E-02   7.782E+01  7.962E+01
   7      3.500E+03  9.031E-02  1.419E-01  -1.260E+01 -1.079E+01
   8      4.000E+03  9.940E-03  1.562E-02   7.377E+01  7.557E+01
   9      4.500E+03  6.992E-02  1.098E-01  -1.618E+01 -1.438E+01
        TOTAL HARMONIC DISTORTION =    4.286648E+01 PERCENT
```

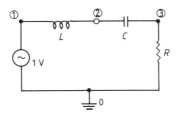

CASE	L	C	R
WIDE BAND	1 MH	1 µF	1 kΩ
NARROW BAND	1.59 MH	2.49 pF	1 kΩ

Figure 7.13 An RLC bandpass filter.

For a unit amplitude the expected DC value is 1/2, the fundamental frequency component is $2/\pi = 0.6366$, and the remaining odd-harmonic components decrease by a factor n relative to the fundamental. These two values match closely to SPICE's output. Also, we can see from the normalized component column of the table that the amplitudes for the odd harmonics are very close to 1/3, 1/5, 1/7, and 1/9 as expected. Theoretically the even-harmonic amplitudes should be zero and therefore 0% normalized, but the actual results show approximately a 1.5% normalized value for each.

7.6 USE OF THE PROBE UTILITY

In several examples in previous sections we have plotted a transient or AC response of a circuit. The use of a .PLOT control line, however, results in a rather primitive display using character rather than bit-mapped graphics. The PSpice version includes a utility program called PROBE, which provides considerably enhanced graphics capability. To take advantage of this capability we simply need to insert the control statement .PROBE into the SPICE input file. When PSpice is run the total circuit response is written to a file PROBE.DAT, which is then available to the PROBE utility.

7.6.1 The bandpass filter

As a first example consider the series RLC circuit shown in figure 7.13. The circuit functions as a bandpass filter, with the centre frequency and bandwidth controlled by the component values.

The SPICE input files are shown below for both the wide-band and narrow-band cases.

USE OF THE PROBE UTILITY 255

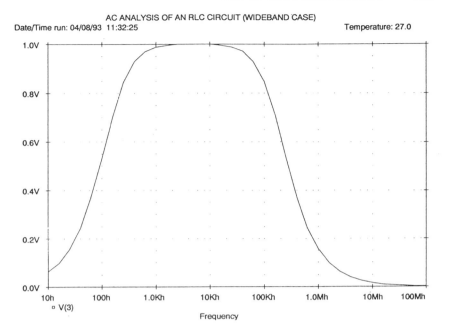

Figure 7.14 AC analysis of an RLC circuit, wide-band case.

```
AC ANALYSIS OF AN RLC CIRCUIT (WIDEBAND CASE)
.WIDTH OUT=80
.OPTIONS NOPAGE
V1 1 0 AC 1V
L 1 2 1MH
C 2 3 1UF
R 3 0 1K
.PLOT AC V(3)
.AC DEC 5 10HZ 100MEGHZ
.PROBE
.END
```

256 DESIGN AIDS—SPICE

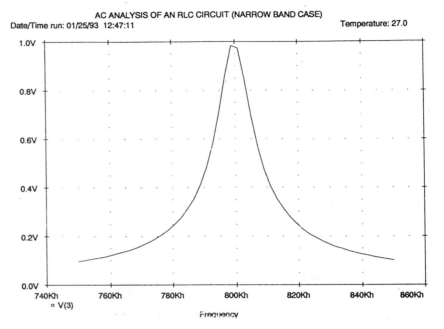

Figure 7.15 AC analysis of an RLC circuit, narrow-band case.

```
AC ANALYSIS OF AN RLC CIRCUIT (NARROW BAND CASE)
.WIDTH OUT=80
.OPTIONS NOPAGE
V1 1 0 AC 1V
L 1 2 15.9MH
C 2 3 2.49PF
R 3 0 1K
.PLOT AC V(3)
.AC LIN 50 750KHZ 850KHZ
.PROBE
.END
```

Figures 7.14 and 7.15 show the resulting graphs produced by the PROBE utility. The wide-band filter response is clearly more meaningful when a logarithmic frequency scale is used, since the passband extends over several decades of frequency. The narrow-band filter illustrates a tuned circuit intended to capture an AM radio frequency of about 800 kHz, with the bandwidth limited to audio frequencies to prevent interference with a nearby AM carrier. For the narrow-band case a linear frequency scale is preferable. Notice for the narrow-band case that only values between 750 kHz and 850 kHz are shown, as a result of the input file's .AC statement. PROBE permits the user to select the ranges for both axes of the plot.

USE OF THE PROBE UTILITY   257

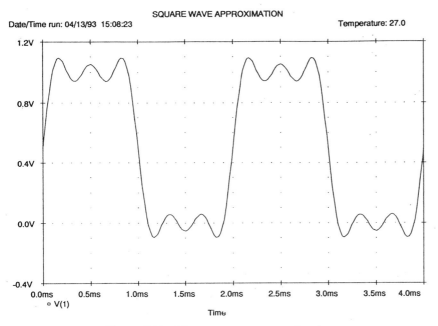

Figure 7.16 The square-wave approximation.

7.6.2 Fourier series—the square-wave approximation

Recall from the previous section that the Fourier series of a periodic square wave consists of odd-harmonic frequencies, along with a DC value if present. The SPICE input file below simulates an approximation to a unit-amplitude 500 Hz square wave, using the fundamental frequency along with the third and fifth harmonics.

```
SQUARE WAVE APPROXIMATION
.WIDTH OUT=80
.OPTIONS NOPAGE
V1  1 2 DC 0.5V
V2  2 3 SIN (0V      0.6366V   0.5KHZ)
V3  3 4 SIN (0V      0.2122V   1.5KHZ)
V4  4 0 SIN (0V      0.1273V   2.5KHZ)
R   1 0 1K
.TRAN  0.02MS   4MS
.PROBE
.END
```

The three sinusoids along with a DC value are implemented as independent voltage sources in series with a resistor of arbitrary value. The resistor voltage is the sum of the source voltages and therefore the square-wave approximation.

258   DESIGN AIDS—SPICE

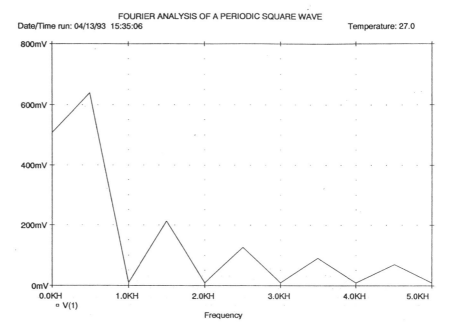

Figure 7.17 Fourier analysis of a periodic square wave.

The transient output from the PROBE utility is shown over two full periods. As more terms are added the approximation obviously improves, although the overshoots and undershoots observed after a discontinuity remain. The Fourier series representation of signals with discontinuities produces this effect, called the Gibbs phenomenon.

From the data file produced by the .PROBE statement a Fourier analysis can also be done. This analysis utilizes the discrete Fourier transform (DFT) built into PROBE. The difference between the DFT and the Fourier series will be discussed in more detail in the next chapter.

Figure 7.17 shows the frequency spectrum produced by the PROBE's DFT for the square wave generated in the example in section 7.5. The DFT, as its name suggests, produces a series of discrete points at multiples of the fundamental frequency. The values therefore correspond to the harmonic amplitudes. The discrete values are joined by straight lines, and we see that the even harmonics at 1 kHz, 2 kHz, 3 kHz, etc are zero. The DC value is indeed 0.5 V, the fundamental amplitude is $2/\pi = 0.633$, and the nth odd harmonic has an amplitude of $1/n$ compared with that of the fundamental.

7.6.3 The FM waveform

The final example illustrates the time and frequency plots of a frequency-modulated (FM) waveform. The input file below uses the built in FM waveform

USE OF THE PROBE UTILITY   259

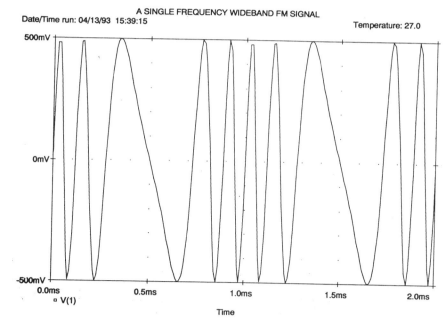

Figure 7.18 A single-frequency wide-band FM signal: the transient response.

generator function for the voltage source.

```
FOURIER ANALYSIS OF A SINGLE FREQUENCY WIDEBAND FM SIGNAL
.WIDTH OUT=80
.OPTIONS NOPAGE
VIN 1 0 SFFM (0V  0.5V    5KHZ  4   1KHZ)
R 1 0 1K
.TRAN 0.01MS 2MS
.PROBE
.END
```

The parameters following SFFM include the waveform's DC and peak voltages, followed by the carrier frequency, modulating index, and modulating frequency. A modulation index of 4 is in the range of wide-band FM, which has several upper and lower sidebands. The time plot of two cycles of the modulating frequency is shown in figure 7.18. The frequency variation is apparent and we can see that at every 1 ms interval the signal is varying at the carrier frequency of 5 kHz.

The frequency spectrum generated by PROBE's DFT, shown in figure 7.19, is centred around the carrier as expected. The first sidebands at 4 kHz and 6 kHz are relatively small but the next three sets of sidebands are quite significant. As the modulation index was decreased we would observe a narrower band,

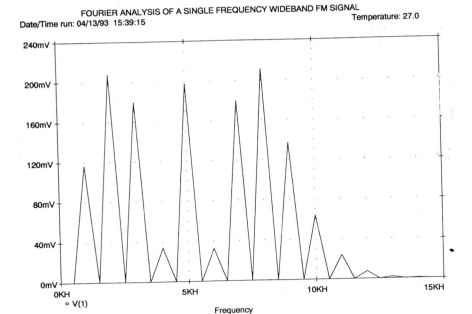

Figure 7.19 A single-frequency wide-band FM signal: the frequency response.

eventually resulting in only one upper and lower sideband. Narrow-band FM is therefore similar in spectral content to an amplitude-modulated (AM) waveform, for which sidebands exist only at the carrier plus and minus the modulating frequency.

REFERENCES

[1] Banzhaf W 1992 *Computer-aided Analysis using PSpice* (Englewood Cliffs, NJ: Prentice-Hall)
[2] Normann R B 1988 *Principles of Bioinstrumentation* (New York, NY: Wiley)
[3] Tuinenga P W 1988 *SPICE—A Guide to Circuit Simulation and Analysis using PSpice* (Englewood Cliffs, NJ: Prentice-Hall)
[4] Byers TJ 1990 Breadboards are giving way to PC-based circuit simulation *Radio-electronics* November, pp 63-8
[5] Mims FW III 1990 Analog Arithmetic *Mod. Electron.* November, pp 57–61

8

Design Aids—MathCAD, MATLAB

8.1 INTRODUCTION

MathCAD®, a product of MathSoft Inc., is a powerful and convenient mathematical computational tool [1, 2]. With MathCAD we can create a worksheet on the screen consisting of text and mathematical equations, and display results individually or with the use of tables or graphs. Unlike a Lotus spreadsheet requiring entries into cells, the MathCAD worksheet employs a 'free format'. That is, a document is created in a format similar to one created manually on paper.

MathCAD uses standard mathematical symbols for algebraic operations, but employs graphics to improve the readability of formulas and permit non-ASCII symbols such as summations and integrals. Graphs are part of the worksheet as well, permitting instant viewing without a command sequence as required by a spreadsheet. MathCAD also has file reading and writing capability permitting the transfer of numerical data between a worksheet and a Lotus spreadsheet. As an example. MATLAB®, an alternative product with similar capabilities as MathCAD, will be discussed in the last section of this chapter.

8.2 ELEMENTARY OPERATIONS

The worksheet in figure 8.1 illustrates some simple MathCAD mathematical operations.

Notice first that the worksheet contains text at the top and down the left-hand side. Each text region is created by pressing the double quotation (") key followed by the text, which may consist of several lines. A text region is left by pressing an arrow key, which moves the cursor away from the text. Throughout the chapter, sequences of keystrokes entered are enclosed in double quotes ("). The quotes themselves are not to be entered.

The calculator operations illustrate that MathCAD displays an immediate result when the equal (=) sign is pressed. For example, typing in "5+3=" results in 8 being displayed. For multiplication the operation symbol entered

```
MathCAD Introductory Examples
-----------------------------------
Calculator:      5 + 3 = 8       5·3 = 15      5
                                               ─ = 1.667      5³ = 125      √5 = 2.236
                                               3

Using variables:       A := 5          B := 3         A
                                                      ─ = 1.667
                                                      B

Summations:       n := 1 ..100           ___                      ___
                                         ╲                  3     ╲    1
                                         ╱   n = 5.05·10          ╱   ─── = 5.187
                                         ‾‾‾                      ‾‾‾  n
                                          n                        n


                          ___
                          ╲   1
                  A :=    ╱  ───        A² = 26.909
                          ‾‾‾ n
                           n


Integral:            ⌠1  2
                     │  x  dx = 0.333
                     ⌡0
```

Figure 8.1 MathCAD operations.

is the asterisk (∗), but the displayed symbol is a dot (.). Likewise, for division a slash (/) is entered but an underscore is displayed. The exponentiation is actually entered as "5^3", and MathCAD formats the exponent as a superscript. The familiar square-root symbol displayed is entered using the backslash (\) preceding the number. The display formatting used by MathCAD improves readability at the expense of extra space used, and increased difficulty in the editing of complex equations.

MathCAD has some characteristics of a programming language. As illustrated by the worksheet, numerical values can be assigned to variables by entering "A:5" for example, which displays as "A:=5". The colon (:) or colon-equal (:=) therefore means 'assigned to' as opposed to 'equal to' (=). Those familiar with a programming language such as Pascal will be comfortable with this distinction. A variable can be set to a range of values as illustrated by the summation example. The summation index is set to range from 1 to 100, in steps of 1 by default, by entering "n:1;100". The semicolon (;) displays as two dots (..) to enhance understanding. If a step size of 0.1, for example, is desired, we would enter "n:1,1.1;100", which is displayed as "n:=1,1.1..100".

For the summation itself first enter the index followed by the dollar sign ($) and then the function to be summed. Thus, to sum all values of n over the selected range we enter "n$n=" and to sum all values of $1/n$ over that range we enter "n$1/n=". Each assignment statement or other non-text statement is a separate object in a worksheet, and can be easily edited. In this way values can be easily changed and the results of new calculations immediately displayed. Thus, MathCAD has the same capability as a spreadsheet in doing 'what if' analyses. As an example, changing the upper limit of n from 100 to 200 results in the summation of n changing from 5050 to 20100, and the summation of $1/n$ from 5.187 to 5.878. Since the change in the summation of $1/n$ for integer n is significant we might conclude that this series is divergent (which it is!)

Finally the integral is constructed by entering the variable name, x for example, followed by the ampersand (&), which sets up an integral template. The function to be integrated can be then be entered. The keystroke sequence for the example shown is then "x&x^2=". If MathCAD attempts to calculate the integral's value an error message will appear, since the integral's limits have not been entered. MathCAD, unlike some other mathematics packages, computes numerical results only, and therefore will not give a closed-form expression for an indefinite integral even if one exists. The integral template displayed permits easy entry of the limits by moving the cursor to designated places highlighted by small blocks.

8.3 GRAPHING WITH MATHCAD

MathCAD places the template for a graph on the screen upon pressing the "@" key. The size as well as other attributes of the graph can be set. A previously defined function can then be plotted over a specified range of the independent variable.

8.3.1 Fourier series—the square-wave approximation

The worksheet in figure 8.2 illustrates MathCAD's 'graphing' (i.e. graph plotting) capability by plotting the Fourier series approximation to a square wave. Recall that the same waveform was generated by SPICE and shown in figure 7.16. The amplitude (A) and period (T) are first assigned their values (1 V and 2 ms respectively) and the fundamental radian frequency is calculated. The subscript used with frequency is part of the variable name and is entered as "w.1". MathCAD also allows subscripts to vary over a range of numbers, enabling convenient reference to array or vector quantities. Range subscripts will be discussed later.

In this example, time (t) is set to vary over a range of 0 to 4 ms giving us a plot of two complete periods of the waveform. The signal itself must be specified as a function of the variable t. The formula displays several pairs of brackets, although entry always requires the use of parentheses.

8.3.2 The transient response of an RC circuit

The worksheet in figure 8.3 illustrates the plotting of the response of a series RC circuit, previously analysed with SPICE in section 7.2.2. The circuit diagram is shown in figure 7.2.

The input pulse to the circuit is generated with MathCAD's Heaviside function, equivalent to the unit step function. The resistor value is set (1 kΩ) as is the pulse duration (0.1 μs). The capacitor value C is set globally to the left of the plot of output voltage. A global assignment, denoted by three bars, can be placed anywhere in the worksheet. In contrast, the ordinary assignment

264   DESIGN AIDS—MATHCAD, MATLAB

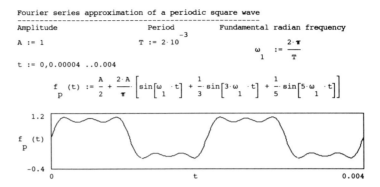

Figure 8.2 Fourier series approximation of a periodic square wave.

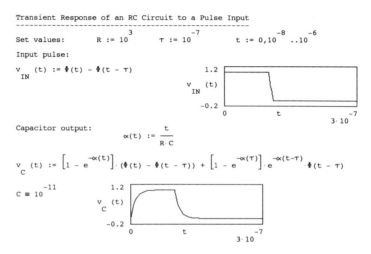

Figure 8.3 Transient response of an RC circuit.

using the colon (:) must always occur before the variable name is used. The use of the global assignment here is only for convenience in viewing the output graph as C is changed.

From the graph of the output shown in figure 8.3 we see that a relatively small value of C produces a sharply rising pulse similar to the input applied. Figure 8.4 shows the effect of increasing C. The first graph for $C = 100$ pF should match the rather primitive plot produced by SPICE (see section 7.2.2). When the value of C is increased by another factor of 10 we have an approximate integrating circuit as discussed in conjunction with SPICE. Notice the ranges for the x- and y-axes of the graphs have been manually set; alternatively, MathCAD will set them for you. For this example we have specified time to a maximum of 1 μs but have only shown a portion of the time axis for better clarity.

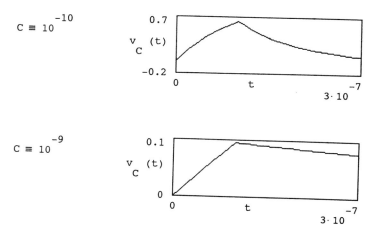

Figure 8.4 Changing capacitor values.

8.3.3 The digital filter approximation

The worksheet in figure 8.5 illustrates the well known bilinear transformation technique to implement a digital filter approximation to a known analogue filter response. The example is shown for a bandpass filter with an analogue transfer function $H_a(s)$ where the Laplace operator s is a function of frequency f. The digital operator z is related to the Laplace operator via

$$z = e^{sT_s}$$

and the transfer function $H_d(z)$ is obtained from $H_a(z)$ by replacing s by

$$s = \frac{2}{T_s} \frac{(z-1)}{(z+1)}.$$

The approximation shown improves as the filter bandwidth is made narrower in comparison to the centre frequency.

8.3.4 Discrete convolution

MathCAD's plots can be also be presented in different formats. For example, a discrete convolution example lends itself to a bar plot. The worksheet in figure 8.6 illustrates the discrete convolution of two pulses. Here the two sequences to be convolved are assigned range subscripted variables x_n and h_n. The entry for x_n is "x[n" as distinguished from "x.n". Notice that n has been assigned a range previously. The values of x_n starting with $n = 0$ for this example are entered by pressing the sequence of keys "x[n:1,1,1,1,1,1,1,0". We see that the values are displayed in tabular form. Although the pulse 'widths' in this example are 7 and 5, N has been chosen as

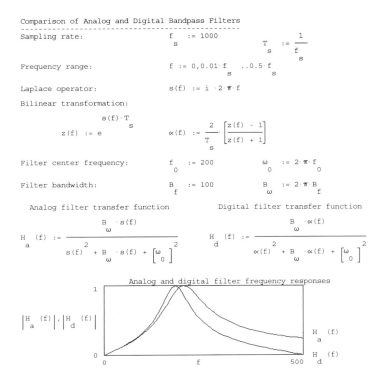

Figure 8.5 Conversion of analogue passband filter to a digital filter.

the next higher power of 2. We will later show convolution using MathCAD's fast Fourier transform (FFT), the algorithm of which requires that the number of samples be a power of 2.

The evaluation of the convolution summation requires a 'trick' since MathCAD has difficulty with negative-integer subscripts. The relational expression $n \geq k$ evaluates to one only if the condition is met, and to zero otherwise. Where n is smaller than k MathCAD doesn't attempt to interpret the subscripts on x and h, but simply evaluates the term of the summation as zero. Notice also that the range of n was increased to include the required $2N$ (rather than N) samples of the output y_n.

8.4 EQUATION SOLVING

MathCAD gives us several techniques for solving mathematical equations. A built-in root function is provided for finding the roots of a single linear or non-linear equation. MathCAD employs the secant numerical method to solve an equation of the form $f(x) = 0$ using an initial guess for x supplied by the user.

EQUATION SOLVING

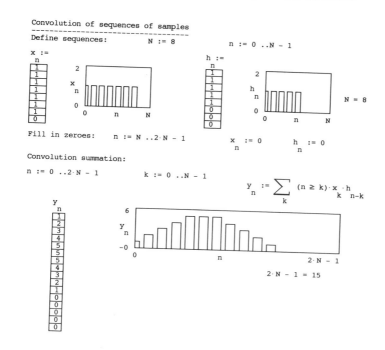

Figure 8.6 Discrete convolution of two pulses.

When $f(x)$ is within a predefined tolerance, the value of the root is returned. As with any numerical method, multiple roots require several trials with somewhat intelligent guesses to be made. Of course the ease of plotting a function with MathCAD is helpful in locating the approximate locations of roots to assist in the use of the root function.

8.4.1 Root finding

The worksheet in figure 8.7 illustrates the locating of the distinct roots of a fourth-order equation. The roots happen to be integers in this case. The plot of $f(x)$ shows that all of the roots are between -10 and 10. Notice that the x-axis is shown by simply including a second function of x, which is always zero. The root function includes the arguments $f(x)$ and x, and will generally find the root closest to the initial guess. Since the first guess is -10 the smallest root $x = -4$ is located first. Choosing the next guess at -2 results in $x = -3$, and the two positive roots are located in similar fashion. MathCAD's 'what if' capability facilitates changing the guesses if previously located roots are again identified.

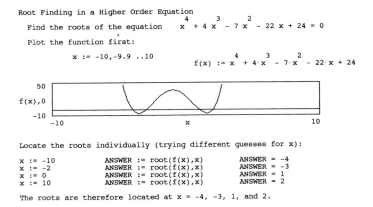

Figure 8.7 Finding roots of an equation.

8.4.2 Linear equations for a differential amplifier

MathCAD's solve block provides the most general approach to solving a set of simultaneous equations. The equations need not be linear, and inequality as well as equality conditions can be handled. For a set of linear simultaneous equations a solution is obtained more directly and rapidly using a matrix approach. To illustrate both of these methods consider the one-op-amp differential amplifier shown in figure 7.5. Using SPICE we verified in section 7.3.5 that the output of this circuit is equal to the input voltage difference amplified by 100. Of course SPICE only requires us to present a circuit description of the amplifier by specifying the components between various pairs of nodes. Since MathCAD has no built-in circuit library we must instead use the fundamental circuit laws to construct a set of simultaneous equations. To accomplish this we use the idealized op amp model shown in figure 7.3 to construct the equivalent circuit shown in figure 8.8.

The resistor branch currents are identified and along with the dependent source voltage V_{IN} we have six unknowns. The DC input voltages have been fixed at $V_1 = 2$ mV and $V_2 = 5$ mV, resulting in an expected output of $100(5 - 2) = 300$ mV.

The worksheet in figure 8.9 below uses a solve block to find the unknowns. The block starts with the word "Given" and ends with a Find function. Prior to entering the block we must guess at a value for each unknown, as is the case when using the root function. A number of constraints equal to the number of unknowns must be entered after "Given". Here we have equality rather than inequality constraints, using a relational equals symbol entered as "<Alt>=". If a solution is found it is returned to the Find function, thus creating a vector. The assignment statement transfers these values to the actual variable names. As an alternative the values could have been assigned to a vector quantity such as X, in which case X_0 is assigned the value of i_1, X_1 the value of i_2, and so

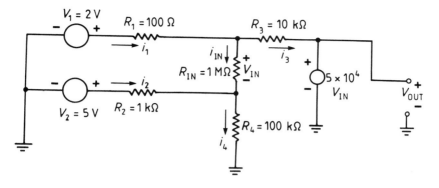

Figure 8.8 Equivalent circuit for a one-op-amp differential amplifier.

on. Finally, the output voltage is calculated, and is very close to the expected value of 300 mV.

The worksheet in figure 8.10 solves the same set of linear simultaneous equations, using matrix inversion. The matrix **B** is created by entering "B:<Alt>M". A message line appears at the top requesting the matrix dimensions (rows, then columns). Enter "6" followed by a space and "6" again. A 6 × 6 matrix appears with placeholders then filled in with the appropriate quantities. Similarly, the column vector **C** is created by entering "C:<Alt>M", "6 1" for the matrix size, and the entries shown. By multiplying the inverse of **B** by **C** we obtain a column vector of the unknowns. Notice that the assignment expression for **X** is entered in the same way for matrix quantities as for scalars. The last unknown V_{IN} corresponds to vector element X_5, and is therefore again used to find the output voltage. The two methods give us identical results.

8.5 FOURIER SERIES AND DISCRETE FOURIER TRANSFORMS

It was shown in sections 7.5 and 7.6.2 that SPICE has built-in Fourier analysis capability. The .FOUR command in SPICE computes the Fourier series coefficients of a continuous periodic signal. With MathCAD we can generate the coefficients from their definitions shown in section 7.5, for a specified periodic function. The worksheet in figure 8.11 generates the DC value and first nine frequency components for the square-wave example (see figure 7.12). The amplitude is unity ($A = 1$) and the period is 2 ms ($T = 0.002$ s). An approximation to the square wave using the computed frequency coefficients is then generated and plotted.

The square wave is defined as a unit-step function using MathCAD's If function, which is of the form

```
if(condition, x, y)
```

DESIGN AIDS—MATHCAD, MATLAB

Solution to the circuit equations for the one op-amp differential amplifier (using Solve block)

Define units: $k \equiv 1000 \qquad meg \equiv 1000000$

Resistor values (ohms):
$R_1 := 100 \qquad R_2 := 1 \cdot k \qquad R_3 := 10 \cdot k$
$R_4 := 100 \cdot k \qquad R_{IN} := 1 \cdot meg$

Source values (mvolts): $V_1 := 2 \qquad V_2 := 5$

Gain: $A := 50 \cdot k$

Guess values:
$i_1 := 0 \qquad i_2 := 0 \qquad i_3 := 0$
$i_4 := 0 \qquad i_{IN} := 0 \qquad v_{IN} := 0$

Given

$$-V_1 + R_1 \cdot i_1 + v_{IN} - R_2 \cdot i_2 + V_2 \approx 0$$

$$-v_{IN} + R_3 \cdot i_3 + A \cdot v_{IN} - R_4 \cdot i_4 \approx 0$$

$$-V_2 + R_2 \cdot i_2 + R_4 \cdot i_4 \approx 0$$

$$v_{IN} \approx R_{IN} \cdot i_{IN}$$

$$i_1 \approx i_{IN} + i_3$$

$$i_2 + i_{IN} \approx i_4$$

$$\begin{bmatrix} i_1 \\ i_2 \\ i_3 \\ i_4 \\ i_{IN} \\ v_{IN} \end{bmatrix} := Find\begin{bmatrix} i_1, i_2, i_3, i_4, i_{IN}, v_{IN} \end{bmatrix} \qquad \begin{bmatrix} i_1 \\ i_2 \\ i_3 \\ i_4 \\ i_{IN} \\ v_{IN} \end{bmatrix} = \begin{bmatrix} -0.03 \\ 4.95 \cdot 10^{-5} \\ -0.03 \\ 4.951 \cdot 10^{-5} \\ 6.012 \cdot 10^{-9} \\ 0.006 \end{bmatrix}$$

Output voltage (mvolts): $\qquad v_{OUT} := A \cdot v_{IN} \qquad v_{OUT} = 300.608$

Figure 8.9 A solve block for the one-op-amp differential amplifier.

The condition is usually an expression with a relational operator, evaluating to either true (1) or false (0). If the condition is true the value of x is returned; for a false condition the y-value is returned. The statement

$$f(t):=if(t>0,1,0)$$

therefore sets $f(t) = 1$ for $t > 0$ and $f(t) = 0$ otherwise. Since the Heaviside function is defined in exactly the same way it could have been used as an alternative. Notice that it is not necessary actually to specify a periodic function here, since the integrations are performed over a single period only.

When MathCAD solves equations or computes integrals and derivatives it uses an internal tolerance (TOL) set by default to 0.001. The tolerance can be changed for all or part of a worksheet by assigning a different value to TOL.

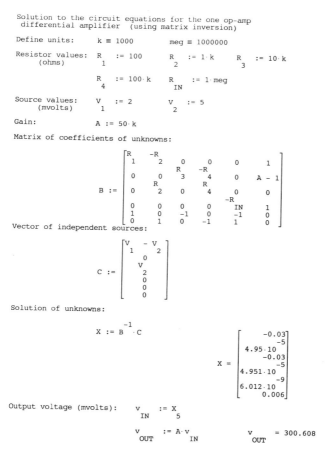

Figure 8.10 A matrix solution for one-op-amp differential amplifier.

Integration is done by approximating the value of the integral until successive approximations differ by less than TOL. For functions with discontinuities or other rapid changes, the solution may be inaccurate or fail to converge at all. The square wave in this example has a discontinuity, and the higher-frequency Fourier coefficients involve integrals of more rapidly varying functions as compared with lower-frequency components.

Running the worksheet using the default value of tolerance produced little error in odd coefficients c_1, c_3, and c_5; however, c_7 was about 10% too low and c_9 more than 30% too high. Also, the DC value was computed as 0.49 rather than 0.5. When running the worksheet with TOL set to 0.0001, the coefficient errors were reduced to 2 to 3% but the DC value remained at 0.49. Setting TOL to 0.00001 as shown in figure 8.11 results in a DC value of 0.497, and less than 1% error in all frequency components. Although decreasing TOL

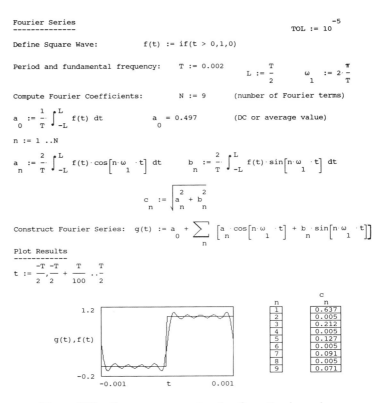

Figure 8.11 Square-wave construction from Fourier series.

improves the accuracy of the results of equation solving and integration, it will obviously result in longer run times as more approximations are required before convergence to an answer is achieved.

The Fourier coefficients a_0, a_n, and b_n are then computed and an approximation to $f(t)$, called $g(t)$, is constructed. The use of L as the half-period ($T/2$) only serves to avoid division operators in the limits of integration, thereby resulting in a more compact display. Both the original and approximated square wave are plotted, and the c_n, the magnitudes of each of the Fourier coefficients, are calculated. As expected, the c_n nearly vanish for even n and have magnitudes of $2/n\pi$ for odd values of n. The values match closely those computed by SPICE in section 7.5.

The Fourier series generates spectral information for a periodic signal, while the Fourier transform performs a similar function for non-periodic waveforms. The discrete Fourier transform (DFT) then transforms a non-periodic discrete time signal, or a sequence of sample values, into frequency components. In contrast to the Fourier series and Fourier transform, the DFT's frequency coefficients are found from a summation rather than from integrations. Since computer evaluation of the DFT is relatively time consuming, a faster computational algorithm called the fast Fourier transform (FFT) is generally used. We will henceforth refer to the DFT as the FFT.

Consider a continuous signal $x(t)$ that has been sampled at uniform intervals $t = 0, T_s, 2T_s, 3T_s, \ldots, (N-1)T_s$. A total of N samples are taken starting at zero time, where the sampling interval or sampling period is T_s. For convenience the discrete signal can be denoted by a sequence of sampled values $x(n)$, where n is an integer ranging from 0 to $N - 1$. The FFT algorithm generally requires that the number of samples N be a power of 2.

The sampled signal $x(n)$ can then be written as

$$x(n) = \frac{1}{N} \sum_{k=0}^{N-1} X(k) e^{i(2\pi/N)nk} \qquad n = 0, 1, 2, \ldots, N-1$$

where the FFT coefficient $X(k)$ is given by

$$X(k) = \sum_{n=0}^{N-1} x(n) e^{-i(2\pi/N)nk} \qquad k = 0, 1, 2, \ldots, N-1.$$

The coefficient $X(k)$ is complex, and is analogous to the Fourier series complex coefficient c_n. The index k represents a normalized digital frequency, and is related to the analogue frequency f of the pre-sampled signal $x(t)$ by

$$\frac{k}{N} = \frac{f}{f_s}$$

where $f_s = 1/T_s$ is the sampling frequency. The nature of sampling is such that the frequency spectrum of a discrete signal is actually periodic in the sampling frequency f_s. Further, it can be shown that each coefficient $X(k)$ has a corresponding complex conjugate coefficient $X(N-k)$. As a result only the coefficients for $k = 0, 1, 2, \ldots, N/2$ are actually independent. Notice that $k = N/2$ corresponds to an analogue frequency of half the sampling frequency ($f = f_s/2$). This correspondence ties into the sampling theorem, which requires that the sampling frequency be at least twice the highest-frequency component of the continuous signal.

MathCAD has a built-in FFT function to compute the coefficients $X(k)$, and an inverse FFT (IFFT) to reconstruct the sampled sequence $x(n)$. MathCAD's slightly different version of the FFT and inverse FFT definition is given below:

$$x(n) = \frac{1}{\sqrt{N}} \sum_{k=0}^{N-1} X(k) e^{-i(2\pi/N)nk} \qquad n = 0, 1, 2, \ldots, N-1$$

where $X(k)$ is given by

$$X(k) = \frac{1}{\sqrt{N}} \sum_{n=0}^{N-1} x(n) e^{i(2\pi/N)nk} \qquad k = 0, 1, 2, \ldots, N-1$$

Convolution using FFTs:

N := 16   X := fft(x)   H := fft(h)   Y := ifft$\left[\sqrt{N} \cdot \overrightarrow{(X \cdot H)}\right]$

$$X = \begin{bmatrix} 1.75 \\ 0.481 + 1.161i \\ -0.177 + 0.177i \\ 0.346 + 0.143i \\ 0.25i \\ 0.154 - 0.064i \\ 0.177 + 0.177i \\ 0.019 - 0.046i \\ 0.25 \end{bmatrix} \quad H = \begin{bmatrix} 1.25 \\ 0.753 + 0.753i \\ 0.604i \\ -0.062 + 0.062i \\ 0.25 \\ 0.209 + 0.209i \\ 0.104i \\ 0.1 - 0.1i \\ 0.25 \end{bmatrix} \quad Y = \begin{bmatrix} 1 \\ 2 \\ 3 \\ 4 \\ 5 \\ 5 \\ 5 \\ 4 \\ 3 \\ 2 \\ 1 \\ 0 \\ 0 \\ 0 \\ 0 \\ 0 \end{bmatrix}$$

Figure 8.12 Convolution using fast Fourier transforms.

Figure 8.12 shows an extension of the worksheet shown in figure 8.6, and illustrates discrete convolution using FFTs. The sequences have been previously defined and the zeros filled in as well for $n = 8$ to $n = 15$ (as in figure 8.6). The value of N is redefined for convenience to reflect the actual vector sizes after zero padding. The FFTs of both pulse sequences are evaluated by the function fft. The resulting vector X consists of the complex $X(k)$-values, and similarly for H. Notice that the display of X and H only shows the first eight values, corresponding to k ranging from 0 to $N/2$. As discussed earlier, the remaining eight $X(k)$-values are complex conjugates of the first eight, and are not even displayed by MathCAD.

It is well known that convolution in the time domain corresponds to multiplication in the frequency domain. The product of the two FFTs results in the FFT of the convolved time sequence. Applying the inverse FFT function, ifft, to the scalar product of X and H gives us the desired result. The difference between MathCAD's FFT definition and the more frequently used one makes necessary the inclusion of the square-root-of-N factor applied to the FFT product. Additionally, a simple scalar product operation on the two vectors does not work since the dimensions are not compatible. MathCAD provides a 'vectorize' operator that facilitates the operation without the need to resort to subscripted variable notation. The vectorize symbol is shown as a 'hooked line' above the product, and is entered after closing the parentheses by pressing "<Alt> -".

We see that the sequence $y(n)$ verifies the result of the convolution summation shown in figure 8.6. MathCAD is normally set to use "i" as the imaginary unit, but can be changed to display "j" via a global format command. Other MathCAD attributes that can be set include the number of decimal places displayed, zero tolerance (the smallest number size below which zero is displayed), and radix or base displayed (octal or hex can be specified instead of the default decimal base).

8.6 SIGNAL DETECTION IN NOISE

MathCAD has a built-in random-number-generating function, similar to that found in several high-level languages. The function generates a predefined sequence of pseudo-random numbers. The seed can be changed by a 'randomize' command, thus producing a different sequence the next time a worksheet is run. A random-number generator permits many interesting simulation problems to be run with MathCAD, although memory capacity and computational speed may be limiting factors. In this section we will show how MathCAD can simulate the recovery of a signal when random noise has been added to it. The two techniques to be demonstrated are filtering and signal averaging (accumulation). The FFT will be used to enable us to specify a filter via its transfer function in the frequency domain, and to permit observation of the frequency content of the signal and noise.

8.6.1 The filtering method

The worksheet in figure 8.13 defines the signal (s) as a unit-amplitude 1 kHz sine wave. A subscripted variable s_n includes 64 uniformly spaced sampled values of the signal over one period (1 ms). Values of random noise at the same intervals are added to the signal. The noise is simulated using the random function (rnd). In MathCAD, rnd(num) yields a uniformly distributed random value between 0 and num, with mean value num/2 as a result. In our worksheet the sampled noise values, noise$_n$, are symmetrically distributed between $-A_{\text{noise}}$ and A_{noise} with zero mean. For illustration, the peak noise amplitude has been set equal to the signal amplitude. The waveform resulting from adding the noise to the signal, denoted by sampled values x_n, shows obvious distortion when plotted.

The FFT of the composite signal plus noise is computed and its amplitude is plotted. Recalling our discussion of the FFT in the previous section, for 64 samples, only Fourier coefficients for k ranging from 0 to 32 are calculated. For clarity we have only plotted the coefficients up to $k = 16$, and compared them to the spectrum of the signal alone. The signal of course contains only one frequency, shown by the line corresponding to $k = 1$. Recall that the ratio of signal frequency to sampling frequency is equal to the ratio of k to the total number of samples. For this example we have sampled one full period 64 times, so the ratio is 1/64 for 64 total samples. In contrast the noise spectrum is spread out with the individual frequency components having relatively small amplitudes compared to that of the signal. A low-pass filter is therefore an obvious choice for removing the higher-frequency noise while retaining the signal.

The filter transfer function H_k is derived from the familiar low-pass Butterworth characteristic

$$|H(i\omega)|^2 = \frac{1}{1 + (\omega/\omega_c)^{2\beta}}$$

where ω_c is the radian cut-off frequency at which the response is down by 3 dB. A higher number of filter stages, β, increases the sharpness of response at the expense of additional components to implement the filter. To retain the signal and eliminate as much noise as possible, it is best to use a filter with a large number of stages and a cut-off frequency slightly above the signal frequency. The worksheet illustrates that a significant improvement occurs even for a two-stage filter, where the attenuation increases only gradually for increasing frequency. The cut-off is chosen to be five times that of the signal frequency (or 5 kHz) to ensure negligible attenuation of the signal. By decreasing α and increasing β on the worksheet you will observe an improvement in signal recovery.

8.6.2 Signal averaging

Signal averaging or accumulation is a common technique for extracting a relatively predictable signal embedded in random or unpredictable noise. The method involves taking several sets of data and averaging the results. The predictable signal values of the different data sets are highly correlated to one another, whereas the noise values are not. It can be shown that correlated signal values increase linearly with the number of signals averaged, but the statistical noise values increase only as the square root of the averaged number. The signal-to-noise ratio thus improves as more signal intervals are used, but at the expense of computational time and memory depending on the algorithm.

The worksheet in figure 8.14 illustrates this accumulation method for the pure 1 kHz sinusoidal signal. The averaging is done on M sets of 64 samples of both signal and noise. Since for this simple example the signal has no randomness associated with it, each set contains identical sampled values. The first part of the worksheet is identical to that using the filtering method, except for the fact that the filter parameter assignments are removed. Only one set of signal and noise values is computed at first, strictly for comparison purposes. MathCAD then performs the summation for all possible values of the two indices m and n. Although the appearance is similar to that of nested loops within a computer program, MathCAD does not perform 'looping' as such. Rather, each operation is performed in sequence for all values of the indices, rather than keeping the indices fixed and performing all operations within the loop. The lack of this programming type of capability does limit MathCAD's versatility.

Averaging over 20 sets of data in this worksheet clearly improves the signal-to-noise ratio. Comparing the two methods for the parameters chosen, we see that the filtering method provides greater attenuation of the higher noise frequencies, while the averaging method did a better job with the lower frequencies near the signal. As a result, the plot of y_n for the averaging method shows the outline of the original signal more clearly than for the filtering method, but the high-frequency 'jitter' is more pronounced with averaging.

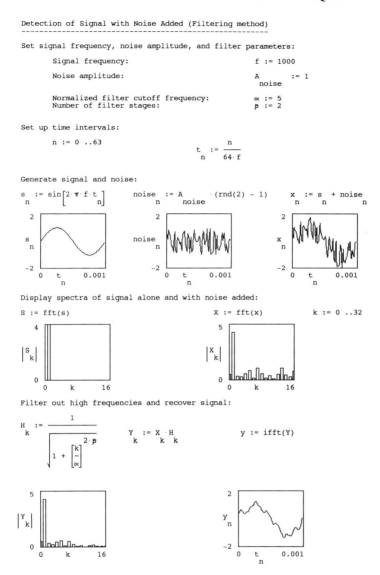

Figure 8.13 Signal detection in noise (the filtering method).

8.7 DATA ANALYSIS TECHNIQUES

In this section we will illustrate how MathCAD can be used to perform numerical data analysis techniques such as sorting, generation of histograms, and curve fitting.

278   DESIGN AIDS—MATHCAD, MATLAB

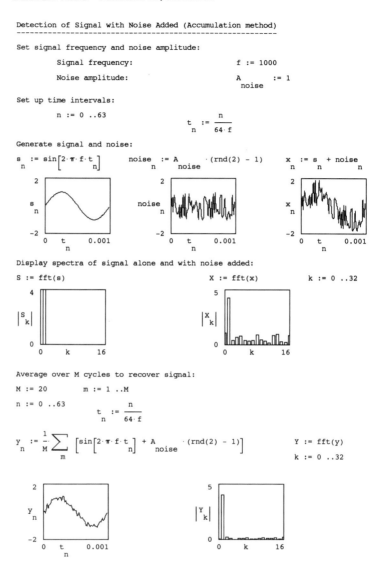

Figure 8.14 Signal detection in noise (the accumulation method).

8.7.1 Data sorting

Various algorithms exist for arranging numerical data in either ascending or descending order. A method that is simple to program but relatively slow is the 'bubble sort'. Bubble sorts can be easily implemented with high-level languages, or in assembly language to increase speed. For example, suppose we have an array of N numerical data values, which have been assigned to a subscripted array x_i, where index i ranges from 1 to N. The objective is to sort the data such that the smallest value is assigned to x_1, the next smallest value to x_2, up to the largest value placed in x_N. A bubble sort algorithm is shown below in both BASIC and C.

```
     BASIC                           C
for i = 1 to N-1              for ( i = 1; i <= N-1; ++i
  for j = i+1 to N              { for ( j = i+1; j <= N; ++j )
    if x(i) > x(j) then           { if ( x[i] > x[j] )
      temp = x(i):                  { temp = x[i];
      x(i) = x(j):                    x[i] = x[j];
      x(j) = temp                     x[j] = temp; }
  next j                          }
next i                          }
```

For both languages a properly dimensioned array variable x is assumed to contain the original data values. With this approach we start with $i = 1$ and compare x_1 successively to $x_2, x_3, \ldots,$ and x_N. Whenever x_1 is larger than any value compared to it, the two values are exchanged. The smallest value has then 'bubbled to the top' and becomes x_1. The process is then repeated for $i = 2$ by comparing x_2 to $x_3, x_4, \ldots,$ and x_N, and swapping values where necessary. The second-smallest number is then in x_2. Each succeeding array variable is then compared to all those with a higher index, with the final comparison being between x_{N-1} and x_N. Notice that two loops are required to keep track of the variables x_i and x_j being compared, where $j > i$ in all cases. A temporary variable (temp) is required if the language does not have a single-exchange or swap operation. Actually most versions of BASIC have a swap command, so the three assignment statements could be replaced by a single operation. A swap does not exist in standard C. The code is written such that if two data values are equal they are simply left in their respective positions.

As discussed previously, MathCAD does not implement the 'for loop' available in BASIC or C. Version 2.5 of MathCAD includes sorting functions that operate on vectors or matrices of data. Vector data can be sorted in either ascending or descending numerical order, and operations on matrices include row reversal as well as sorting of values in a specified row or column.

A ranking method is an alternative 'data sort' algorithm, which can be implemented within the structure of earlier versions of MathCAD [3]. The worksheet in figure 8.15 illustrates this technique on a small array of integers (although the numbers could be real as well). The indices i and j are both set to range from 1 to N, as for the bubble sort approach. The array variable x is then assigned the N arbitrary data values. A second array variable, RANK, produces integer values corresponding to the rank of each data value. The ranks range from 1 for the smallest value to N for the largest. The *greater than* relational expression adds one to the rank of each x_i for each smaller data value found.

280 DESIGN AIDS—MATHCAD, MATLAB

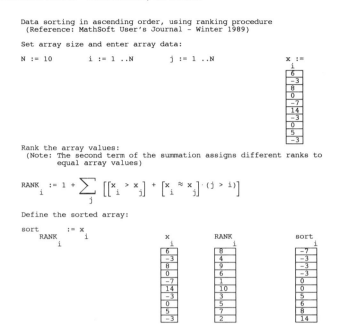

Figure 8.15 Data sorting by ranking.

The *equal* relational expression arbitrarily ranks equal data values, since they must be assigned different ranks.

Recall that each MathCAD statement is evaluated for all values of the indices before going on to the next statement (unlike nested BASIC or C loops for bubble sorts). Finally a new data array (SORT) is formed using the rank corresponding to each original data value as a subscript to the new array.

8.7.2 Statistical analysis of simulated temperatures

MathCAD has a number of additional built-in functions useful for data analysis. These functions calculate the mean, variance, standard deviation, and the minimum and maximum values of a data vector. A histogram function facilitates the tabulation and plotting of the frequency of distribution of the data values.

The worksheet in figure 8.16 analyses a simulated distribution of room temperatures in degrees Fahrenheit. Before running this worksheet the subscript origin, zero by default, must be set to one with a MathCAD command. The array Temp is assigned N values of temperature, where each value is computed by adding ten random numbers, each between 0 and 6, to a fixed or base value of 40 Fahrenheit degrees. The range of temperature values is therefore between 40 and 100 Fahrenheit degrees, although values close to those extremes are highly unlikely. The addition of ten uniformly distributed random numbers results in a reasonable approximation to the normal bell-shaped curve. In general if m

independent random numbers between 0 and A are added to a base value, it can be shown that for a large number of trials the mean and variance approach the following values:

$$\text{mean} = \text{base} + mA/2 \qquad \text{var} = mA^2/12.$$

For our example we have base $= 40$, $m = 10$, and $A = 6$. The expected mean temperature is then 70 Fahrenheit degrees with a variance of 30 Fahrenheit degrees. The expected standard deviation, defined as the square root of the variance, is therefore 5.477 for this example.

For our worksheet, the 1000 values of temperature range from a low of 55.889 °F to a high of 87.026 °F. The mean of 70.051 °F, variance (var) of 30.242 °F, and standard deviation (stdev) of 5.499 °F are close to the expected values. A plot of only the first 100 values is shown with a bar-graph format. The histogram function (hist) requires that we specify a vector of intervals, or 'bins', into which the data values are placed. Arbitrarily we have specified a bin size of 5, thereby setting up the intervals 40–45, 45–50, 50–55, ..., 95–100. The value of each bin_j is the lower limit of the corresponding interval. Notice that the hist function requires num+1 boundaries for the num intervals, thereby necessitating use of the two indices j and k. The table of values for freq, the vector generated by the histogram function, verifies that there are no temperatures below 55 °F or above 90 °F. Of the 1000 simulated readings, 37 are between 55 and 60, 145 between 60 and 65, 316 between 65 and 70, and so on. The frequency distribution plot, even for a relatively small number of intervals, resembles a normal or bell-shaped curve.

8.7.3 Linear regression applied to population estimation

Fitting a curve to a set of data points is an important data analysis technique, permitting us either to interpolate or to extrapolate to data values not provided. If data are accumulated as a function of time, future values can be predicted by extrapolation from the fitted curve. We will focus our discussion on the fitting of a straight line to a set of data points, or a linear regression analysis. The fitting of higher-order equations may improve the predictability of future values. In particular, MathCAD has a built-in cubic spline function, which fits a third-order polynomial to the data points.

To understand linear regression consider figure 8.17, which shows N pairs of datum points denoted by $\{x_1, y_1\}$, $\{x_2, y_2\}$, $\{x_3, y_3\}$, ..., $\{x_N, y_N\}$. A straight line fitted to the data can be written in equation form as

$$Y = mX + b$$

where m is the slope and b the y-intercept. At each x_i the vertical deviation or 'error' between the line and the data point is

$$e_i = mx_i + b - y_i.$$

282 DESIGN AIDS—MATHCAD, MATLAB

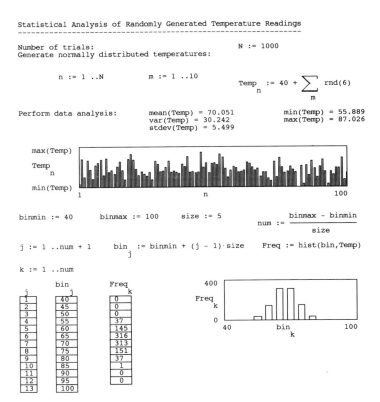

Figure 8.16 Statistical analysis of simulated temperature readings.

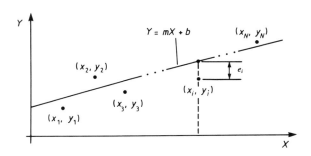

Figure 8.17 Linear curve fitting.

The usual criterion for the 'best' fit is that of least squares, where the sum of the squares of the errors at each point is minimized. Therefore, the objective is to choose m and b to minimize the total error given by

DATA ANALYSIS TECHNIQUES

$$\text{Err} = \sum_{i=1}^{N} e_i^2 = \sum_{i=1}^{N}(mx_i + b - y_i)^2.$$

For minimum error, the partial derivatives of Err with respect to both m and b must be zero. These conditions result in the following two linear equations in the unknowns m and b:

$$a_{11}b + a_{12}m = c_1$$
$$a_{21}b + a_{22}m = c_2$$

where the equation coefficients are given by the following summations:

$$a_{11} = \sum_{i=1}^{N} 1 = N \qquad a_{12} = a_{21} = \sum_{i=1}^{N} x_i \qquad a_{22} = \sum_{i=1}^{N} x_i^2$$

$$c_1 = \sum_{i=1}^{N} y_i \qquad c_2 = \sum_{i=1}^{N} x_i y_i.$$

MathCAD has built-in slope and intercept functions, which compute the values of m and b for minimum total error. The worksheet shown in figure 8.18 applies linear regression to the prediction of population growth based upon past statistics. Three methods are shown to find the slope and y-intercept to satisfy the minimum-error criterion. A solve block is first used to find m and b explicitly from the two linear equations above. The results are then verified by the slope and intercept functions, and again using MathCAD's solve block with a minimum-error function (Minerr) in place of the Find function.

This worksheet requires that the subscript origin be set to 1. The population data are based on the official United States 20th-century census figures shown below.

Year	US population (millions)
1900	76.2
1910	92.2
1920	106.0
1930	123.2
1940	132.2
1950	151.3
1960	179.3
1970	203.3
1980	226.5

The x_i array values are assigned as 0, 10, 20, 30, ..., 80 and represent the number of years from the starting year of 1900. The y_i-values are the population figures in millions. After calculating m and b using each technique, the straight line is extended to predict population for the years 1990 and 2000. Notice that the equation of the straight line uses upper-case X as a non-subscripted variable.

284 DESIGN AIDS—MATHCAD, MATLAB

Estimating U.S. Population using Linear Curve Fitting:
```
Enter data:    N := 9    i := 1 ..N    x_i := 10·(i - 1)    y_i :=
```
$$\begin{array}{|c|} \hline 76.2 \\ \hline 92.2 \\ \hline 106.0 \\ \hline 123.2 \\ \hline 132.2 \\ \hline 151.3 \\ \hline 179.3 \\ \hline 203.3 \\ \hline 226.5 \\ \hline \end{array}$$

Analytical least square curve fit:

$$a_{11} := N \qquad a_{12} := \sum_i x_i \qquad c_1 := \sum_i y_i$$

$$a_{21} := a_{12} \qquad a_{22} := \sum_i x_i^2 \qquad c_2 := \sum_i [x_i \cdot y_i]$$

Guess values: b := 0   m := 0

Given

$$a_{11} \cdot b + a_{12} \cdot m \approx c_1$$

$$a_{21} \cdot b + a_{22} \cdot m \approx c_2$$

$$\begin{bmatrix} m \\ b \end{bmatrix} := Find(m,b) \qquad \begin{array}{l} m = 1.849 \\ b = 69.409 \end{array}$$

Y1(X) := m·X + b   X := 90   Y1(X) = 235.789   (1990 population)
                    X := 100   Y1(X) = 254.276   (2000 population)

Use of slope and intercept functions:

m := slope(x,y)   m = 1.849
b := intercept(x,y)   b = 69.409

Y2(X) := m·X + b   X := 90   Y2(X) = 235.789   (1990 population)
                    X := 100   Y2(X) = 254.276   (2000 population)

Use of Minerr function:

F(X,m,b) := m·X + b

$$Err(m,b) := \sum_i \left[y_i - F[x_i,m,b] \right]^2$$

Guess values: m := 0   b := 0

Given
    Err(m,b) ≈ 0   1 ≈ 1

$$\begin{bmatrix} m \\ b \end{bmatrix} := Minerr(m,b) \qquad m = 1.851 \qquad b = 69.301$$

Y3(X) := m·X + b   X := 90   Y3(X) = 235.875   (1990 population)
                    X := 100   Y3(X) = 254.383   (2000 population)

Plot data points and "optimum" linear curve:

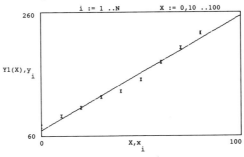

Figure 8.18 A linear regression example.

Assigning values of 90 and 100 to X gives us the 1990 and 2000 predictions respectively.

Explicit solution of the two linear equations uses the Given and Find functions shown in previous examples. Notice that the slope and intercept functions require the two data vectors, x and y, as arguments. In using the Minerr function the quantity to be minimized (Err) is first defined as a function of the variables m and b. Initial guesses are assigned as usual before entering the solve block. Notice that the first relational constraint sets Err to zero. Unlike the Find function which attempts to solve an equation as an equality, the Minerr function will find m and b for which Err is closest to zero. Since the straight-line fit is not perfect in general, the Find function used here would give us a 'solution not found' error. Notice that a second relational constraint is necessary since the solve block requires as many equations as unknowns. On employing the 'dummy' constraint $1 = 1$ which is always satisfied, the solution is not affected by it.

Interestingly, the optimum straight-line fit appears to predict a rather conservative population growth of less than 10 million from 1980 to 1990. Although the population growth for several previous decades was well over 20 million, we see from the plot that the line tries to fit itself to the smaller growth rates of the earlier decades as well. As a result the 1980 population from the straight-line fit is only about 217.4 million as compared with the actual 226.5 million data value. The calculated slope $m = 1.851$ translates into 18.5 million per decade, a reasonable growth rate based on the statistics. The point is that this example illustrates the limitations imposed by a straight-line fit, where only two parameters (slope and intercept) can be chosen independently.

As stated earlier, a better fit may be achieved using a higher-order polynomial. For example, for a parabolic curve fit the straight-line equation

$$Y(X) = mX + b$$

can be replaced by

$$Y(X) = aX^2 + bX + c.$$

Minimizing total error in this case results in three linear equations from which the parameters a, b, and c have been found. The worksheet in figure 8.19 uses the Minerr function again to find the three parameters to satisfy the minimum-error criterion. The optimum parabola fitted to the US population data is found to satisfy the equation

$$Y(X) = 0.010\,X^2 + 1.040\,X + 78.809$$

where again X is years since 1900, and Y is population in millions. From this equation the 1980 population is calculated as 226.9 million, which is very close to the actual value of 226.5 million. As shown on the worksheet, the predicted populations are then 254.4 million for 1990 and 284.0 million at the turn of the

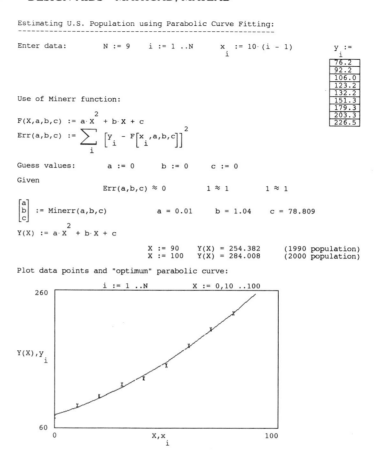

Figure 8.19 Parabolic curve fitting.

century. The total error, Err, is found to decrease by about a factor of 5 when the parabolic curve is used in place of the straight-line fit. A better fit is evident from the plot in figure 8.19.

It should be noted that the official 1990 census count was close to 249 million, but in fact the actual population was estimated to be close to 253 million. Thus, the parabolic curve fit proves to be a much better predictor than the straight line fit, at least for this example.

8.8 FILE TRANSFER BETWEEN MATHCAD AND A SPREADSHEET

This presentation of MathCAD will be concluded with a simple example of how MathCAD can complement the features of other application programs such as

FILE TRANSFER BETWEEN MATHCAD AND A SPREADSHEET 287

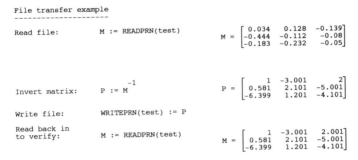

Figure 8.20 File reading and writing.

Lotus (to be discussed in chapter 10). MathCAD has the ability to read numerical data into a worksheet from a file, and similarly to write data to a file. Since Lotus has the same capability, a Lotus spreadsheet can take advantage of some of MathCAD's extensive mathematical capabilities.

In this example we will outline the steps to transfer a small matrix of data from Lotus to MathCAD, perform an operation on the matrix, and then return the new data to Lotus. We will assume for the example that the data file, consisting of a 3 × 3 matrix, is saved to the PC's drive A with the file name "test". The matrix will be inverted in MathCAD. The worksheet in figure 8.20 verifies the file reading and writing from MathCAD's point of view.

- Step 1. In Lotus enter a 3 × 3 matrix starting at an arbitrary cell D4.
- Step 2. Enter "/P F" (Lotus print file command) and "a:test" as the text file name. If the file already exists, enter "R" (replace).
- Step 3. Enter "R" (range) and select the range of the array as "D4..F6". Enter "G" (go) and "Q" (quit). The file "test.prn" is written to drive A.
- Step 4. Exit Lotus and enter MathCAD.
- Step 5. Press "<F10> F F" (file filename commands). Enter "test" as the file variable, and change "filename=test" to "filename=a:test".
- Step 6. Enter the MathCAD statement "M:READPRN(test)", which assigns the values from the file "a:test.prn" to a matrix variable **M**. To verify, enter "M=" to display the matrix.
- Step 7. Invert the matrix and assign the new matrix to **P** by entering the expression "P:M^-1". To verify, enter "P=". Enter the MathCAD statement "WRITEPRN(test):P" to save the new values.
- Step 8. Exit MathCAD, re-enter Lotus, and set to the desired spreadsheet cell to store the new matrix.
- Step 9. Enter "/ F I N" (Lotus file import numbers command) and "A:test" as the name of the file to import. The new matrix values appear on the Lotus spreadsheet.

8.9 AN INTRODUCTION TO MATLAB

An alternative product to MathCAD is MATLAB, a product of The MathWorks, Inc [4, 5, 6]. MATLAB is short for Matrix Laboratory, reflecting the fact that it deals with matrix objects. However, single-dimension arrays can be easily handled as single-row or column matrices or vectors. Likewise, a 1×1 matrix is treated as a scalar. MATLAB is available in both a 'Professional' and a 'Student' version. Optional toolboxes facilitate digital signal processing, control system design, and other areas. In comparison to MathCAD, MATLAB along with its toolboxes has more power and graphing capability. However, MathCAD appears to be more user interactive.

8.9.1 The MATLAB command mode

Starting MATLAB places you into a command mode with the MATLAB prompt appearing as ">>". A command is of the form

$$\text{variable} = \text{expression}.$$

The expression produces a matrix, which is displayed on the screen and assigned to the variable name. If the command is terminated with a semicolon (;) the display is suppressed. If an expression alone is entered the variable ans (short for answer) is created. The sequence of commands shown below illustrate the creation of scalar, vector, and matrix variables. A matrix element can be referenced by the variable name with row and column numbers in parentheses.

```
>> A = 5/2
 A =
     2.5000
>> B = [ 1  -2  3  -4 ]
 B =
     1  -2  3  -4
>> C = [ 1  -2  3 ; -4  5  -6 ; 7  -8  9 ]
 C =
     1  -2  3
    -4   5  -6
     7  -8  9
>> C(2,3)
   ans =
       -6
```

MATLAB also features a convenient way to handle range variables by creating a row vector containing all the values. For example,

$$x = \text{first} : \text{incr} : \text{last}$$

assigns to x values ranging from first to last with a constant increment (incr). An example is shown below.

```
>> x = 2 : 0.5 : 4
x =
    2.0000    2.5000    3.0000    3.5000    4.0000
```

Expressions can include arithmetic operators and a variety of functions. An example of a useful function from a MATLAB toolbox allows us to find the roots of an nth-order polynomial. The commands shown below solve the fourth-order equation entered into the MathCAD worksheet in figure 8.7.

```
>> p = [ 1  4  -7  -22  24 ];
>> r = roots(p)
r =
   -4.0000
    2.0000
   -3.0000
    1.0000
```

Other functions to be used in our examples facilitate graphing, convolution, random-number generation, and statistical analysis of data.

8.9.2 MATLAB analysis and graphing examples

Although MATLAB does not have a built-in editor, a sequence of commands can be created using an external editor and saved in a file of the form filename.m. Upon the command prompt, the filename is entered (as if it were an expression) resulting in the sequence of commands being executed. For the remaining examples, the files are shown along with the resulting output.

The file below calculates and plots the first few terms of the Fourier series of a periodic square wave, as done in the MathCAD worksheet in figure 8.2. Notice that comments can be inserted into a MATLAB file by preceding them with the per cent sign (%); also, more than one statement can be placed on a line.

```
% FOURIER SERIES APPROXIMATION OF A PERIODIC SQUARE WAVE
A=1; T=0.002; w=2*pi/T;    % Amplitude, period, fund. radian freq.
t=0:0.00004:0.004;
f = A/2 + 2*A/pi * ( sin(w*t) + sin(3*w*t)/3 + sin(5*w*t)/5 );
plot (t,f)
title ('Fourier Series Approximation of a Periodic Square Wave')
xlabel ('t')
ylabel ('f(t)')
```

This example features the sine function which operates as if it were a scalar function, as well as the graphing functions plot, title, xlabel, and ylabel. The special value pi is also built in. The resulting plot is shown in figure 8.21.

290 DESIGN AIDS—MATHCAD, MATLAB

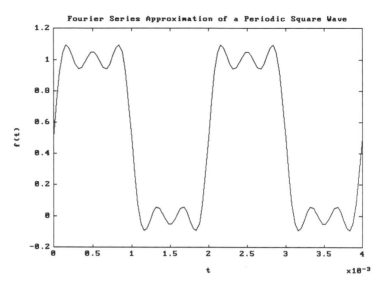

Figure 8.21 Fourier series approximation of a periodic function.

The next file replicates the one-op-amp differential amplifier (see figures 7.5 and 8.10). Notice that the first row and column are identified in MATLAB by subscript 1, whereas the default origin in MathCAD is 0.

```
% MATRIX SOLUTION FOR THE ONE OP-AMP DIFFERENTIAL AMPLIFIER
R1=100; R2=1E3; R3=10E3; R4=100E3; Rin=1E6;     % Resistors (ohms)
V1=2; V2=5;                            % Source values (mvolts)
A=50E3;                                % Gain
B = [ R1 -R2 0 0 0 1; 0 0 R3 -R4 0 A-1; 0 R2 0 R4 0 0; ...
      0 0 0 0 -Rin 1; 1 0 -1 0 -1 0; 0 1 0 -1 1 0 ];
C = [ V1-V2 0 V2 0 0 0];
X = inv(B) * C'                        % Solution of unknowns
Vin = X(6);   Vout = A * Vin           % Output voltage (mvolts)
```

The resulting output to the screen is shown below.

```
X =
   -0.0296
    0.0000
   -0.0296
    0.0000
    0.0000
    0.0060
Vout =
  300.6079
```

As done with MathCAD, we now show the convolution of two pulses

AN INTRODUCTION TO MATLAB 291

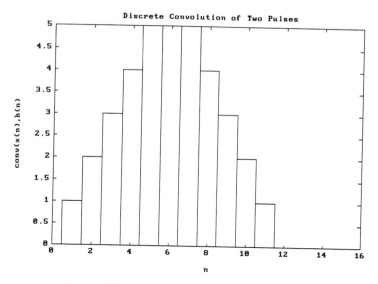

Figure 8.22 Discrete convolution of two pulses.

directly, and indirectly using FFTs (see figures 8.6 and 8.12). The file for direct convolution is shown below.

```
% DISCRETE CONVOLUTION OF TWO PULSES
x = [1 1 1 1 1 1 1 0];
h = [1 1 1 1 1 0 0 0];
y = conv(x,h)
pause
bar(y)
```

The convolved output is shown below, and the resulting bar graph in figure 8.22. The pause function keeps the display on the screen until any key is pressed. The bar function plots a bar graph of the specified vector as a function of its row index.

```
y =
  Columns 1 through 12
   1   2   3   4   5   5   5   4   3   2   1   0
  Columns 13 through 15
   0   0   0
```

The file for convolution using FFTs is next shown. Notice that the fft function specifies the size (previously called N). Also, the MATLAB's FFT conforms to the more standard one discussed in section 8.5 (as opposed to MathCAD's modified definition). Finally, notice the use of the modified multiply operator (.*) in the inverse fft. Ordinary multiplication is not permitted since both **X** and **H** are row vectors. The special operator results in a same-size vector with corresponding elements equal to the product of the individual elements.

```
% DISCRETE CONVOLUTION USING FFT AND INVERSE FFT
x = [1 1 1 1 1 1 1 0];
h = [1 1 1 1 1 0 0 0];
X = fft(x,16)
pause
H = fft(h,16)
pause
y = ifft(X .* H)
pause
bar(y)
```

A portion of the resulting screen output is shown below.

```
X =
  Columns 1 through 4
    7.0000                    1.9239 - 4.6447i       .......
H =
  Columns 1 through 4
    5.0000                    3.0137 - 3.0137i       .......
y =
  Columns 1 through 4
    1.0000                    2.0000 + 0.0000i       .......
```

The values of **X** and **H** differ from those calculated by MathCAD (see figure 8.12) by a constant factor of 4, as a result of the factor $1/\sqrt{N}$ term in MathCAD's FFT definition. The inverse FFT as expected produces the same result as the direct convolution, resulting in the same bar plot as in figure 8.22.

The next example uses random numbers to simulate signal averaging in noise, introduced in section 8.6.2. MATLAB's rand function generates a vector or matrix of uniformly distributed random numbers between 0 and 1. A single argument for rand can be either a constant or a variable, resulting in a vector of random numbers of the size of the constant or variable vector. Thus, in this example rand(signal) generates 64 values, corresponding to the size of the signal vector which is a function of n. Averaging of the noise is done with the aid of the sum function and the two-argument rand which generates M independent sets of 64 values (with M set to 20 for the example).

AN INTRODUCTION TO MATLAB

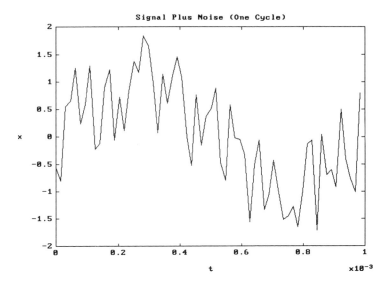

Figure 8.23 Signal plus noise (one cycle).

```
% SIGNAL DETECTION IN NOISE (ACCUMULATION METHOD)
f=1000; A=1;        % Signal frequency, noise amplitude
M=20; m=1:M;        % Average over M cycles
n=0:63; t=n/(64*f); % Time intervals (64 per cycle)
signal=sin(2*pi*f*t);
noise=A*(2*rand(signal)-1);
x=signal+noise;
plot(t,x)
title ('Signal Plus Noise (One Cycle)')
xlabel ('t')
ylabel ('x')
y=signal+sum(A*(2*rand(M,64)-1))/M;
pause
plot(t,y)
title ('Signal Plus Noise (Averaged Over 20 Cycles)')
xlabel ('t')
ylabel ('y')
```

The resulting plots before and after averaging are shown in figures 8.23 and 8.24.

MATLAB's statistical functions are next used to repeat the simulated temperature example of section 8.7.2.

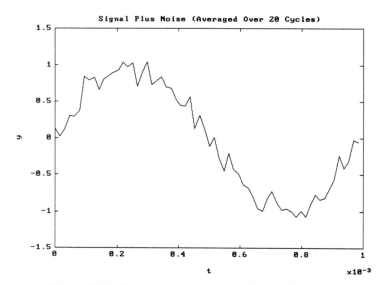

Figure 8.24 Signal plus noise (averaged over M cycles).

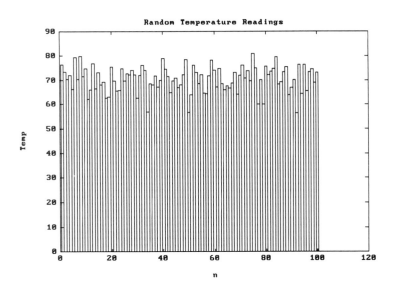

Figure 8.25 Simulated temperature readings.

```
% STATISTICAL ANALYSIS OF RANDOMLY GENERATED TEMPERATURE READINGS
n=1:1000;      % Number of trials
rand ('normal')
Temp=70+sqrt(30)*rand(n);
Min=min(Temp)
Max=max(Temp)
Avg=mean(Temp)
StDev=std(Temp)
pause
bar(n(1:100),Temp(1:100))
title ('Random Temperature Readings')
xlabel ('n')
ylabel ('Temp')
pause
bins=42.5:5:97.5;
[m]=hist(Temp,bins)
pause
hist(Temp,bins)
title ('Distribution of Temperature Readings')
xlabel ('bins')
ylabel ('frequency')
```

Recall that to approximate a normal distribution with MathCAD we added ten uniformly distributed random numbers. With MATLAB the command rand('normal') switches the default uniform distribution to a normal one. Using similar statistical functions available with MathCAD, MATLAB finds the minimum and maximum values, as well as the mean and standard deviation. A bar graph of the first 100 simulated temperature readings is shown in figure 8.25, and is similar to the graph in the MathCAD worksheet shown in figure 8.16.

A histogram is also generated for specified bin sizes. A difference from MathCAD is that MathCAD's bin vector requires the smallest values of each bin whereas with MATLAB the centre of each bin is specified. For our example the first bin collects temperature values between 40 and 45 °F. The first element of the bin vector is therefore 40 °F for MathCAD and 42.5 °F for MATLAB. The hist function calculates the number of readings in each bin. This vector can be assigned to a variable as shown, or plotted as a bar graph if no variable is specified. The numerical results of a sample run are shown below, and the histogram is shown in figure 8.26.

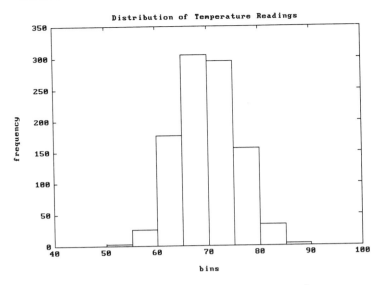

Figure 8.26 The distribution of temperature readings.

```
Min =
    55.3325
Max =
    90.9271
Avg =
    69.9682
Stdev =
    5.5147
m =
    0   0   0   36  152  323  318  132  35  2   2   0
```

In section 8.7.3 we showed how MathCAD can perform linear regression on data in various ways. MATLAB's toolboxes provide a convenient function called polyfit, which finds a polynomial curve fit of a specified order to minimize mean squared error. The example below applies this function to the population prediction problem shown in the MathCAD worksheets in figures 8.18 and 8.19.

```
% ESTIMATING U.S. POPULATION USING CURVE FITTING
x = 0:10:80;
y = [ 76.2    92.2   106.0   123.2   132.2 ...  % Population data
      151.3  179.3   203.3   226.5 ];            % (1900 - 1980)
c = polyfit (x, y, 1)                            % Straight line fit
q = polyfit (x, y, 2)                            % Parabolic fit
X = 90:10:100;
LINEAR = c(1) * X      + c(2)                    % Linear curve
                                                 % predictions
PARAB = q(1) * X .^ 2 + q(2) * X + q(3)          % Parabolic curve
                                                 % predictions
```

Both first- and second-order curves (linear and parabolic) are fitted to the data, and the estimates of the US population in 1990 and 2000 are found for both fits. The output shown below matches the expected results.

```
c =
    1.8487   69.4089
q =
    0.0103    1.0251   79.0170
LINEAR =
  235.7889  254.2756
PARAB =
  254.6619  284.4724
```

REFERENCES

[1] Anderson R B 1989 *The Student Edition of MathCAD Version 2.0* (New York, NY: Addison-Wesley and Benjamin/Cummings)
[2] Wieder S 1992 *Introduction to MathCAD for Scientists and Engineers* (New York, NY: McGraw-Hill)
[3] MathSoft Inc. 1989 *MathSoft User's J.* **3** Winter 3
[4] The MathWorks Inc. (Natick, MA) 1985–91 *386-MATLAB User's Guide*
[5] The MathWorks Inc. (Natick, MA) 1992 *The Student Edition of MATLAB* (Englewood Cliffs, NJ: Prentice-Hall)
[6] Saadat H 1993 *Computational Aids in Control Systems Using MATLAB* (New York, NY: McGraw-Hill)

9

Design Aids—DSPlay

9.1 INTRODUCTION

DSPlay, a product of Burr Brown Corp., is a digital signal-processing (DSP) simulator developed for the PC [1, 2]. Data-processing systems are simulated by creating and executing block diagrams or 'flowgrams'. A wide range of function blocks are available for creating flowgrams. To facilitate signal-processing simulations, single-function blocks exist that perform such operations as convolution, filtering, and FFT generation. Input waveforms such as DC, pulse, sinusoidal, and random can be simulated, as can an arbitrary sequence of sample values. Waveforms can be differentiated or integrated, and can be combined using arithmetic add, subtract, multiply, or divide function blocks. Other trigonometric, non-linear, and windowing functions can be used as well. For advanced applications, user-defined functions can be created in Turbo Pascal source code. Several analogue input and output boards are supported for signal capture and display.

9.2 DSPLAY FEATURES

When DSPlay is run, the initial screen appears as in figure 9.1.

At the top of the screen is a command line permitting us to save flowgrams in files. Additionally, flowgrams can be loaded in, edited, executed, and have their results displayed or printed. The flowgram workspace is blank, except for one empty function block placed at an arbitrary starting location. Status and keystroke prompt lines are found below the workspace. Notice that the size of a function block is such that a flowgram of any complexity can easily overfill the screen. Unlike a spreadsheet, which may be much larger than what is visible on the screen, a DSPlay flowgram must fit onto one screen. To overcome the size problem, however, we can place a number of function blocks into a single subgram. Like subroutines in a program, subgrams can be nested in other subgrams. As a result, the size of a flowgram is actually limited by available memory rather than by screen size.

DSPLAY FEATURES 299

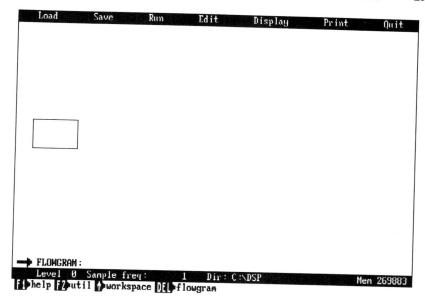

Figure 9.1 DSPlay's initial screen.

At any given time, DSPlay is either in the flowgram or block mode. The difference is that any operation is performed on the entire flowgram or on an individual block selected. As examples, we must be in flowgram mode to set the sampling frequency, save or load flowgram files, and execute the complete flowgram. Block mode permits us to create or edit individual function blocks, and display their outputs after flowgram execution.

Figure 9.2 shows the overall flowgram parameters that may be edited, as well as the utility programs available with DSPlay.

In addition to editing the sampling frequency parameter, the flowgram can be set to display automatically up to six data outputs after executing the flowgram. Calling for displays of individual blocks, when in block mode, has the advantage of full screen utilization for each display and the inclusion of axis coordinate information. For examples in this chapter, outputs from automatic display will be presented only when plot size and numerical detail are not important.

The utilities menu, invoked by pressing function key ⟨F2⟩, enables us to examine our directory of flowgram files or to invoke other DOS commands without leaving DSPlay (via the OS shell). The filter design utility is particularly useful. DSPlay includes function blocks for finite-impulse-response (FIR) or recursive-infinite-impulse-response (IIR) digital filters. These blocks can only be used, however, if the filter coefficients are previously known. The filter design utility invokes the bilinear transformation to design a digital filter approximation to a standard filter such as a Butterworth or Chebyshev type. The utility provides

300 DESIGN AIDS—DSPLAY

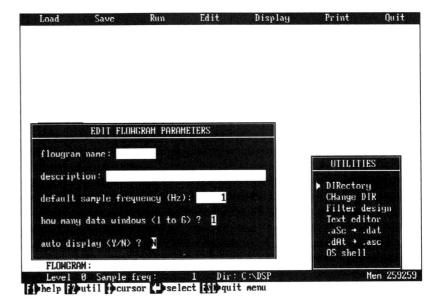

Figure 9.2 Flowgram parameters and utilities.

such characteristics as low-pass, high-pass, or bandpass with specified cut-off frequency or bandwidth, and number of filter stages. The calculated filter coefficients can be saved in a file and then imported into a function block in the flowgram.

9.3 SPECTRAL ANALYSIS OF A SIMPLE SIGNAL

After some practice, the creation of DSPlay flowgrams is relatively straightforward. Since the keystroke prompt line below the flowgram space is nearly self-explanatory, we will not detail the sequence of keys required to create and edit flowgrams, or to display results. In block mode a main function menu can be brought up, from which submenus are invoked for each main function. For example, selecting the main function "Input" gives a choice of input waveforms such as sine, cosine, DC, or random. From the "Arith" submenu we can choose one of the four arithmetic operations as well as absolute value, accumulate, or quantize. "Filter" functions include the FIR and IIR digital filters as well as the convolution function. From "Spectrum" we can select the FFT or inverse FFT functions to convert between time and frequency domains. For most functions various parameters need to be set. For example, if a sine or cosine generator input block is selected, the amplitude, frequency, and phase must be specified.

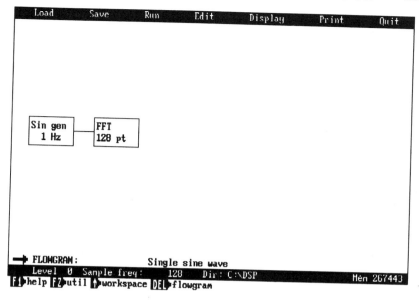

Figure 9.3 The frequency spectrum of a single sinusoid.

9.3.1 The spectrum of a sine wave

The flowgram in figure 9.3 simply connects an FFT block to a sine-wave generator. The sine wave is set to 1 V amplitude, 1 Hz frequency, and zero phase. The number of FFT points, N, is set to 128, and the sampling frequency is chosen to be 128 Hz. Multiplying N by the sampling interval gives us the total record length, which is 1 s for this example. Since DSPlay time plots extend over one record length, one full cycle of the 1 Hz sampled signal is shown.

When the flowgram is run, the resulting output of the sine generator is shown in figure 9.4. The plot format shows the individual sampled values as vertical lines ("bars"), clearly outlining the 1 Hz sine wave. A line-plot format would result in straight lines connecting successive sample values. For a sufficiently high sample rate, the line plot appears to be a continuous waveform. In subsequent examples we will, for clarity, generally use the line plot for time functions and bar plots for frequency spectra. Recall, however, that DSPlay is always processing sampled values rather than a continuous signal.

Figure 9.5 shows the resulting frequency spectrum in magnitude/phase (polar) format. Although the spectrum can be displayed in rectangular format, the polar form is generally more useful. In our examples, we will be concerned mainly with the magnitude of the complex FFT coefficients.

Recalling the discussion in section 8.4 of the FFT, the FFT index k is related to the analogue frequency f by

302   DESIGN AIDS—DSPLAY

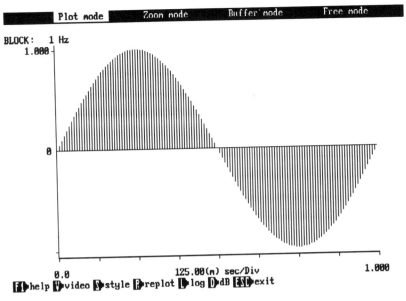

Figure 9.4 The output of the sine-generator block.

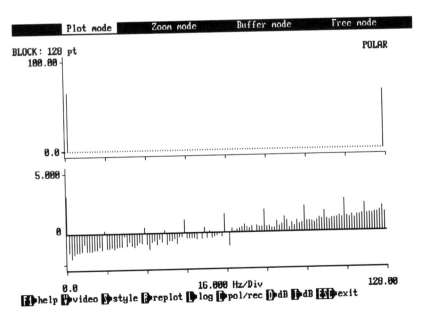

Figure 9.5 The frequency spectrum of a single sinusoid.

SPECTRAL ANALYSIS OF A SIMPLE SIGNAL

$$\frac{k}{N} = \frac{f}{f_s}$$

where

N = total number of samples
f_s = sampling frequency.

The FFT coefficients, $X(k)$, are calculated only for integer values of k from 0 to $N-1$. Therefore, a given analogue frequency will transform into a single spectral line only if Nf/f_s is in fact an integer. Since Nf/f_s is actually the total record length divided by the period of the analogue signal, an integer value corresponds to a record length equal to an integral number of periods or cycles.

For this example, $Nf/f_s = 1$, and we therefore observe a line corresponding to $k=1$. Notice that an identical line exists for the complex conjugate coefficient $X(N-k)$, or $X(127)$. It can be shown that for the case where Nf/f_s is an integer, the magnitude of the non-zero FFT coefficients is $N/2$. Since $N=128$ for this example, the line heights of 64 are as expected.

9.3.2 Spectral leakage

For analogue frequencies resulting in a non-integer value for Nf/f_s, spectral 'leakage' occurs. This manifests itself in multiple non-zero spectral components, with the major components corresponding to k index values close to Nf/f_s. Spectral leakage is a problem inherent in digital signal processing. For a single-frequency signal, as in our example, N and f_s can easily be adjusted to avoid leakage. In practice, however, we often digitize analogue signals consisting of multiple frequencies. It is usually inevitable that an FFT will produce spectral leakage at least for some analogue frequencies.

To illustrate the effect of spectral leakage for a single sine wave, let us simply change the sampling rate in our example to 100 Hz. When the flowgram of figure 9.3 is run again, we get the time plot shown in figure 9.6 and the frequency spectrum of figure 9.7.

Notice from the time plot that 128 samples are again shown, but that the sampling rate has been reduced to 100 samples s^{-1}. The record length has been therefore extended to 1.28 s and more than one cycle of the 1 Hz signal is displayed. For this case,

$$Nf/f_s = 128(1)/100 = 1.28$$

corresponding to a record length of 1.28 periods of the sine wave. The resulting non-integer value causes the spectral leakage observed in figure 9.7. As expected, the lines cluster near $k=1$ (the closest integer to 1.28). The amplitudes of the first four FFT coefficients, rounded off to the nearest integer, are

$$X(0) = 9 \quad X(1) = 57 \quad X(2) = 22 \quad X(3) = 9.$$

304 DESIGN AIDS—DSPLAY

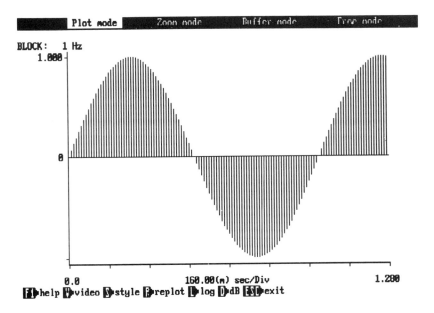

Figure 9.6 The sine wave sampled at 100 Hz.

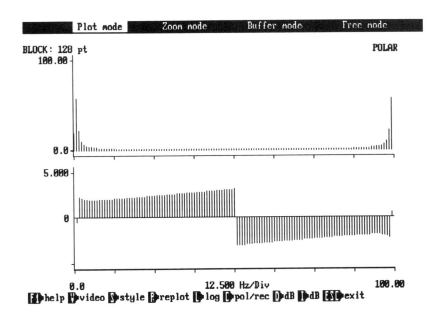

Figure 9.7 An illustration of spectral leakage.

ADDING AND MULTIPLYING SIGNALS   305

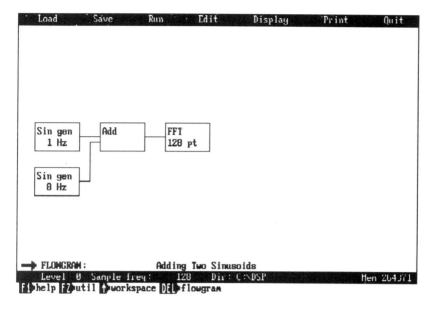

Figure 9.8 Adding two sinusoids.

Notice that the amplitude of $X(0)$ appears to be twice the value shown, but this is a result of periodicity causing the identical value for $X(N)$ to be folded back and added to $X(0)$.

In contrast, for our initial example where $f_s = 128$ Hz,

$$X(0) = 0 \quad X(1) = 64 \quad X(2) = 0 \quad X(3) = 0.$$

The signal energy at each frequency is roughly proportional to the square of the amplitude of the corresponding coefficient. The effect, therefore, of sampling over a non-integral rather than an integral number of periods is to 'spread out' the spectrum. The total energy is not affected, however.

9.4 ADDING AND MULTIPLYING SIGNALS

Figure 9.8 shows how we can simulate a higher-frequency (8 Hz) sine wave added to the original 1 Hz signal. Both sine waves have unit amplitude. The outputs of the two sine-wave-generator blocks are connected to the inputs of an adder, chosen from the arithmetic menu. The sampling rate has been reset to 128 Hz for a 1 s record length.

The composite signal at the adder output is shown in figure 9.9, and the resulting spectrum in figure 9.10. The record length of 1 s corresponds to one cycle of the lower frequency and eight cycles of the 8 Hz wave. Since the

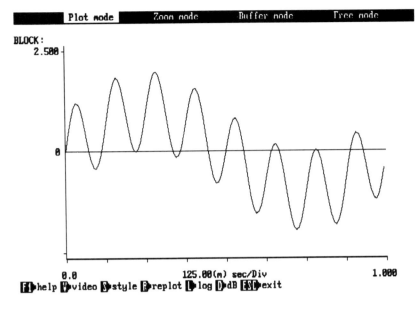

Figure 9.9 Addition of two sinusoids.

record length is an integral number of cycles of both waveforms, no spectral leakage should occur. We therefore see lines of amplitude $N/2 = 64$ for $X(1)$ and $X(8)$, along with the complex conjugate lines corresponding to $X(128-1)$ and $X(128-8)$.

9.4.1 AM signals

A simple change in the flowgram allows us to simulate a double-sideband AM signal, minus the carrier. In the flowgram of figure 9.11, the adder block previously used is replaced by a multiply block.

Multiplying unit-amplitude sine waves at frequencies f_1 and f_2 results in a composite waveform given by

$$\sin(2\pi f_1 t)\sin(2\pi f_2 t) = \frac{1}{2}\cos[2\pi(f_1 - f_2)t] - \frac{1}{2}\cos[2\pi(f_1 + f_2)t].$$

If f_1 represents the carrier frequency, we now have components at frequencies $f_1 - f_2$ (lower sideband) and $f_1 + f_2$ (upper sideband). Each sideband has an amplitude of 1/2. The composite waveform is shown in figure 9.12 and the resulting spectrum in figure 9.13. For a carrier frequency of 8 Hz and a lower frequency of 1 Hz the sidebands appear at 7 and 9 Hz as expected. The time plot shows the classical AM signal with the lower frequency forming a 1 Hz envelope for the 8 Hz carrier.

ADDING AND MULTIPLYING SIGNALS 307

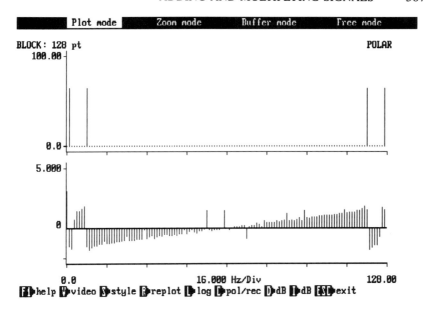

Figure 9.10 The spectrum of added sinusoids.

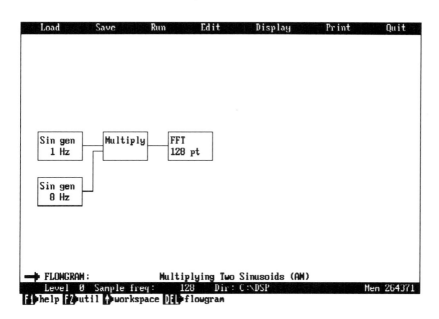

Figure 9.11 Multiplying two sinusoids (double-sideband AM).

308   DESIGN AIDS—DSPLAY

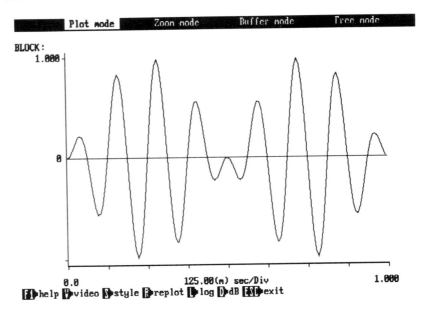

Figure 9.12   A double-sideband AM signal.

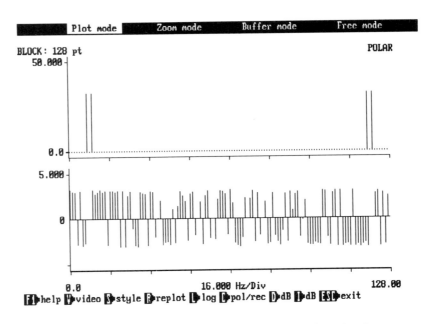

Figure 9.13   The spectrum of a double-sideband AM signal.

ADDING AND MULTIPLYING SIGNALS

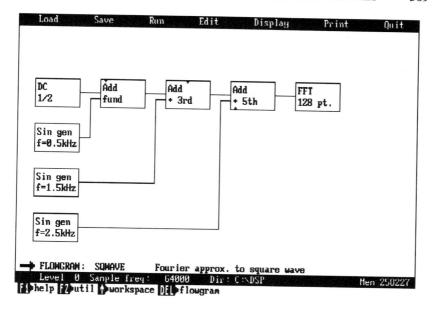

Figure 9.14 A flowgram to approximate a periodic square wave.

9.4.2 The square-wave approximation

The flowgram of figure 9.14 approximates the square wave shown in the example in section 7.5. The flowgram implements the DC and first three frequency terms of the Fourier series, as given below:

$$f_p(t) = \frac{A}{2} + 2\frac{A}{\pi}[\sin(\omega_1 t) + \tfrac{1}{3}\sin(3\omega_1 t) + \tfrac{1}{5}\sin(5\omega_1 t)]$$

where $A = 1$, $\omega_1 = 2\pi f_1$, and the fundamental frequency is $f_1 = 500$ Hz. Notice that a DC input block along with three sine-generator input blocks are used. The amplitude and frequency parameters are set to the required values. Since adder blocks, along with all other DSPlay function blocks, are limited to two inputs at most, a total of three adders are required. To avoid spectral leakage and to display only one cycle of the fundamental frequency, we choose

$$N = 128 \quad \text{(number of samples)}$$
$$f_s = 128(500) = 64000 \text{ Hz} \quad \text{(sampling rate).}$$

Figure 9.15 shows the three adder output waveforms, obtained using the automatic display feature. We see the fundamental frequency alone offset by the DC value, the addition of the third harmonic, and finally the addition of the fifth harmonic.

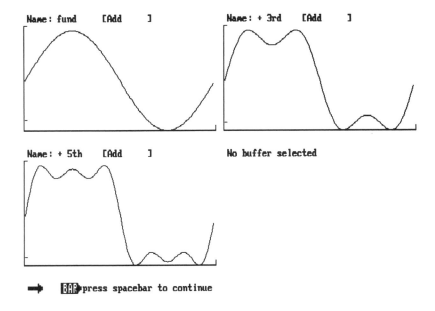

Figure 9.15 Square-wave-approximation waveforms.

The FFT output is shown in figure 9.16, after using the zoom feature to focus on the frequencies of interest. As expected, the amplitude of the third harmonic is approximately a third of that of the fundamental, and the ratio for the fifth harmonic is 1/5. The DC value is again twice the value expected, as a result of the fold-over problem.

9.4.3 The D/A converter

The flowgram in figure 9.17 simulates a D/A converter using a summing circuit. The illustrate the concept with a flowgram fitting on a screen, we have limited to D/A size to four bits. The summing circuit was discussed in section 7.3.7, and for the 4-bit case the output analogue voltage is given by

$$V_{\text{out}} = -\frac{1}{2^4}(2^3 V_3 + 2^2 V_2 + 2^1 V_1 + 2^0 V_0)$$

and therefore

$$|V_{\text{out}}| = \tfrac{1}{2}V_3 + \tfrac{1}{4}V_2 + \tfrac{1}{8}V_1 + \tfrac{1}{16}V_0.$$

The fractional weighting factors are applied with the aid of DC inputs and multiplier blocks. Alternatively, a weight block from the arithmetic menu could be used in place of each DC and multiply block.

The digital binary inputs V_3 to V_0 will be simulated as either 0 or 1 V. To illustrate the D/A conversion over the range of values of the 4-bit numbers,

ADDING AND MULTIPLYING SIGNALS 311

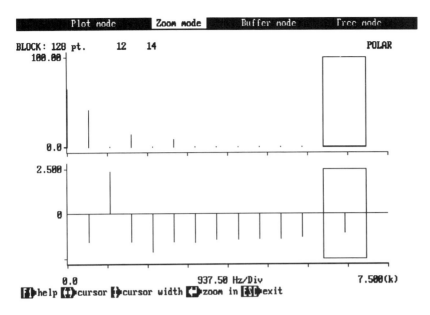

Figure 9.16 The spectrum of a square-wave approximation.

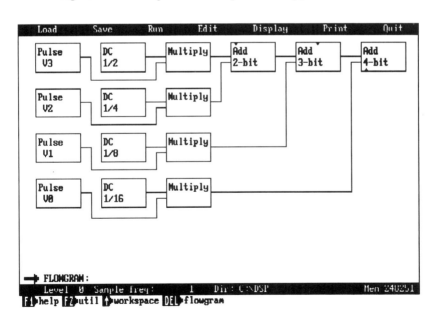

Figure 9.17 4-bit D/A simulation.

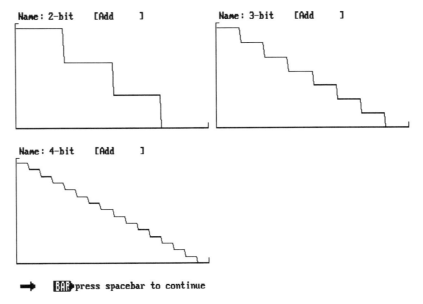

Figure 9.18 The analogue output from a D/A converter.

the four input voltages are created as unit-amplitude pulses with different frequencies. For a unit sampling frequency, the time base is set to 128 s by default. The pulse input V_3 is then set to a frequency of $\frac{1}{128}$ Hz, resulting in one full cycle over the time base. Input V_2 is set to $\frac{1}{64}$ Hz, V_1 to $\frac{1}{32}$ Hz, and V_0 to $\frac{1}{16}$ Hz. Each pulse starts at 1 V and has a 50% duty cycle. With V_3 as the most significant bit, the result is a downward binary sequence of 1111, 1110, 1101, 1100, ..., 0001, 0000 over the time base.

Figure 9.18 shows the automatic display of the adder outputs after running the D/A flowgram. The final adder output shows the 4-bit analogue output, but the intermediate adders simulate 2-bit and 3-bit conversions. On each plot the voltage ranges from 0 to $1 - 2^{-N}$, where N is the number of bits. As expected, the adder outputs are monotonically decreasing with a step size of $1/2^N$.

9.5 CONVOLUTION AND FILTERING

Since convolution is an often-used operation in digital signal-processing applications, DSPlay has provided a single function block for this purpose. The convolution block is one of the choices provided by the filter menu, and its use is illustrated with the flowgram in figure 9.19. The waveform inputs are set up to implement the sequences

$$\{1111111000\ldots\} \quad \text{and} \quad \{11111000\ldots\}$$

CONVOLUTION AND FILTERING   313

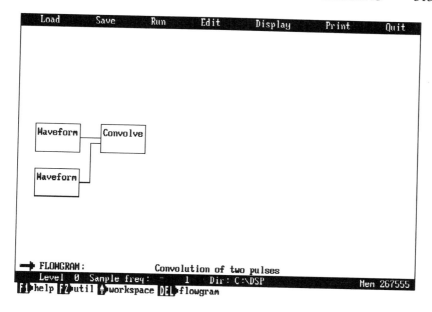

Figure 9.19 Convolution of two pulses.

used previously in the convolution example in section 8.3.4.

Notice that the inputs represent sampled values of unit-amplitude pulses of width 7 and 5 respectively. DSPlay's pulse input block could have been used as an alternative to the waveform block, to generate the signals. As shown in figure 9.20, the output of the convolution block is the sequence

$$\{12345554321000\ldots\}$$

verifying the result produced by MathCAD in section 8.3.4.

The flowgram in figure 9.21 illustrates the use of DSPlay's built-in filter design utility. The flowgram simulates power supply circuitry which converts standard line frequency AC to a DC signal. The sine-wave input is set to 110 V (RMS) amplitude and 60 Hz. The actual amplitude entered is the peak value of the sinusoid, which is $110\sqrt{2} = 155.56$ V. The arithmetic menu's weight block implements a stepdown transformer, simply reducing the input amplitude by an appropriate amount. The value is chosen to give us a 5 V DC output from the converter. An absolute-value function, also taken from the arithmetic menu, simulates a full-wave rectifier.

A Fourier analysis of a full-wave rectified sinusoid results in a DC value of $2/\pi$ times the peak amplitude, and frequency components at even harmonics of the original 60 Hz. Using DSPlay's filter design utility, a low-pass Butterworth filter is inserted to 'smooth out' the waveform. The parameters selected for the filter design are a cut-off frequency of 20 Hz, unity passband gain, and two

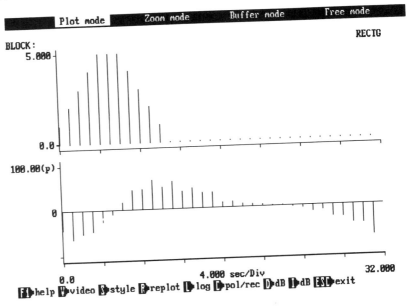

Figure 9.20 The output of a convolve block.

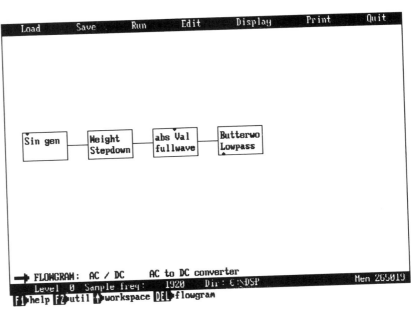

Figure 9.21 An AC-to-DC converter (flowgram).

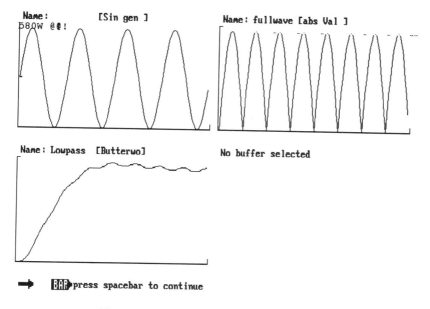

Figure 9.22 An AC-to-DC converter (outputs).

poles or stages. Although a two-stage filter does not have a particularly sharp response, the cut-off frequency is well below the first AC component of 120 Hz.

A sampling frequency of 1920 Hz results in displaying four cycles of the original 60 Hz AC signal. The display shown in figure 9.22 confirms that the filter output rises to a nearly constant value during the first two cycles. Notice the rippling caused by the filtering, an effect inherent in real power supplies also. The scales for the displayed waveforms are different of course. Whereas the AC signal ranges from $-110\sqrt{2}$ to $+110\sqrt{2}$ V, the full-wave rectified signal ranges from 0 to 7.85 V. The filtered output is rippling around 5 V DC, and the ratio of DC voltage to peak rectified value is confirmed to be very close to $2/\pi$.

9.6 SIGNAL DETECTION IN NOISE

Recall from section 8.6 that MathCAD's random-number-generating function enabled us to simulate a noise environment. DSPlay provides two input blocks that generate a random set of sampled values. The blocks give us a choice of either a 'random' or gaussian distribution. We will use the slightly more flexible gaussian function block, which permits us to choose the mean and variance. We will set the mean value to zero for the examples to be presented. The variance is comparable to the noise amplitude, A_{noise}, used in the MathCAD examples in sections 8.6.1 and 8.6.2. As for the MathCAD signal detection examples, the signal (a 1 kHz sine wave) and noise amplitudes will both be set to unity, to simulate a relatively significant amount of noise added to the signal. As with MathCAD or spreadsheets, it is relatively easy to perform 'what if' analyses with DSPlay. Parameters such as noise variance can be changed without altering the flowgram structure, and the flowgram can be rerun as often as is desired.

9.6.1 Signal filtering and averaging

The flowgram in figure 9.23 illustrates the detection of analogue signals with noise added. Of course, signals generated by DSPlay are digital in the form of sequences of numbers; however, we are reserving the term digital for a subsequent example involving binary detection using phase shift keying (PSK). The flowgram combines the techniques of filtering and accumulation, discussed in sections 8.6.1 and 8.6.2, into a single example. A low-pass Butterworth filter is created with the filter design utility, to match the transfer function used in the MathCAD example. The filter is chosen to have two stages and a cut-off frequency of 5 kHz.

As a comparison, a digital finite-impulse-response (FIR) filter is provided as well. The coefficients are chosen to give us a 'moving-average filter', one whose current output is an average of the most recent sampled inputs. Specifically, for an input sequence x_i, the current input x_n and output y_n are related by

$$y_n = \frac{1}{L}(x_n + x_{n-1} + x_{n-2} + \ldots + x_{n-L+1}).$$

The FIR coefficients for the moving-average filter are all equal to $1/L$, where L is the number of most recent samples to be averaged. These coefficients need to be entered as parameters to the FIR block. For the example we have chosen $L = 20$. A spectral analysis of the moving-average filter shows that it is another form of a low-pass filter, and therefore should be effective in eliminating the higher-frequency components of noise.

The accumulate function block, selected from the arithmetic menu, results in the flowgram being run a specified number of times. The resultant output at each sampling time is then a total of the corresponding outputs for each run. Alternatively, by setting the appropriate parameter of the accumulate block, the total output at each sampling time is divided by the number of runs. This gives us the average value rather than the total at each interval. For the nth input x_n, the output y_n is calculated as

$$y_n = \frac{1}{M} \sum_{j=1}^{M} x_{jn}$$

where x_{jn} is the nth input value for the jth run. Recall that the MathCAD worksheet in figure 8.14 performs the equivalent calculation built into DSPlay's

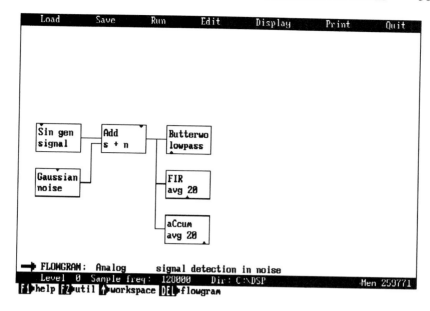

Figure 9.23 Analogue signal detection in noise.

accumulate block. The number of runs, M, is set to 20 to match the value used in the MathCAD example.

The displayed output of all six function blocks is shown in figure 9.24. As for some previous examples, the sampling frequency is chosen to display one cycle of the signal. It is clear from the plots that the signal has been corrupted considerably by the noise, and that each detection technique does a reasonable job in eliminating the noise. As for the MathCAD examples in sections 8.6.1 and 8.6.2, we see that the accumulate output contains more high-frequency components, or signal 'jitter', as compared with the output of the Butterworth filter. The averaging FIR filter also succeeds in reproducing the signal with only a little more jitter than the Butterworth's output, confirming its low-pass function. An FFT analysis of the outputs confirms the conclusions drawn here.

9.6.2 Binary signal detection

Transmission of digital information over telephone lines requires that the binary signals be converted to audio tones. Usually the binary levels are distinguished by using two separate frequencies, or by using a single frequency with separate phases. The latter method, called phase shift keying (PSK), can be used with single bits or with groups of successive bits. When used on single bits the phases are chosen 180 degrees apart, to provide the greatest possible separation between the two signals.

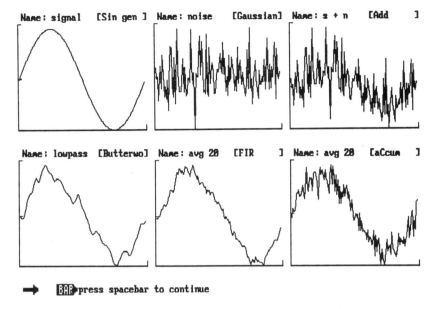

Figure 9.24 Analogue signal detection by filtering and averaging methods.

The flowgram in figure 9.25 implements a matched filter detector for binary PSK signals. The signal (s) is either s_1 or s_2, corresponding to bit values 0 and 1. The signal with noise added is multiplied by reference signal s_1 and then integrated. The parameters are again chosen to sample the signal over one full cycle, and the final integrator output value is used to make an optimum decision as to which binary level occurred.

If the received signal at the integrator output is denoted by r, we have

$$r = \int_0^T s_1(s+n)\,dt$$

where s is either s_1 or s_2, and n represents the noise. Signal and noise are of course functions of time, with the noise being 'random'. If the signals are at the same frequency and opposite in phase, then assuming unit amplitude we have

$$s_1 = \sin(2\pi f t) \qquad s_2 = \sin(2\pi f t + \pi) = -\sin(2\pi f t) = -s_1.$$

If the integration is performed over one period, then $T = 1/f$. The received signals for s_1 and $s_2 = -s_1$ are then

$$r_1 = \int_0^T s_1(s_1+n)\,dt = \int_0^T \sin^2\left(2\pi \frac{t}{T}\right) dt + \int_0^T s_1 n\,dt$$

$$r_2 = \int_0^T s_1(s_2+n)\,dt = -\int_0^T \sin^2\left(2\pi \frac{t}{T}\right) dt + \int_0^T s_1 n\,dt.$$

SIGNAL DETECTION IN NOISE 319

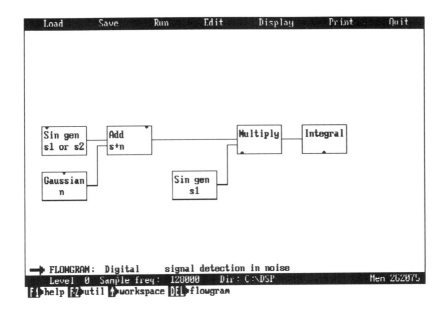

Figure 9.25 Matched filter detection of a PSK digital signal.

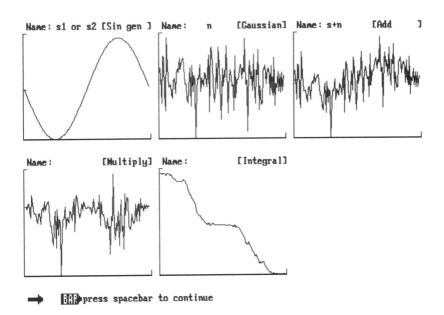

Figure 9.26 Detection of the signal s_1.

320   DESIGN AIDS—DSPLAY

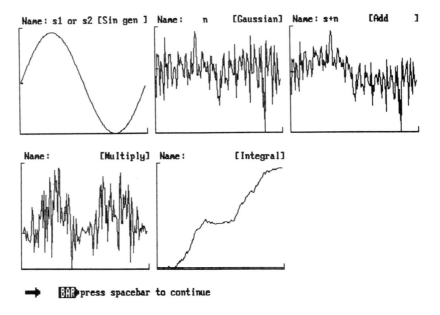

Figure 9.27 Detection of the signal s_2.

The first integral evaluates to $T/2$ for r_1 and $-T/2$ for r_2. The integral involving noise can be positive or negative. However, if zero mean value of noise is assumed, the average value of the integral of n is zero. Therefore,

$$\text{avg}(r_1) = \frac{T}{2} \qquad \text{avg}(r_2) = -\frac{T}{2}.$$

The optimum decision rule is to establish a threshold level half-way between $\text{avg}(r_1)$ and $\text{avg}(r_2)$, at zero in this case. If the integrator output at $t = T$ exceeds the threshold, assume that the signal is s_1, otherwise assume s_2. The rule is simply to choose

$$\begin{array}{ll} s_1 & \text{if } r > 0 \\ s_2 & \text{if } r < 0. \end{array}$$

As in the previous example we set the signal to a 1 kHz unit-amplitude sine wave, and the gaussian noise amplitude (variance) to unity also. The displayed outputs for signal s_1 are shown in figure 9.26, and for s_2 in figure 9.27. The output of the integrator starts at zero in both cases. For s_1 the output clearly rises to a positive value, and for s_2 it descends to a final negative value. On observing the output, therefore, it becomes clear which signal is present.

If this flowgram is run repeatedly the output waveform will exhibit minor differences in shape, due to the randomness in the noise. It is unlikely though that the final output will be on the 'wrong side' of the zero threshold value. The flowgram was also modified by increasing the noise amplitude from unity to 10, and then to 100 and 1000. Several runs were carried out for each case, for both s_1 and s_2. The ability to predict the correct signal remained high for noise of amplitude 10 and 100, but diminished considerably when the noise reached the 1000 level.

REFERENCES

[1] Kamas A and Lee E A 1989 *Digital Signal Processing Experiments* (Englewood Cliffs, NJ: Prentice-Hall)
[2] Burr-Brown Corp. (Tucson, AZ) 1988 *DSPlay Users Guide*

10

Spreadsheets: Lotus 1-2-3

10.1 INTRODUCTION

Development of automated spreadsheet programs was motivated by the need for them in business applications. Lotus 1-2-3 and a competing program, Symphony, were patterned after an earlier program called VisiCalc. Newer spreadsheets such as Quattro Pro and Windows-based Excel are gaining in popularity, while Lotus Development Corp. has continued to provide enhancements to 1-2-3 through releases 2.0, 2.2, and 3.0. Scientists and engineers have found the extensive features of Lotus to be useful in performing numerical calculations for their own applications. The majority of Lotus 1-2-3 books on the market stress business applications, with a few exceptions (see [1–3]).

Lotus 1-2-3 is actually a combination worksheet, graphics, and data-base program. Separate utility programs facilitate the printing of graphics, the exchange of files with other programs, and 1-2-3 installation. A worksheet format consists of rows and columns of cells, which may be identified individually or as a group of cells within a specified range. When starting 1-2-3 a blank worksheet appears on the screen with row and column identifiers as shown below.

```
A1:
    A    B    C    D    E    F    G    H
1
2
3
:
:
19
20
```

The rows are identified by number and the columns by letter. The cell in row 1 and column A is initially highlighted and referenced as A1 in the upper left-hand corner of the screen. The cursor, to be referred to as a cell pointer, can be easily moved around the worksheet. Only a small portion of the worksheet (20 rows and 8 columns) is visible at a time, but the entire worksheet size is

8K (8192) rows by 256 columns. Of course most applications will only utilize a small fraction of the worksheet. Additionally, the width of each column can be adjusted to accommodate as many as 240 characters in each cell.

A cell entry can be either text (label), a number, or a formula from which a numerical value is computed. As formulas reference cells or cell ranges, changing values in referenced cells results in recomputation of values in other cells. With Lotus we can easily observe the effect of changes in the worksheet, thereby performing a 'what if' analysis. A range of cells is identified by its first row and column (upper left-hand side) and last row and column (lower right-hand side). For example, the range C5..F12 identifies the cells in rows 5 to 12 and columns C to F (8 rows by 4 columns, or 32 total cells).

10.2 LOTUS 1-2-3 FEATURES

Lotus has a wide variety of commands that facilitate file manipulation, data entry into the worksheet, and analysis and graphing of the data. An initial command menu can be invoked by pressing the "/" key, from which submenus can be accessed by moving the cell pointer to a menu item and pressing ⟨Enter⟩. A faster way is simply to type the first letter of the menu name, in which case ⟨Enter⟩ is not needed. Pressing the ⟨Esc⟩ key will take you back to the previous menu or 'exit' you from the initial command menu. The menus are displayed immediately above the column labels. When "/" is first pressed the initial command menu appears on the first line, the first choice is highlighted, and the corresponding submenu is displayed on the line below it. The two lines shown below are initially displayed, with the second line being the Worksheet submenu:

Worksheet Range Copy Move File Print Graph Data System Quit
Global,Insert,Delete,Column,Erase,Titles,Window,Status,Page

An overview of many of the commands is given below. Some of the commands will be illustrated by the examples in the remaining sections of this chapter.

- Worksheet:
    Global settings
        Number format, such as number of decimal places displayed
        Label alignment (left, centre, or right justified)
        Column width
    Insert blank rows or columns
    Delete rows or columns
    Erase entire worksheet
    Set split screen (Window)
- Range:
    Format cell or cell range (features similar to global format)

Label alignment for cell or cell range
Erase cell or cell range
Create, delete, or modify range names
Adjust width of a column of labels
- Copy:
    Copy cell or cell range
- Move:
    Move cell or cell range
- File:
    Retrieve file
    Save file
- Print:
    Output range to printer or to file
- Graph:
    Create graph
    Set as many as six data ranges
    View current graph
    Save graph in file for later printing
    Options
        Insert data range legends
        Format graphs with lines between data points
        Enter graph title
        Insert horizontal/vertical grid lines
        Scaling options (auto, manual, format)
- Data:
    Fill range with numbers of fixed increment
    Sort data records
    Find data records satisfying given criteria
    Create frequency distribution for a range
    Matrix operations
        Invert
        Multiply
    Linear regression
- System:
    Enter DOS shell

One of the more powerful features of Lotus 1-2-3 offered is the ability to automate a worksheet by creating and invoking special routines called macros. Keyboard macros permit the storage of sequences of keystrokes to be executed with a single keystroke. In addition, programming macros give 1-2-3 many features of a programming language, including the capability of performing conditional tests (IF), looping, assignment of values or labels to cells, and the calling of subroutines.

With 1-2-3 we can create mathematical formulas utilizing a wide variety of operators and functions, examples of which are shown below. For each example, the worksheet entry is shown first followed by a description. Notice the use of # preceding certain logical operators and @ preceding all function names prevents confusion with label entries into the worksheet.

- Mathematical operators:

+	Add
-	Subtract, negate
*	Multiply
/	Divide
^	Exponentiate

- Logical operators:

=	Equal
<	Less than
>=	Greater than or equal
#NOT#	Logical NOT
#AND#	Logical AND

- Mathematical functions:

@SIN(x)	Value of sine of angle x
@ACOS(x)	Angle whose cosine is x
@PI	Value of pi (3.14159...)
@EXP(x)	Value of e^x
@LOG(x)	Value of $\log_{10}(x)$
@ABS(x)	Absolute value of x
@SQRT(x)	Value of square root of x
@RAND	Random number between 0 and 1

- Logical functions:

@TRUE	Value of logical TRUE (1)
@FALSE	Value of logical FALSE (0)
@IF(c,x,y)	Value of x if condition c is TRUE
	Value of y if condition c is FALSE

- Statistical functions:

@MAX(list)	Maximum value in list
@MIN(list)	Minimum value in list
@AVG(list)	Average of values in list
@VAR(list)	Variance of values in list
@STD(list)	Standard deviation of values in list

10.3 TABULATING AND GRAPHING A FORMULA

As a first worksheet example, consider a simple equation $y = x^2$ where x ranges from 0 to 1. Suppose that we wish to tabulate values of x and y over the range of x, with a step size of 0.1. The desired form of the worksheet is shown below, with a title and table headings included.

	A	B	C	D
1	Table of values for the equation y = x^2			
2				
3	x	y		
4	0	0		
5	0.1	0.01		
6	0.2	0.04		
7	0.3	0.09		
8	0.4	0.16		
9	0.5	0.25		
10	0.6	0.36		
11	0.7	0.49		
12	0.8	0.64		
13	0.9	0.81		
14	1	1		

The title can be entered character by character in cell A1, the table heading x in cell A3 and y in A4. As an option, we can centre the characters within the cell by entering ^x and ^y respectively. The values of x then occupy the cell range A4..A14, and the calculated values of y are in the range B4..B14. The values of x can be entered in one of three ways. Obviously they could be entered manually one at a time; although it is not especially tedious for only eleven entries, we would clearly need to re-enter each value if we later decided to change the step size.

A second approach is to enter the first value (0) in cell A4 and a formula into cell A5. The formula can be entered as +A4+0.1 or as 0.1+A4; notice, however, that A4+0.1 without the leading plus sign would be interpreted as a label rather than a formula. When a formula is entered into a cell the computation is done immediately and the resulting value is displayed in the cell position. In this case the value is the content of A4 (0) added to 0.1, resulting in 0.1. All formulas become part of the worksheet and are displayed in the upper left-hand corner of the screen when the cell pointer is at that cell. The advantage of this approach is that the formula can now be copied to the remaining cells below it. Invoke the copy command by pressing / to display the main command menu followed by c (to select 'copy').

When selecting copy you are prompted to enter a range to copy from (the single-cell range A5..A5 in this case) and a range to copy to (A6..A14). The range is entered by typing the first cell, a single dot (.) which is displayed as two dots (..), and the last cell. Alternatively, position the cursor on the first cell, type a dot (.), and then move the cursor to the last cell. The range will be highlighted on the screen and displayed as well. The entry is completed by pressing the ⟨Enter⟩ key.

When a formula is copied, the cell names are by default relative rather than absolute references. Thus, a formula such as +A4+0.1 in cell A5 is copied as +A5+0.1 to cell A6, as +A6+0.1 to A7, etc. The result in each cell is then the value of the previous cell incremented by 0.1, as desired. Incidentally we can copy a formula without changing cell references by preceding the row and/or column identifier by $. That is, a formula such as +A$4+0.1 is copied as +A4+0.1 to all cells in column A. Retaining the absolute cell reference would be useful if the cell contents were to be treated as a constant. An example using absolute cell references will be shown later.

The third and easiest approach to entering the x-values in this example is to press /df to invoke the 'data fill' command prior to entering any values. Enter the requested range as A4..A14 and the starting value as 0. The default step value of 1 should be changed to 0.1, and the stop value of 8191 need not be changed (since the value of x is well below that value within the specified range). When using the data fill command the cell entries are values rather than formulas, thereby equivalent to manual entry of the values.

The y-values can be easily computed using the 'formula copy' approach. Since the first value of x is in cell A4, enter the formula +A4^2 into B4 to evaluate the first value of y. Then press /c and copy from range B4..B4 to range B5..B14. The formula is copied as +A5^2 into B5, +A6^2 into B6, etc. The complete worksheet is then displayed.

To compute the values of y it was necessary to enter the formula +A4^2. In developing larger worksheets with numerous formulas, it would be desirable to be able to reference variables by name rather than by cell location. Fortunately 1-2-3 permits us to name a range. For the first example, assume that we have entered only the title and table headings. We then identify cell A4 with the name "x" by pointing to that cell, and invoking the 'range name create' command by pressing /rnc. Enter "x" when prompted to enter a name. The range A4..A4 is shown, and is accepted by simply pressing ⟨Enter⟩. The name "x" can now be used in place of A4 anywhere in the worksheet.

Now the formula +x+0.1, rather than +A4+0.1, can be entered into cell A5 and copied to the cell range A6..A14 as before. You will notice that the copied formulas reference the cell locations as before, and produce the correct table of x-values. In a similar manner, the table of y-values can be created by entering the formula +x^2 into cell B4 and copying to B5..B14.

Now that we have created a table of values of x and y, let us plot the values as x–y coordinates. Invoke the graphing command by pressing /g. Depending on the version of Lotus 1-2-3 being used, a new screen may or may not appear with a graph menu line on top. If it does, the current graph attributes are then displayed on the rest of the screen. Alternatively, the worksheet may remain displayed with the graph menu line above it. The graphing menu items allow you to choose the type of graph you want (line, bar, or pie for example), set the "X" data range, and set up to six "Y" data ranges referred to as A to F. You can choose to view the current graph on the screen, as well as to save it in a file for later printing. If Options is selected you can control legends, format, titles, grid lines, and scaling of your graph.

The graph in figure 10.1 results from setting the type to line graph, the "X"

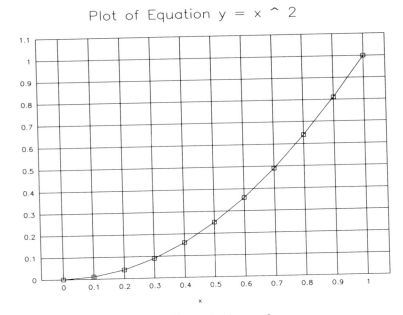

Figure 10.1 Graphing a simple equation.

data range to A4..A14, and the "A" data range to B4..B14. The remaining data ranges (B to F) are not used. The graph title was entered by selecting Options (a two-line title is permitted as well), the axes were labelled (x and y), and grid lines were requested for both axes. Automatic scaling results by default, and generally appropriate minimum and maximum values are chosen to cover the range of data values. In this case the values on both axes are obviously between 0 and 1.

10.3.1 Fourier series—the square-wave approximation

As a second example of tabulating and graphing a formula, consider the Fourier series approximation to a unit square wave. For a period T the square wave is defined within a period starting at $t = 0$ as

$$f(t) = 1 \quad (0 < t < T/2)$$
$$f(t) = 0 \quad (T/2 < t < T)$$

with a fundamental angular frequency of $\omega = 2\pi/T$. Recall that section 7.5 defined the general Fourier series representation of a periodic function. It was shown that the Fourier series of the square wave with unit amplitude ($A = 1$) is given by

$$f(t) = \frac{1}{2} + \frac{2}{\pi}[\sin(\omega t) + \tfrac{1}{3}\sin(3\omega t) + \tfrac{1}{5}\sin(5\omega t) + \ldots].$$

TABULATING AND GRAPHING A FORMULA

A portion of a Lotus worksheet to sum the values of the DC term, the fundamental frequency term, and the third- and fifth-harmonic components, is shown below.

	A	B	C	D	E
1	t (ms.)	wt	f1(t)	f3(t)	f5(t)
2	0	0.000	0.500	0.500	0.500
3	0.025	0.157	0.600	0.696	0.786
4	0.05	0.314	0.697	0.868	0.996
5	0.075	0.471	0.789	0.999	1.089
:	:	:	:	:	:
:	:	:	:	:	:
41	0.975	6.126	0.400	0.304	0.214
42	1	6.283	0.500	0.500	0.500

The headings in row 1 have been centred by preceding them with ^ in each case. The A2 cell entry is 0, and this cell is identified with the name t by invoking the 'range name create' command (/rnc). The A3 entry is then the formula +t+1/40, which establishes a step size of 0.025 for time t. The formula is then copied to the remainder of column A. Column B calculates the values of wt for convenience only. The formula 2*@PI*t is entered into cell B2. The period is assumed to be unity here, but if the times in column A are in units of milliseconds (ms) as suggested by the column heading, the period of 1 ms corresponds to a 1 kHz fundamental frequency. The remaining formulas entered are

```
0.5+2/@PI*@SIN(B2)          into cell C2
+C2+2/@PI*@SIN(3*B2)/3      into cell D2
+D2+2/@PI*@SIN(5*B2)/5      into cell E2.
```

Column C therefore contains the DC plus fundamental term, and columns D and E add to these terms the third and fifth harmonics respectively. The formulas in cells B2, C2, D2, and E2 are copied to the remaining cells in their respective columns, to complete the worksheet. Notice that the worksheet number format has been set to display three decimal places for all numbers.

Figure 10.2 illustrates the plotting of more than one set of data points. The X data range is A2..A42, and the Y data ranges are C2..C42 (the A range) and E2..E42 (the B range). We are therefore superimposing plots of signal f1(t), the fundamental-frequency sine wave plus DC, along with f5(t), the composite signal with the third and fifth harmonics included. As expected, the composite signal resembles the square wave more closely when the harmonics are added. The same waveform was shown in figure 7.16, using PSpice's PROBE utility. Although scaling of both the X- and Y-axes has been done automatically, the scaling option called Skip has been invoked. On entering a 'skip factor' of 4 the displayed values of time on the X-axis are spaced apart. Otherwise all time values 0, 0.025, 0.05, 0.075, etc are displayed and the result has a jumbled

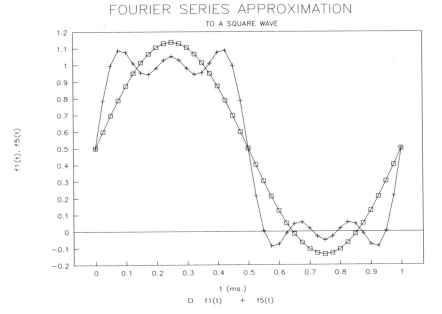

Figure 10.2 Fourier series approximation to a square wave.

appearance. Notice that we can easily replace the plot of f5(t) by f3(t) if the B data range is changed to D2..D42. In this way we can see the effect of eliminating the fifth harmonic.

10.4 DATA ANALYSIS TECHNIQUES

As indicated earlier, Lotus has built in statistical functions which calculate the mean, variance, and standard deviation of a set of data values. Minimum and maximum values can be extracted as well. In addition the data commands allow us to sort the data values in ascending or descending order, and perform an optimum linear curve fit (linear regression). To illustrate the statistical functions and data sort command, the worksheet below is created with ten integer values and appropriate headings. The objective is to tabulate the sorted data in ascending order in column C, and calculate the mean, variance, and standard deviation.

	A	B	C	D	E	F	G	H
1	Example of Use of Statistical Functions and Sort Data Command							
2								
3	Unsorted Data		Sorted Data					
4	6							
5	-3							
6	8							
7	0							
8	-7							
9	14							
10	-3							
11	0							
12	5							
13	-3							
14								
15	Average							
16	Variance							
17	Standard Dev.							

The data values in column A could be sorted directly, but we will copy them to another column to preserve the original ordering. Invoke the copy command (/c) and copy from range A4..A13 to C4..C13. Then invoke the data sort command (/ds), press d to select Data Range from the data sort menu, and enter the range C4..C13. Next press p to select the Primary Sort Key option and position the cell pointer on C4, and press ⟨Enter⟩. The primary key indicates the field by which the data are to be sorted when data records are composed of separate columns of fields. Next enter A (ascending) for the Sort order and then G (go). The sorted data will be placed in column C. Finally, the formulas below are entered:

@AVG(A4..A13) into cell C15

@VAR(A4..A13) into cell C16

@STD(A4..A13) into cell C17.

The data range in each formula could of course be replaced by C4..C13, since the mean, variance, and standard deviation don't change as a result of sorting the data. The final worksheet values are shown below. Notice that the variance is the square of the standard deviation, as expected.

	A	B	C	D	E	F	G	H
1	Example of Use of Statistical Functions and Sort Data Command
2
3	Unsorted Data			Sorted Data
4	6			-7
5	-3			-3
6	8			-3
7	0			-3
8	-7			0
9	14			0
10	-3			5
11	0			6
12	5			8
13	-3			14
14
15	Average			1.7
16	Variance			36.81
17	Standard Dev.			6.067124

10.4.1 Linear regression applied to population estimation

To illustrate linear regression with 1-2-3 consider the worksheet below with official US Census population data from the start of the 20th century. The objective is to use linear regression to predict the population in 1990 and 2000. Section 8.7.3 presents the theory behind optimum linear curve fitting, and the same population example is solved in various ways using MathCAD.

	A	B	C	D	E	F	G	H
1	Estimating U.S. Population using Linear Curve Fitting
2
3	Year	Population (millions)
4		Actual	Predicted
5	1900	76.2
6	1910	92.2
7	1920	106.0
8	1930	123.2
9	1940	132.2
10	1950	151.3
11	1960	179.3
12	1970	203.3
13	1980	226.5
14	1990
15	2000

The years can be entered individually, or, as illustrated previously for data with a fixed increment, the entries can be made with the aid of a 'data fill' command or by entering and copying a formula. The optimum straight-line fit to the population data is obtained by invoking the 'data regression' command (/dr). From the 'data regression' menu enter A5..A13 for the X-range and

B5..B13 for the Y-range. For the output range we must enter the upper left-hand cell of an area to display the results of the regression analysis. This area requires nine rows and four columns and should be placed in a blank area of the worksheet. For our example we enter E4, and then select Go to perform the analysis. The updated worksheet entries are shown below.

```
         A        B          C        D       E         F         G         H
 1    Estimating U.S. Population using Linear Curve Fitting
 2
 3         Year   Population (millions)
 4                Actual    Predicted         Regression Output:
 5         1900    76.2                       Constant              -3443.05
 6         1910    92.2                       Std Err of Y Est       7.643218
 7         1920   106.0                       R Squared              0.980447
 8         1930   123.2                       No. of Observations           9
 9         1940   132.2                       Degrees of Freedom            7
10         1950   151.3
11         1960   179.3                       X Coefficient(s)       1.848666
12         1970   203.3                       Std Err of Coef.       0.098673
13         1980   226.5
14         1990
15         2000
```

The straight lines being fitted to the data are of the form

$$Y = mX + b$$

where the slope m is given by the X-coefficient (cell G11) and the y-intercept by the constant (cell H5). For these data the resulting equation is then

$$Y = 1.848\,666X - 3443.05.$$

We have reserved column C to allow us to tabulate the results of this equation for each census year, and use it to predict the population in the years 1990 and 2000. As in previous examples, we can enter a formula into the first cell (C5) and copy it to the remaining cells in the column. The X-values are the years specified in column A, while the slope and intercept values are constants. Therefore, we enter the formula

  +G$11*A5+H$5 into cell C5

and copy to the range C6..C15. Recall that a row or column identifier preceded by a $ fixes that row or column reference when the formula is copied. As a result, the formulas in subsequent cells become

  C6: +G$11*A6+H$5
  C7: +G$11*A7+H$5
     ⋮
  C15: +G$11*A15+H$5

as only the X-values are changed. The final worksheet is shown below, with the predicted values in column C formatted to one decimal place.

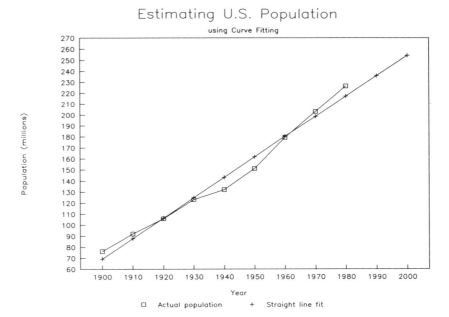

Figure 10.3 Population data and an optimum linear curve fit.

```
          A      B         C       D      E        F         G       H
1    Estimating U.S. Population using Linear Curve Fitting
2
3         Year  Population (millions)
4               Actual   Predicted           Regression Output:
5         1900   76.2     69.4      Constant                 -3443.05
6         1910   92.2     87.9      Std Err of Y Est         7.643218
7         1920  106.0    106.4      R Squared                0.980447
8         1930  123.2    124.9      No. of Observations             9
9         1940  132.2    143.4      Degrees of Freedom              7
10        1950  151.3    161.8
11        1960  179.3    180.3      X Coefficient(s)  1.848666
12        1970  203.3    198.8      Std Err of Coef.  0.098673
13        1980  226.5    217.3
14        1990           235.8
15        2000           254.3
```

Now it is a simple matter to plot the actual and predicted data. The X data range is A5..A15, and the Y data ranges are B5..B13 (the A range) and C5..C15 (The B range). The actual population and best straight-line fit are shown superimposed in figure 10.3.

10.5 SIGNAL DETECTION IN NOISE

The mathematics functions of 1-2-3 include @RAND, which returns a pseudo-random number between 0 and 1. Noise can be simulated by choosing a succession of random numbers representing voltage amplitudes at equally spaced sampling intervals (as discussed in section 8.6.1). To simulate a zero average random value between $-A$ and $+A$, we can use the formula

A*(2*@RAND-1)

The worksheet below simulates a single-frequency signal (1 kHz) sampled 64 times per cycle, with random noise superimposed. The data are all formatted to three decimal places, except for the times t in column B for which five places are used.

	A	B	C	D	E	F	G	H
1	Signal	and	Noise					
2								
3	freq =	1000		Anoise =	1			
4								
5	n	t	signal	noise	signal+noise			
6	0	0.00000	0.000	0.261	0.261			
7	1	0.00002	0.098	0.453	0.551			
8	2	0.00003	0.195	-0.234	-0.038			
:								
68	62	0.00097	-0.195	0.489	0.294			
69	63	0.00098	-0.098	-0.603	-0.701			

The signal frequency (f) is placed in cell B3 and the noise amplitude (Anoise) in cell E3. Using the 'range name create' (/rnc) command we identify the variable names with the corresponding cells. In the same manner we identify cell A6 with the sample number (n) and cell B6 with time (t). To create the remainder of the table we insert the titles and the following cell entries:

0	into cell A6
+N+1	into cell A7
+N/(64*F)	into cell B6
@SIN(2*@PI*F*T)	into cell C6
+Anoise*(2*@RAND-1)	into cell D6
+C6+D6	into cell E6.

The formulas in cells A7, B6, C6, D6, and E6 are then copied to the cells below in the respective columns. The sampled values of signal and noise can be written in algebraic form as

$$\text{signal}(n) = \sin(2\pi f t_n) \qquad \text{noise}(n) = A_{\text{noise}}(2r - 1)$$

where

$$t_n = n/(64f)$$

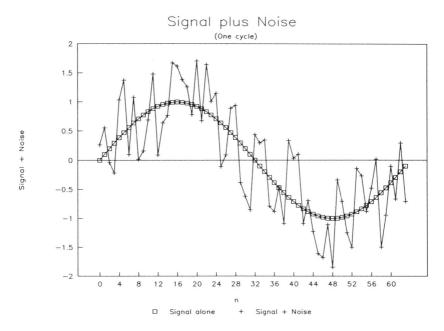

Figure 10.4 The signal plus random noise (one cycle).

with n ranging from 0 to 63, and r being a random number between 0 and 1.

Signal alone as well as signal plus noise are both plotted in figure 10.4. On setting the X data range to A6..A69 the values are plotted with respect to the sample number (n). Alternatively, for a plot versus time values we would set the X-range to B6..B69. The A and B data ranges are C6..C69 and E6..E69 respectively. As expected, the signal is considerably corrupted by the noise, since both signal and noise amplitudes are set to unity.

10.5.1 Signal averaging

Signal averaging or accumulation is a method of detection resulting in an increased signal-to-noise ratio. As discussed in section 8.6.2, the method involves averaging signal and noise values over a longer time interval, to take advantage of the relative predictability of the signal in comparison to the noise. If we average the signal plus noise over M cycles, we can write the sampled values of signal plus noise as

$$\frac{1}{M}\sum_{m=1}^{M}[\text{signal}(n) + \text{noise}(n)] = \frac{1}{M}\sum_{m=1}^{M}[\sin(2\pi f t_n) + A_{\text{noise}}(2r - 1)].$$

Since the signal values are the same for each cycle, the averaging over any number of integral cycles gives us the same values as for a single cycle. The

SIGNAL DETECTION IN NOISE 337

signal-plus-noise expression can be rewritten as

$$\frac{1}{M}\sum_{m=1}^{M}[\text{signal}(n) + \text{noise}(n)] = \sin(2\pi f t_n) + \frac{1}{M} A_{\text{noise}} \left(2\sum_{m=1}^{M}(r) - M\right).$$

The problem with implementing this expression in our worksheet is that we need to sum the @RAND function M times. Notice that simply multiplying M by @RAND does not produce an equivalent result, as this would result in the adding of one random value to itself repeatedly. The formulas for the noise component are shown below for small values of M:

+Anoise*(2*(@RAND+@RAND)-2)/2	for $M = 2$
+Anoise*(2*(@RAND+@RAND+@RAND)-3)/3	for $M = 3$
+Anoise*(2*(@RAND+@RAND+@RAND+@RAND)-4)/4	for $M = 4$
+Anoise*(2*(@RAND+@RAND+@RAND +@RAND+@RAND)-5)/5	for $M = 5$.

We see that not only does the formula get cumbersome as M gets larger, but also it must be edited every time we want to change M to perform a 'what if' analysis. We will use this example shortly to introduce a programming macro to resolve the problem.

As an illustration of improving the extraction of the signal from the noise before resorting to macros, let us arbitrarily choose five cycles for accumulation. We can then set up another column in the previous worksheet to display averaged signal plus noise. After inserting an appropriate column heading, the formula

    +C6+Anoise*(2*(@RAND+@RAND+@RAND+@RAND+@RAND)-5)/5

is then entered into cell G6. As was the case for the other columns, the formula is then copied down to G7..G69. The final worksheet is shown below.

	A	B	C	D	E	F	G	H
1	Signal and Noise							
2								
3	freq =	1000		Anoise =		1		
4								averaged
5	n	t	signal	noise	signal+noise		signal+noise	
6	0	0.00000	0.000	0.261	0.261		-0.078	
7	1	0.00002	0.098	0.453	0.551		0.001	
8	2	0.00003	0.195	-0.234	-0.038		-0.077	
⋮								
68	62	0.00097	-0.195	0.489	0.294		-0.106	
69	63	0.00098	-0.098	-0.603	-0.701		-0.214	

Figure 10.5 shows the signal plus noise averaged over five cycles, superimposed on the plot of signal alone. The same graph parameters as for figure 10.4 are used except for a different title, and a change in the B data

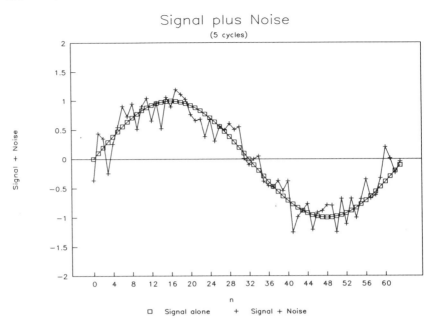

Figure 10.5 The signal plus random noise (averaged over five cycles).

range from column E (E6..E69) to column G (G6..G69). The improvement in signal-to-noise ratio is evident when the two graphs are compared.

10.5.2 Use of macros

As mentioned in section 10.2, we can employ features of a programming language through the use of Lotus macros. Macros contain a combination of keystrokes and advanced 'programming-like' commands. We will not attempt to overview all of the macro commands available, but will simply show examples of macros providing looping operations. The purpose of the programming loop is to average M random values, thus implementing the noise component of the example just presented. Consider the sequence of statements below:

```
         {goto}TOTAL~
         {for I,1,M,1,LOOP}~
LOOP     {calc}(TOTAL+Anoise*(2*@RAND-1)/M)~
```

Each statement defines an operation and ends with "~", the key equivalent of ⟨Enter⟩. Macro execution proceeds with the statements down a column until it reaches a blank cell. Alternatively, a macro can be terminated by either a {quit} or a {return} command. To understand the operation of the macro, assume first that TOTAL references a cell into which an initial value of zero has been entered. When the macro is executed the {goto} command positions the cursor at TOTAL. The {for} command is a loop command as in the BASIC programming language, repeatedly executing a sequence of instructions (a sort of a 'sub-macro') a specified number of times. In this example, an index I varies from 1 to M in steps of 1, and the sequence identified by LOOP is executed repeatedly.

The names I and M reference cells, with cell M initialized to the desired integer value. No value need be entered for I as the {for} command initializes it and then repeatedly increments it by the step size. The name LOOP references the cell containing the statement starting with calc. In this case the 'sub-macro' executed repeatedly consists of a single statement only. The {calc} command recalculates the worksheet values, resulting in adding a new random value to the previous TOTAL. The noise amplitude, Anoise, also identifies a cell as in the previous worksheet. Dividing each random value by M ensures that the value of TOTAL after M loops is in fact the average we seek.

We now turn to the question of entering the macro into the spreadsheet and executing it. The spreadsheet below contains the macro and appropriate documentation to identify the cells of importance.

	A	B	C	D	E	F	G	H
1	\X		{goto}TOTAL~					
2			{for I,1,M,1,LOOP}~					
3								
4		LOOP	{calc}(TOTAL+Anoise*(2*@RAND-1)/M)~					
5								
6							Anoise	1
7							M	20
8								
9		Range Name Table					I	
10								
11		Anoise	H6				TOTAL	0
12		I	H9					
13		M	H7					
14		LOOP	C4					
15		TOTAL	H11					
16		\X	C1					

The range names are created in the usual manner by invoking the range name create command (/rnc) and entering each variable name and its associated range. Each name is identified by a single cell and documented in the range name table entered as worksheet text. The cells in column H are identified with the parameters Anoise, M, I, and TOTAL. The variable names themselves are entered as text into the respective rows of column G for ease of identification. The desired values of A_{noise} and M are entered into the appropriate cells, and TOTAL is initialized to zero to meet the requirements of the macro. Since the {calc} statement in the macro has been placed in cell C4, the name LOOP is associated with that cell. Inclusion of the word LOOP itself in cell B4 is not necessary, but obviously enhances the readability of the macro.

The macro itself must be associated with a single lettered key. In our case we have chosen "X" arbitrarily. The letter is then identified with the cell containing the first macro statement by again invoking the range name create command (/rnc). To distinguish the macro key identifier from a variable name, however, the letter must be preceded by a backslash. By entering "\X" as the name and C1..C1 as the range, we thereby identify "X" as a key associated with the macro starting with the {goto} statement in cell C1. Incidentally, Lotus 1-2-3 release 2.2 (or later versions) permits a macro to be identified with a name having as many as 15 characters.

To execute a macro we simply need to press the ⟨Alt⟩ key along with the key identifying that macro. In our example the initial position of the cell pointer is unimportant since the first macro command repositions it at TOTAL. Therefore we simply press ⟨Alt⟩X. Depending on the speed of your machine, you may or may not be able to observe the looping action by watching the index I and the sum of random values in TOTAL repeatedly changing. The resulting spreadsheet after macro execution is shown below.

	A	B	C	D	E	F	G	H
1	\X		{goto}TOTAL~					
2			{for I,1,M,1,LOOP}~					
3								
4		LOOP	{calc}(TOTAL+Anoise*(2*@RAND-1)/M)~					
5								
6							Anoise	1
7							M	20
8								
9		Range Name Table					I	21
10								
11		Anoise	H6				TOTAL	-0.05484
12		I	H9					
13		M	H7					
14		LOOP	C4					
15		TOTAL	H11					
16		\X	C1					

As indicated by the value of I in cell H9, the {for} command terminated when the index exceeded the upper limit M. Since TOTAL was initialized to zero, the final value of -0.05484 is the average of 20 random numbers, each between $-A_{\text{noise}}$ and A_{noise}. The macro can be rerun as many times as desired by reinitializing TOTAL to zero manually (cell H11) and pressing ⟨Alt⟩X again. Repeated runs of this worksheet showed that the final value of TOTAL was generally between -0.1 and 0.1.

Several additional runs were performed after changing M from 20 to 100 (cell H7). The macro execution time increased by a factor of about five as expected.

SIGNAL DETECTION IN NOISE   341

The value of TOTAL decreases by about the same factor, with the absolute value staying below 0.02 for most runs. These results reinforce the notion that signal averaging over a longer time interval does indeed enhance signal-to-noise ratio, by reducing the average of the random or noise component.

We will conclude this section by incorporating the macro just discussed into the worksheet simulating sampled signal plus noise. Consider the worksheet outline and separate range name table shown below.

```
            A       B       C           D           E       F       G           H
1   Signal and Noise (Accumulation Method - using Macro)
2
3   \X              {goto}TEMP~
4                   {for I,1,64,1,LOOP1}~
5                                                                   f =      1000
6           LOOP1   {let TEMP,0}~                               Anoise =         1
7                   {for J,1,M,1,LOOP2}~                            M =        10
8                   {put ACCUM,0,+I-1,TEMP/M}~                      I =         0
9                                                                   J =         0
10          LOOP2   {calc}(TEMP+Anoise*(2*@RAND-1))~             TEMP =         0
11
12                              single  cycle                   averaged
13          n       t           signal  noise   s + n           noise   s+n
14
15
16
:
76
77

                    Range Name Table
                    ACCUM       G14..G77
                    Anoise      H6
                    F           H5
                    I           H8
                    J           H9
                    LOOP            C10
                    M           H7
                    N           A14
                    START       C6
                    T           B14
                    TEMP            H10
                    \X          C3
```

To the original worksheet in this section we have added the macro shown along with the list of parameters and their initial values (see rows 3 to 10). Also, headings are set up for a table of averaged values of noise and signal plus noise, to be placed below the parameters in columns G and H. We see that the looping macro defined earlier, which averages M random noise values between $-A_{noise}$ and A_{noise}, is placed inside another loop. The averaging of M values is performed a total of 64 times, recalling that 64 is the number of sampled values

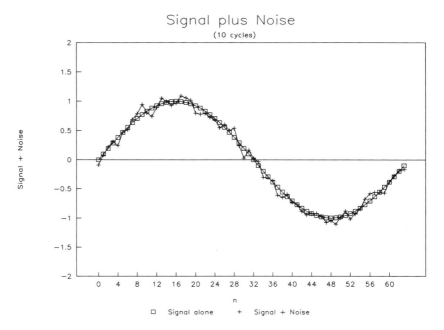

Figure 10.6 The signal plus random noise (averaged over ten cycles).

taken over a full cycle of the signal.

Automation of this worksheet is facilitated by use of the {let} and {put} macro commands. The function and format of these commands are given below:

Command	Function	Format
{let}	Set single cell contents	{let cell-location,entry}
{put}	Set range contents	{put range,column,row,entry}

In our macro, the cell identified with TEMP is used to accumulate the sum of the M random values at each sampling time. The average, computed by dividing the sum by M, is then inserted in the table of averaged noise values (column G identified by ACCUM). The {let} command simply reinitializes TEMP to zero to prepare it for the next sum. The {put} command places TEMP/M, or the average, into the appropriate row of ACCUM. Notice that the column and row identifiers in the {put} command are numbers relative to the start of the range. The first cell of ACCUM, namely G14, corresponds to column 0 and row 0. Since the range consists of a single column, the column identifier is fixed at 0. Since the loop index I varies from 1 to 64, the row identifier $I - 1$ ranges from 0 to 63 as desired.

Aside from the averaged noise values, the remaining table is created as with previous examples by entering the formulas below, and copying them to the remaining cells in each column:

0	into cell A14
+N+1	into cell A15
+N/(64*F)	into cell B14
@SIN(2*@PI*F*T)	into cell C14
+Anoise*(2*@RAND-1)	into cell D14
+C14+D14	into cell E14
+C14+G14	into cell H14 .

After the worksheet entries are completed, the macro is executed by pressing ⟨Alt⟩X. The final worksheet is shown below.

	A	B	C	D	E	F	G	H
1	Signal and Noise (Accumulation Method - using Macro)							
2								
3	\X		{goto}TEMP~					
4			{for I,1,64,1,LOOP1}~					
5						f =		1000
6		LOOP1	{let TEMP,0}~			Anoise =		1
7			{for J,1,M,1,LOOP2}~			M =		10
8			{put ACCUM,0,+I-1,TEMP/M}~			I =		0
9						J =		0
10		LOOP2	{calc}(TEMP+Anoise*(2*@RAND-1))~			TEMP =		0
11								
12				single	cycle		averaged	
13		n	t	signal	noise	s + n	noise	s+n
14	0	0.00000	0.000	-0.867	-0.867		-0.117	0.117
15	1	0.00002	0.098	-0.049	0.049		0.067	0.165
16	2	0.00003	0.195	-0.010	0.185		0.044	0.239
:								
76	62	0.00097	-0.195	-0.126	-0.321		-0.140	-0.335
77	63	0.00098	-0.098	-0.197	-0.295		0.077	-0.021

A plot of averaged signal plus noise over ten cycles is shown in figure 10.6. The X-range is set to A14..A77 and the A data range to H14..H77. Comparing figures 10.5 and 10.6 for five- and ten-cycle averaging, respectively, reinforces the fact that the signal-to-noise ratio continues to improve as the total interval increases.

REFERENCES

[1] Etter D M 1992 *Lotus 123—a Software Tool for Engineers* (New York, NY: Addison-Wesley and Benjamin/Cummings)
[2] Mezei L M 1989 *Laboratory LOTUS—a Complete Guide to Instrument Interfacing* (Englewood Cliffs, NJ: Prentice-Hall)
[3] Orvis W J 1991 *1-2-3 for Scientists and Engineers* (Alameda, CA: SYBEX)

11

A Graphical User Interface Development Tool: LabWindows

11.1 INTRODUCTION

LabWindows®, a product of National Instruments, is a software development system for creating data-acquisition and instrument control applications. Programs can be created, compiled, and debugged in an interactive environment, using LabWindows subsets of either C language or QuickBASIC. A set of library functions are available for acquisition, analysis, and presentation of data. Special interfaces called function panels facilitate generation of the code for library functions [1, 2, 3].

For data acquisition, high-level functions are provided for controlling National Instruments data-acquisition boards. Included are analogue and digital I/O, counter/timer, and waveform generation functions. A collection of instrument drivers are also provided in a special instrument library. Tools are provided for creating your own instrument drivers for single or multiple instruments, or for a virtual instrument where no physical instrument exists. Additional libraries provide functions for handling addressing and bus management protocol for GPIB, IEEE 488.2, and RS-232 standards. An optional library is available for the newer VXI instrument standard.

The data analysis library includes functions that perform arithmetic operations on arrays and complex numbers. Statistical functions calculate the mean and standard deviation of a data array, create a histogram, or perform numerical sorting. Vector and matrix operations include dot product, matrix multiplication, matrix inversion, transpose, and evaluation of a determinant. An optional advanced analysis library includes digital processing functions such as FFT, inverse FFT, convolution, power spectrum, FIR and IIR digital filter design, and windowing. Other features include noise generation, curve fitting, and additional matrix, vector, and complex operations.

For data presentation, the user interface and graphics libraries add objects such as menu bars, instrument panels, controls, and graph pop-up panels to an application program. A formatting and I/O library includes various file I/O, string

DATA ANALYSIS AND DATA PRESENTATION EXAMPLES

manipulation, and data formatting functions.

11.2 DATA ANALYSIS AND DATA PRESENTATION EXAMPLES

In this section we will show five C language programs, each created in the LabWindows environment and supplemented by functions from the standard data analysis and user interface libraries. In all but the first program input data values are simulated with the aid of a built-in pseudo-random-number generating function.

11.2.1 Linear equations for a differential amplifier

LabWindows example 1 uses matrix manipulation functions from the data analysis library to solve for the unknowns of a set of six linear equations. The equations represent circuit laws applied to the equivalent circuit of a single-op-amp differential amplifier. For a fuller description of this circuit see sections 7.3.5 and 8.4.2, where SPICE and MathCAD are used to solve the equations for the same circuit.

```
/* LabWindows Example 1 - Solution to the circuit equations */
/*    for the one op-amp differential amplifier              */
/*    (using matrix inversion)                               */
# define K     1e3
# define MEG   1e6
main()
{
  int n;
  double r1, r2, r3, r4, rin, v1, v2, a, vin, vout;
  double b [6] [6], invb [6] [6], c [6] [1], x [6] [1];
  cls();
/* Resistor, source, and gain values                         */
/*    (voltages in millivolts)                              */
  r1 = 100;
  r2 = 1*K;
  r3 = 10*K;
  r4 = 100*K;
  rin = 1*MEG;
  v1 = 2;
  v2 = 5;
  a = 50*K;
/* Matrix of coefficients of unknowns                        */
/*    (all other values default to zero)                    */
  b [0] [0] = r1;
  b [0] [1] = -r2;
  b [0] [5] = 1;
  b [1] [2] = r3;
```

```
    b [1] [3] = -r4;
    b [1] [5] = a-1;
    b [2] [1] = r2;
    b [2] [3] = r4;
    b [3] [4] = -rin;
    b [3] [5] = 1;
    b [4] [0] = 1;
    b [4] [2] = -1;
    b [4] [4] = -1;
    b [5] [1] = 1;
    b [5] [3] = -1;
    b [5] [4] = 1;
/*  Independent source vector                              */
    c [0] [0] = v1 - v2;
    c [2] [0] = v2;
/*  Use matrix operations from analysis library            */
/*     to solve for unknowns                               */
    InvMatrix (b, 6, invb);
    MatrixMul (invb, c, 6, 6, 1, x);
    FmtOut ("\n Solution of unknowns:");
    for (n = 0; n <= 5; ++n)
        FmtOut ("\n %f", x [n] [0]);
    vin = x[5] [0];
    vout = a * vin;
    FmtOut ("\n\n Output voltage = %f   mV.", vout);
}
```

The set of linear equations is shown on the MathCAD worksheet in figure 8.9, and the matrix solution on the worksheet in figure 8.10. The 6×6 matrix **b** contains the coefficients of the unknown currents and voltages, represented by the 6×1 vector x. We use c for the 6×1 independent source vector. Only the non-zero values of **b** and c have been assigned, as the remaining entries default to zero anyway. Since the equations to be solved are in matrix form $\mathbf{b}x = c$, the solution is given by

$$x = \mathbf{b}^{-1}c.$$

Therefore, the inverse of matrix **b** must first be computed and then multiplied by vector c. The data analysis library functions `InvMatrix` and `MatrixMul` accomplish these tasks. The other functions, `cls` (clear screen) and `FmtOut` (formatted print), are part of the LabWindows C system library. The system library also includes other LabWindows C functions as well as many from the Microsoft C library.

After compiling and running the program the screen displays the following:

```
Solution of unknowns:
-0.029565
4.9499e-005
-0.029565
```

DATA ANALYSIS AND DATA PRESENTATION EXAMPLES 347

```
4.950501e-005
6.012158e-009
0.006012


Output voltage = 300.607889  mV.
```

As explained in section 7.3.5, the differential amplifier output voltage is the difference between the two inputs v_2 and v_1 multiplied by the ratio R_3/R_1. For the numbers selected the differential input 3 mV is multiplied by 100, giving us close to 300 mV.

11.2.2 Signal detection in noise

LabWindows example 2 simulates signal detection in noise, using an accumulation method, which averages a number of sets of received data. MathCAD and DSPlay are also used to simulate the same problem (see sections 8.6.2 and 9.6.1). The program replicates the MathCAD worksheet in figure 8.14 by creating a single-frequency signal added to 'random' noise. The signal plus noise is 'sampled' 64 times per signal period, and corresponding sampled values over M cycles are averaged. The system library's rand function generates a random integer within the range 0 to 32K; dividing it by 32K gives us a uniform distribution between 0 and 1.

Noise generated with positive and negative values no greater in magnitude than A_{noise} is added to the value to the sine-wave signal. The array y accumulates the M sets of values at each sample time in the cycle, and averages them by simply dividing the sum by M. For comparison, an array x is also generated for a single period of the signal. In this example the peak signal and noise values are set to unity, and averaging is done over 20 data sets.

```
/*   LabWindows Example 2 - Signal Detection in Noise    */
/*      (Accumulation method)                            */
# define PI   3.14159
main()
{
   int      k, M, n;
   double   f, Anoise;
   double   t[64], x[64], y[64];
/* Initialize parameters                                 */
   cls();
   f = 1000;
   Anoise = 1;
   M = 20;
/* Simulate signal plus noise                            */
   for (n = 0; n <= 63; ++n)
```

```
     { t[n] = n / (64 * f);
       x[n] = sin(2*PI*f*t[n]) + Anoise*(2*(rand()/32767.0) - 1);
       y[n] = 0;
       for (k = 1; k <= M; ++k)
           y[n] = y[n] + sin(2*PI*f*t[n]) + Anoise *
 (2*(rand()/32767.0) - 1);
       y[n] = y[n] / M;
       FmtOut("%d %f %f \n", n, x[n], y[n]);
     }
 /*  Graph results with user interface library           */
 /*    (X-Y Graph Popup function)                        */
    XYGraphPopup (t, x, 64, 4, 4);
    XYGraphPopup (t, y, 64, 4, 4);
 }
```

Although the program displays the 64 array values of *x* and *y*, graphing these values on a time base gives us a more meaningful data presentation. The function XYGraphPopup is generated by selecting the Pop-up Panels option of the user interface library. The graphs produced by the program are shown in figures 11.1 and 11.2. It is clear from figure 11.2, which shows array *y*, that averaging has enhanced the signal with respect to the noise. In the absence of noise we would see one cycle of a pure 1 kHz sine wave with unit amplitude. Notice that the graphing function has automatically scaled the axes in an appropriate fashion.

11.2.3 Statistical analysis of simulated temperatures

LabWindows example 3 analyses a simulated distribution of room temperatures in degrees Fahrenheit. The same simulation is performed using MathCAD in section 8.7.2 (see the worksheet in figure 8.16). The program approximates a normal distribution by adding ten uniformly distributed random numbers, and storing the result in sum. The values are then scaled to produce an array temp, consisting of values normally distributed about 70 °F. Although a true normal distribution has an infinite range of values, the approximate distribution is limited to a range between 40 and 100 °F with a variance of 30 °F. The square root of the variance, the standard deviation, is then about 5.61. For a large number of sample values the calculated mean and standard deviation should be close to the expected values.

```
 /*  LabWindows Example 3 - Statistical Analysis of Simulated  */
 /*  Temperature Readings (using Analysis Library Functions)   */
 main()
 {
   int n, intervals, i, j, key, hist[12];
   double tempbase, temptop, sum, avg, sDev;
   double temp[1000], tempsort[1000], axis[12];
```

DATA ANALYSIS AND DATA PRESENTATION EXAMPLES 349

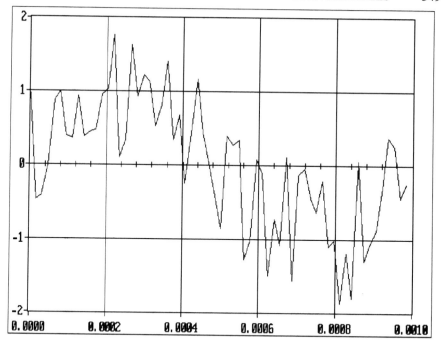

Figure 11.1 Random noise added to the signal.

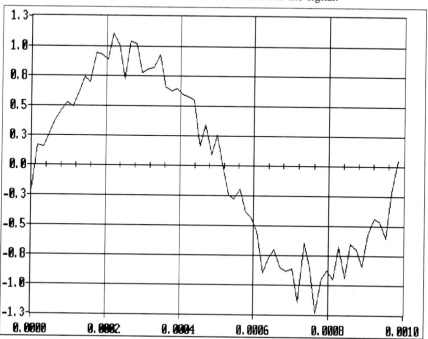

Figure 11.2 Detection of a signal in noise using accumulation.

```
/* Initialize parameters                                              */
  cls();
  n = 1000;
  intervals = 12;
  tempbase = 40.0;
  temptop = 100.0;
/* Generate normally distributed temperatures                         */
  for (i = 0; i <= n-1; ++i)
    { sum = 0;
      for (j = 1; j <= 10; ++j)
        sum = sum + rand()/32767.0;
      temp[i] = tempbase + sum * (temptop - tempbase) / 10.0;
    }
/* Use library functions to find mean and standard deviation */
/*   Sort the data and generate a histogram                  */
  StdDev (temp, n, &avg, &sDev);
  Sort (temp, n, 0, tempsort);
  Histogram (tempsort,n,tempbase,temptop,hist,axis,intervals);
/* Display results with statistical analysis library functions */
  FmtOut ("Mean = %f, Standard deviation = %f", avg, sDev);
  FmtOut ("\n min. temp. = %f, max. temp. = %f", tempsort [0],
  tempsort[n-1]);
  FmtOut ("\n \n Histogram results");
  FmtOut ("\n \n Mid.         Freq.");
  for (i = 0; i <= intervals - 1; ++i)
    FmtOut ("\n %f       %d", axis[i], hist[i]);
  FmtOut ("\n \n Press any key to continue");
  getkey();
/* Graph unsorted and sorted temperatures with user interface */
/*    library functions                                       */
  YGraphPopup (temp, n, 4);
  YGraphPopup (tempsort, n, 4);
}
```

Three statistical functions from the data analysis library are used. First the function StdDev is used to calculate both the mean and standard deviation of the 1000 simulated temperatures. The values are then sorted in ascending order with the Sort function, resulting in the array tempsort. The Histogram function produces an array, hist, consisting of the number of readings within prescribed intervals. Notice that Histogram requires as inputs the array of values (either temp or tempsort could have been used), the number of values ($n = 1000$), the number of intervals (intervals = 12), and the extreme values expected from the distribution (tempbase = 40, temptop = 100). In addition to the array hist, a corresponding array of interval midpoints is generated (axis).

Using the printing function (FmtOut) the mean and standard deviation are displayed, followed by the histogram results presented in tabular form. The display from a sample run is shown below.

```
Mean = 70.244191, Standard deviation = 5.510338
min. temp. = 51.758477, max. temp. = 88.524064
Histogram results
Mid.           Freq.
42.5            0
47.5            0
52.5            4
57.5           30
62.5          140
67.5          315
72.5          312
77.5          157
82.5           39
87.5            3
92.5            0
97.5            0
Press any key to continue
```

We see that the calculated mean differs from the expected value of 70 °F by less than 0.5%, while the standard deviation is about 2% lower than its expected value of 5.61. After the data have been sorted, the minimum and maximum simulated readings are displayed by simply specifying the first and last item of the tempsort array. It should not be surprising that no values close to 40 and 100 °F are generated in a sample of this size, as these values are more than five standard deviations from the mean. Since the range from 40 to 100 °F has been divided into 12 intervals for the histogram, each interval width is 5. Therefore, the first interval ranges from 40 to 45, with the midpoint shown as 42.5. The remaining intervals are 45–50, 50–55, ... , 95–100. If the histogram were plotted in bar-graph form instead of being tabulated, its outline would bear a slight resemblance to a normal curve. A larger number of intervals and total sample size would be required to produce the familiar bell shape of the normal distribution.

Notice that the program includes a prompt message to press a key to continue. The system function getkey waits for the user to press a key and returns the ASCII code as an integer. Since the value in this case is not assigned and tested, pressing any key will advance the program to the next line. Without the getkey function the graphs will be plotted immediately, not allowing time to observe the data display on the screen.

In contrast to the procedure in example 2, where XY graphs were presented on a time base, we use the YGraphPopup functions here to display the array values alone. The X-axis markings simply indicate the sample numbers. Since the array temp contains the unordered values, figure 11.1 has an appearance similar to normally distributed noise. When the ordered values of tempsort are plotted in figure 11.2 the resulting curve increases monotonically as expected.

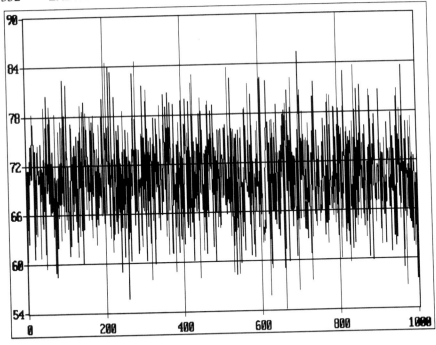

Figure 11.3 Simulated unsorted temperature readings.

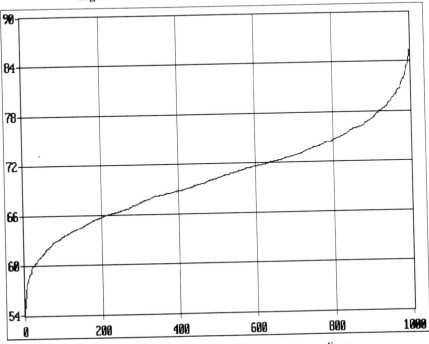

Figure 11.4 Simulated sorted temperature readings.

11.2.4 Simulated EMG muscle contraction

A bioinstrumentation application is presented in LabWindows example 4, which simulates an electromyograph (EMG) recording produced by a voluntary muscle contraction. The period of contraction is represented by a random waveform with amplitudes uniformly distributed between -1 and 1. The recorder output is reduced by a factor of 10 both before and after contraction, to account for low-level muscle action or stray pick-up. The array x stores 128 values, with the middle third of them representing the muscle contraction.

An EMG is often converted to an absolute integral before being displayed on an oscilloscope. The system function fabs is first used to convert x to an array xabs of positive values. Then the integration is approximated by taking a moving average of the absolute values of the most recent N sample values, with N set to 10. The moving average, stored in array avgN, is actually an implementation of a finite-impulse-response (FIR) digital filter (see section 9.6.1 for a further discussion).

```
/* LabWindows Example 4 - Simulation of Muscle Contraction */
main ()
{
  int i, j, p, N;
  double x[128], xabs[128], sumN[128], avgN[128];
/* Simulate low level, then muscle contraction,           */
/*     then low level again                               */
  cls();
  for (i = 0; i <= 42; ++i)
    x[i] = 0.1 * (2 * (rand()/32767.0) - 1);
  for (i = 43; i<= 85; ++i)
    x[i] = 2 * (rand()/32767.0) - 1;
  for (i = 86; i <= 127; ++i)
    x[i] = 0.1 * (2 * (rand()/32767.0) - 1);
/* Full wave rectify and take moving average              */
/*    N = number of most recent samples to average       */
  N = 10;
  for (i = 0; i <= 127; ++i)
    { xabs[i] = fabs(x[i]);
      sumN[i] = 0;
      p =i - N + 1;
      if (p < 0) p = 0;
      for (j =p; j <= i; ++j)
        sumN[i] = sumN[i] + xabs[j];
      avgN[i] = sumN[i] / N;
    }
/* Graph results with user interface library function    */
  YGraphPopup (x, 128, 4);
  YGraphPopup (xabs, 128, 4);
  YGraphPopup (avgN, 128, 4); }
```

354 LABWINDOWS

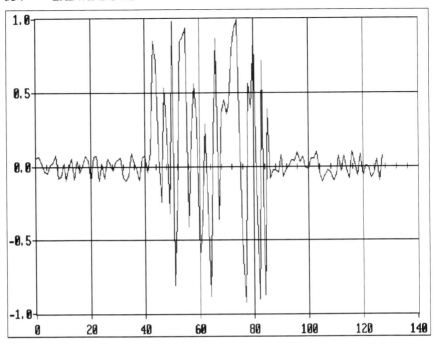

Figure 11.5 Original muscle contraction.

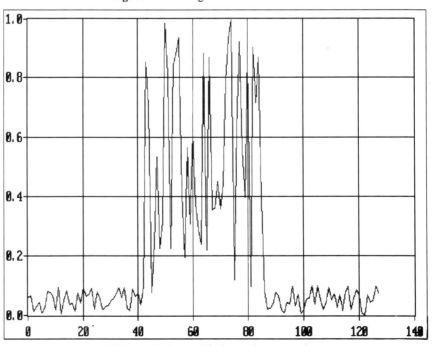

Figure 11.6 Rectified muscle contraction.

As in previous examples, the pop-up panels from the user interface library create the code for graphing the data. Figures 11.5, 11.6, and 11.7 show the original data (x), rectified values (xabs), and integrated or averaged values (avgN), respectively.

11.2.5 Instrument behaviour shaping

LabWindows example 5 simulates an example of instrument behaviour shaping, presented in chapter 12 of 'Digital Biosignal Processing' [4]. The objective is to stimulate a subject going through an evolutionary learning process, to achieve time-varying target behaviour or goals. The behaviour (response) is tracked over a period of time, while it is continually compared to the desired target (criterion). Stimulation is provided by reward and reinforcement. After each tracking interval behaviour shaping occurs, resulting in possible modification of the goals depending on the reinforcement level. Shaping serves to ease the task if subject behaviour does not respond sufficiently to reinforcement, or to make the goal more difficult to achieve if the task proves too 'easy'.

The routines for tracking and shaping behaviour are incorporated in user-defined functions (track, shaping). Notice that, since parameters must be passed between these functions and the main function, C requires that variable addresses be referenced by main. A random-number generator is again used as in previous examples. Here the generated random number is used to determine whether the subject's response is correct. A seed function (srand) has been included for flexibility. Changing the seed produces a different set of random values. As the subject learns, the likelihood of a correct response (probability) should increase.

Flow diagrams of the tracking and shaping procedures are shown in figure 11.8. The tracking and shaping procedures can be repeated as many times as desired, with the program terminating only upon pressing "x" (generating ASCII code hex 78).

```
/* LabWindows Example 5 - Simulation of instrument learning  */
/*   (Ref. Silverman and Dworkin, "Instrumental behaviour    */
/*    shaping automata" (Digital Biosignal Processing - Ch.12) */
main()
{
  int seed, response, reinforce, reward, criterion, key;
  double probability;
  seed = 12345;
  srand(seed);
  cls();
  response = 5;
  probability = 0.25;
  reinforce = 0;
  reward = 0;
```

```
    criterion = 5;
    FmtOut ("\n Press return to continue, 'x' to stop");
    FmtOut ("\n response  probability  reward  criterion reinforce");
/*  Track behaviour                                                    */
    while (key != 0x78)
    { track (&response,&reinforce,&reward,&criterion,&probability);
/*  Observe results                                                    */
        FmtOut ("\n    %d         %f        %d        %d
%d", response, probability, reward, criterion, reinforce);
/*  Perform behaviour shaping                                          */
       shaping (&response, &reinforce, &reward, &criterion,
&probability);
/*  Another iteration ?                                                */
      key = getkey();
    }
}
/*  Simulate tracking                                                  */
int track (resp, rein, rew, crit, probability)
int *resp, *rein, *rew, *crit;
double *probability;
{
  int index;
  double prob, ranumbr;
  prob = *probability;
  for (index=1; index <= 50; ++index)
  { ranumbr = rand() / 32767.0;
    if (ranumbr < prob)
    { ++ *resp;
      if (*resp > 100) *resp = 100;
    }
    else
    { -- *resp;
      if (*resp < 0) *resp = 0;
    }
    if (*resp >= *crit)
    { ++ *rew;
      ++ *crit;
      if (*rew > 30) *rew = 30;
      if (*crit > 100) *crit = 100;
    }
    else
    { if (*rew > 0) --*rew;
      else
      { ++ *rein;
        prob = prob + 0.01;
        -- *crit;
        if (*crit < 0) *crit = 0;
```

DATA ANALYSIS AND DATA PRESENTATION EXAMPLES

```
      }
    }
  }
  *probability = prob;
}
/*  Simulate shaping                                        */
int shaping (resp, rein, rew, crit, probability)
int *resp, *rein, *rew, *crit;
double *probability;
{
  if (*rein < 20)
  { *resp = *resp - 10;
    *probability = *probability - 0.1;
    if (*probability < 0.0) *probability = 0.0;
  }
  else
  { if (*rein > 40)
    { *resp = *resp + 10;
      *probability = *probability + 0.1;
      if (*probability > 1.0) *probability = 1.0;
    }
  }
  *rein = 0;
}
```

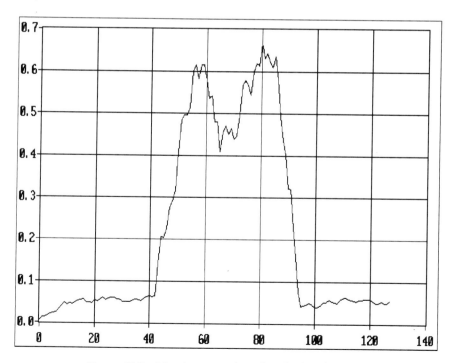

Figure 11.7 Muscle contraction after absolute integration.

Running this program resulted in the following output for the first 20 tracking intervals. Notice from the program that a tracking interval consists of 50 repetitions (using index) of the procedure shown on the flow diagram. The outputs are then printed and the behaviour shaping procedure is carried out before tracking resumes. Both the response and criterion have been normalized to a scale between 0 and 100, and initialized to 5. The results indicate that learning has indeed occurred, resulting in a response that tracks increasing the criterion as the task is made more difficult. After 20 tracking intervals and beyond, both the response and criterion are hovering near their maximum values of 100. This does not imply an 100% correct response as noted by the other results. The probability of a correct response, initialized to 0.25, eventually settles within the range 0.6 to 0.8. Both reward and reinforcement continue to be given and taken away, as their values are not approaching a steady state.

```
Press return to continue, 'x' to stop
response  probability  reward  criterion  reinforce
2         0.43         2       4          18
5         0.48         1       6          15
7         0.54         1       7          16
6         0.59         0       9          15
14        0.64         7       15         15
11        0.71         0       11         17
23        0.72         9       24         11
27        0.76         9       28         14
43        0.74         17      45         8
59        0.68         9       60         4
65        0.69         4       66         11
71        0.7          7       76         11
77        0.75         8       79         15
83        0.76         3       85         11
93        0.79         12      95         13
100       0.75         6       100        6
100       0.78         9       100        13
99        0.74         7       100        6
99        0.69         12      100        5
96        0.69         8       99         15
```

11.3 DATA-ACQUISITION EXAMPLES

This section will present an overview of a data-acquisition board, and show how to program it within the LabWindows environment to input, process, and display data. The use of panels, created with the aid of the User Interface Editor, is explored as well.

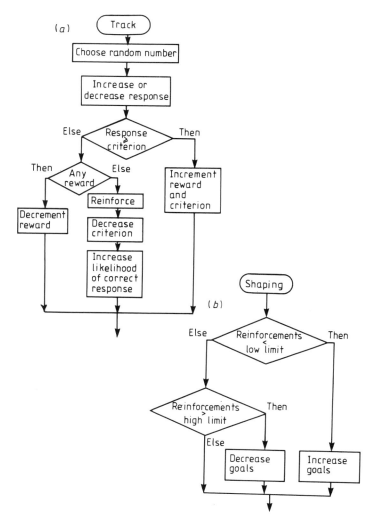

Figure 11.8 Flow diagrams of (a) the tracking procedure and (b) the shaping procedure.

11.3.1 Data-acquisition hardware

The DAS-16 PC expansion board, manufactured by MetraByte Corp., has the following features:

- Sixteen multiplexed channel inputs (used as 16 single-ended or 8 differential channels) to a 12-bit A/D converter.
- Switch-selectable unipolar or bipolar voltage ranges. Maximum ranges are +10 V (unipolar) or −10 V to +10 V (bipolar).

- A/D conversion initiated by software, an internal programmable interval timer (counter), or an external trigger. Converted data transfer by program transfer, interrupt, or DMA.
- Two independent 12-bit D/A channel outputs, using an on-board or external reference voltage.
- 4-bit digital input and 4-bit digital output.

The A/D conversion time of 12 μs results in a maximum throughput of 50 kHz in DMA mode (100 kHz for the faster DAS-16F board). The I/O board contains address decoding, data buffers, and interrupt/DMA control logic interfacing to the PC's bus. The DAS-16 requires 16 successive address locations in the PC's expansion I/O space (Hex 200–3FF). To avoid conflict with the PC's standard I/O devices, the base address was set at hex 330 via an on-board base address DIP switch.

The portion of the A/D address map relevant to our examples is shown below.

Location	Function	Type
Base addr.+0	A/D low byte	In
	Start A/D conversion	Out
Base addr.+1	A/D high byte	In
Base addr.+2	Mux scan control	In/Out
Base addr.+3	Digital I/O	In/Out

The channel 0 and channel 1 D/A outputs use the next four locations, with the remaining addresses reserved for status, control, and counters. The 12-bit digital data format used in conjunction with the A/D or D/A converter is shown below:

```
-----------------------------------------------------------
 D11 D10 D9 D8 D7 D6 D5 D4    D3 D2 D1 D0  x x x x
-----------------------------------------------------------
     Base addr.+1 (high byte)     Base addr.+0 (low byte)
```

The low 4-bits (xxxx) are arbitrary when writing to the D/A converter, but identify the channel number (0 to 15) when read from the A/D converter.

11.3.2 Programming the data-acquisition board

Using the addressing and data format information, we can easily communicate with the DAS-16 by using the 8-bit input and output instructions available in BASIC or from the standard C library. For example, to output an analogue voltage through the channel 0 D/A converter requires only three lines of code. In interpretive BASIC the code below outputs a voltage near the middle of the preset range.

```
10 BASADR% = &H330          ' Set base address
20 OUT BASADR%+5, &H7F      ' Write high byte to D/A
30 OUT BASADR%+4, &HF0      ' Write low byte to D/A
```

DATA-ACQUISITION EXAMPLES   361

Note: &H is a hex value; and % shows an integer variable.

The 12-bit digital value is hex 7FF, where the maximum analogue voltage corresponds to hex FFF.

Converting analogue input to digital requires two steps after setting the base address, before the A/D output is read. Since the A/D channels are multiplexed, a scan register at base addr.+2 must be written to set the channel scan limits. For example, if hex 84 is written, channels 4 to 8 are scanned in succession. In our examples we will only be inputting data through channel 0, so we will write all zeros to the scan register. A write of any number to base addr.+0 then initiates the A/D conversion. The following lines of BASIC code read a single converted analogue value from channel 0.

```
10  BASADR% = &H330         ' Set base address
20  OUT BASADR%+2, 0        ' Scan channel 0 only
30  OUT BASADR%, 0          ' Start A/D conversion
40  HI% = INP(BASADR%+1)    ' Read high byte
50  LO% = INP(BASADR%)      ' Read low byte
```

The program below takes the high and low bytes read in and converts them to the actual voltage within the preset range. The incoming signal is then plotted on the screen continuously.

```
10  BASADR% = &H330            ' Set base address
20  OUT BASADR%+2, 0           ' Scan channel 0 only
30  SCREEN 2                   ' 640 x 200 graphics mode
40  CLS                        ' Clear screen
50  LINE(0,100)-(639,100)      ' Draw time axis
60  FOR X=0 TO 639             ' Go across screen
70     OUT BASADR%,0           ' Start A/D conversion
80     HI% = INP(BASADR%+1)    ' Read high byte
90     LO% = INP(BASADR%)      ' Read low byte
100    NUM%= (256*HI% + LO%)/16 ' Convert to single value
110    V = (NUM%/2^12)*10 - 5  ' Scale to -5v to +5v range
120    PSET (X,20*(5-V))       ' Plot one point
130 NEXT X
140 GOTO 40                    ' Erase screen and re-plot
```

Within the program loop starting at line 60, the two bytes are brought in (lines 70 to 90) and converted to an unsigned 12-bit value (line 100). The high and low bytes are joined by multiplying HI% by 256 and adding LO%. Dividing the result by 16 then shifts the result four bits to the right, thus discarding the four bits identifying the channel number and leaving the correct 12-bit data value. Since HI% is treated as an unsigned number, dividing it by 2^{12} and multiplying

362 LABWINDOWS

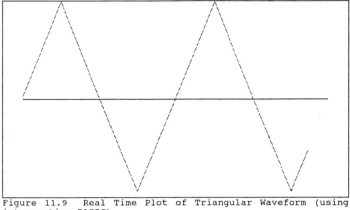

Figure 11.9 A real-time plot of a triangular waveform (using Interpretive BASIC).

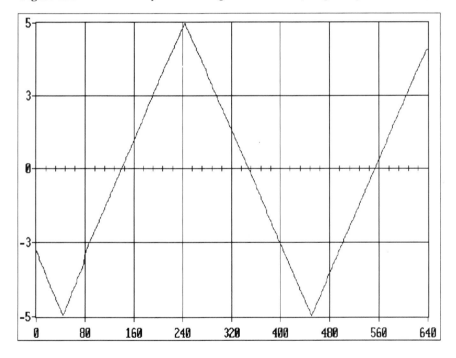

Figure 11.10 A triangular waveform plot (using LabWindows graphing).

it by 10 (line 110) scales the value to the range 0 to 10 V. Subtracting 5 then adjusts the scale to the bipolar range desired.

The PSET statement (line 120) sets a single pixel in PC BASIC. Since the high-resolution graphics mode has been selected, the voltage values need to be converted to vertical pixel positions 0 to 199 from top to bottom of the screen. The calculation 20*(5-V) plots large positive voltages near the top of the screen and large negative values at the bottom, to create a realistic looking graph of the waveform.

To test this program and subsequent ones described in this section, the output of a standard three-function signal generator was connected to the DAS-16 board through input channel 0. A choice of sine, triangular, or square-wave function is provided. The board was installed in an 80486-based PC running at 25 MHz. Despite the power of the CPU, the execution time of the BASIC interpreter limits the signal frequency to about 10–20 Hz. For a compiled language the A/D conversion time of 12 μs may exceed instruction times. In this case the following line of code should be inserted:

```
75 IF INP(BASADR%+8) >= &H80 GOTO 75
```

The most significant bit in a status register at base addr.+8 is an end-of-conversion (EOC) indicator. If EOC = 1 the A/D converter is busy, and when EOC = 0 the converted data are valid and the next conversion can be initiated. The line of code results in waiting until EOC = 0 before proceeding to the next instruction.

Figure 11.9 shows a PC screen capture while the test program is running for a triangular waveform at approximately 10 Hz. The amplitude and DC offset of the signal generator were set to give us a triangle between about −5 V and +5 V, thus utilizing the full screen for the display.

A similar C language program to input and display channel 0 data has been created within the LabWindows environment. The program shown in LabWindows example 6 differs from the data-acquisition program written in interpretive BASIC in one important respect. In place of a continuously running real-time plot, here the data values read in are saved in an array. The YGraphPopup function used in previous examples replaces BASIC's screen mode set (SCREEN) and point set (PSET) statements, and plots the array values all at once. The graph produced by the program listing below is shown in figure 11.10.

```
/* LabWindows Example 6 - Inputting data from DAS16 board    */
main()
{
   int     basadr, n, i, lo, hi;
   double  num, voltage[640];
   cls();
   basadr = 0x330;
   outp (basadr+2, 0);
   n = 640;
   for (i = 0; i <= n-1; ++i)
      { outp (basadr, 0);
```

```
            hi = inp(basadr+1);
            lo = inp(basadr);
            num = (256.0 * hi + lo) / 16.0;
            voltage[i] = (num / 4096) * 10.0 - 5.0;
         }
   /* Graphing the waveform                                          */
      YGraphPopup (voltage, n, 4);
   }
```

Saving the voltage values in an array has the advantage of allowing us to process the data without interrupting the collection of those values. For example, the discrete Fourier transform (DFT) of the waveform can be computed. LabWindows example 7 illustrates the capturing of the data and calculation and display of the magnitudes of the DFT frequency components. For a full discussion of the DFT see section 8.5.

```
/* LabWindows Example 7- DFT on Data Input from DAS16 board
*/
main()
{
   int     basadr, n, i, lo, hi, k;
   double  num, pi, real, imag, voltage[640], dftmagn[640];
   cls();
   basadr = 0x330;
   outp (basadr+2, 0);
   n = 380;
   for (i = 0; i <= n-1; ++i)
      { outp (basadr, 0);
         hi = inp(basadr+1);
         lo = inp(basadr);
         num = (256.0 * hi + lo) / 16.0;
         voltage[i] = (num / 4096) * 10.0 - 5.0;
      }
/* Calculating and plotting the DFT magnitude                    */
   pi = 3.14159;
   for (k = 0; k <= 16; ++k)
      { real = 0;
         imag = 0;
         for (i = 0; i <= n-1; ++i)
            { real = real + voltage [i] * cos (2*pi*i*k/n);
               imag = imag + voltage [i] * sin (2*pi*i*k/n);
            }
         dftmagn [k] = sqrt (real * real + imag * imag);
      }
   YGraphPopup (voltage, n, 4);
   YGraphPopup (dftmagn, 16, 4);
}
%
```

Since the DFT is a complex operation, the real and imaginary parts are computed separately and then used to find the magnitude. The value of n has been reduced to give us enough samples for approximately one period of the waveform (the frequency was left unchanged from the previous example). In this way spectral 'leakage', as described in section 9.3.2, is avoided. Also, only the first 16 frequency harmonics are calculated to reduce the computation time and enhance the plot. Higher-frequency harmonics have very small amplitudes anyway.

Figures 11.11 and 11.12 show the waveform and magnitude spectrum for a square-wave input from the signal generator. As expected, the odd harmonics are approximately in the ratio 1, 1/3, 1/5, 1/7, ... and the even harmonics vanish.

11.3.3 Strip-chart recording of simulated temperatures

As we have seen from numerous examples, the pop-up panels available with the User Interface Library of LabWindows give us a simple tool for graphing array data. The predefined pop-up panels, however, restrict us to line plots and have no provision for labelling the axes or providing a title for the graph. Further, they are not intended for real-time data presentation. Pop-up panels are only one of many components of a graphical user interface used with an application program. With the aid of the User Interface Editor a resource file can be created with such objects as panels, controls, and menu bars. Examples of controls include push buttons, LEDs, graphs, and strip charts.

LabWindows example 8 uses the code from example 3 to approximate a normal distribution of 1000 temperature readings with a mean of 70 °F. The statistical analysis and graphing done in example 3, however, are replaced by code that displays a strip-chart recording of the data.

```
/* LabWindows Example 8 - Strip Chart of Simulated      */
/*    Temperature Readings                              */
main()
{
  int n, i, j, handle, strip, k;
  double tempbase, temptop, sum;
  double temp[1000];
/* Initialize parameters                                */
  cls();
  n = 1000;
  tempbase = 40.0;
  temptop = 100.0;
/* Generate normally distributed temperatures           */
  for (i = 0; i <= n-1; ++i)
    { sum = 0;
      for (j = 1; j <= 10; ++j)
        sum = sum + rand()/32767.0;
```

```
            temp[i] = tempbase + sum * (temptop - tempbase) / 10.0;
        }                                                                       */
   /* Plot strip chart
        handle = LoadPanel ("test.uir", strip);
        DisplayPanel (handle);
        PlotStripChart (handle, strip, temp, n, 0, 0, 4);
        getkey();
   }
```

The strip chart is a control object contained in a region of the screen called a panel. Separate from the program, the Create menu of the User Interface Editor is invoked and the panel and strip chart are set up and saved in the resource file "test.uir". The LoadPanel function loads the panel from the specified resource file and returns an integer "handle" used to reference the panel in subsequent functions. The integer variable "strip" is a defined constant assigned to the panel in the User Interface Editor.

When the program runs, the panel is displayed and the strip chart plotted as a result of the next two function calls. The n points of the array of temperatures (temp) are plotted. The two parameters following n in the PlotStripChart function enable one to start the plot with any array index and plot non-successive values. Since the starting and skip values are both zero, all of the data points are plotted. The getkey function freezes the action after all the data are displayed, until a key is pressed to terminate the program.

Although this example demonstrates a strip-chart recording, the program clearly arrays all the data first before presenting them. Capturing and presenting data 'on the fly' will be treated in the next example. Figure 11.13 is a 'snapshot' of the strip chart after the program has filled the array and displayed all the values, just before pressing any key to terminate the program. In contrast to figure 11.3 where all the values were displayed via a pop-up panel, the strip chart only displays a preset number of data points at any one time (100 for this example). The strip chart scrolls across the screen, a fact not apparent when viewing a single snapshot.

11.3.4 A control panel for data acquisition and presentation

LabWindows example 9 combines data acquisition, real-time strip-chart display, and DFT computation and graphing. Control buttons are provided to start and stop data acquisition, to calculate and display the DFT of the most recently acquired data samples, and to exit the program. Similarly to what was done in example 8, a panel was created with the User Interface Editor and saved in the file "p1.uir". The panel, shown in figure 11.14, contains six control objects—a strip chart, graph, and the four control buttons placed near the bottom of the panel. Notice that the graph is a panel object here, in contrast to a separate pop-up panel as used in previous examples.

DATA-ACQUISITION EXAMPLES 367

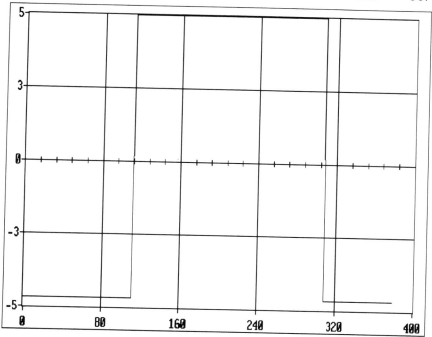

Figure 11.11 A plot of one period of a square waveform.

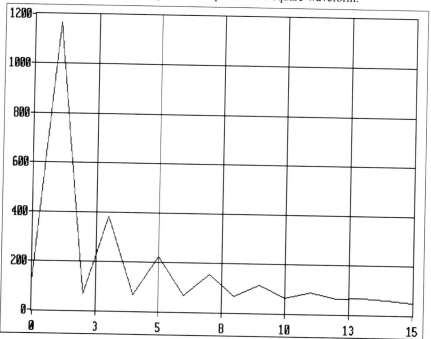

Figure 11.12 The DFT magnitude for a square waveform.

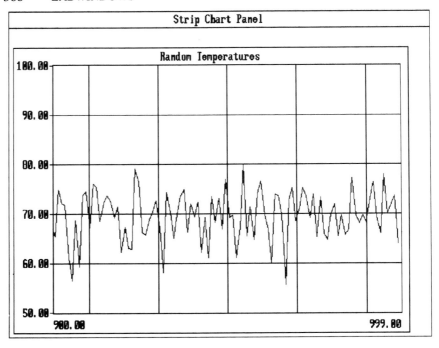

Figure 11.13 A strip-chart recording of simulated temperature readings.

```
/* LabWindows Example 9 - Panel with Strip Chart, Graph,   */
/*      and Button Controls                                */
#define  p1            0
#define  p1_sc1        0
#define  p1_START      1
#define  p1_STOP       2
#define  p1_quit       3
#define  p1_DFT        4
#define  p1_dftgraf    5
#define  pi            3.14159
main ()
{
    double voltage[640], dftmagn[64];
    double begintim, num, real, imag;
    int ctl, handle, basadr, n, STOPflag;
    int hi, lo, i, j, k;
/*  Panel Display and Initialization                       */
    ctl = -1;
    handle = LoadPanel ("p1.uir", p1);
    DisplayPanel (handle);
    SetInputMode (handle, p1_STOP, 0);
    cls();
```

DATA-ACQUISITION EXAMPLES   369

```
      basadr = 0x330;
      outp (basadr + 2, 0);
      n = 100;
/* Continue until QUIT button                                     */
      STOPflag = 0;
      while (ctl != p1_quit)
      { GetUserEvent (0, &handle, &ctl);
/* Run strip chart on START button                                */
        if (ctl == p1_START)
        { while (STOPflag == 0)
/* Enable STOP and disable other buttons                          */
          { SetInputMode (handle, p1_STOP, 1);
            SetInputMode (handle, p1_START, 0);
            SetInputMode (handle, p1_quit, 0);
            SetInputMode (handle, p1_DFT, 0);
/* Read and plot n data values at rate of 100 per second          */
/* STOP only after multiple of n values                           */
            for (i=0; i<=n-1; ++i)
            { GetUserEvent (0, &handle, &ctl);
              if (ctl == p1_STOP)  STOPflag = 1;
              begintim = timer();
              outp (basadr, 0);
              hi = inp (basadr + 1);
              lo = inp (basadr);
              lo &= 0xf0;
              num = (256.0 * hi + lo) / 16.0;
              voltage[i] = (num / 4096) * 10.0 - 5.0;
              PlotStripChart (handle, p1_sc1, voltage, 1, i, 0, 4);
              syncwait (begintim, 0.01);
            }
          }
/* After STOP, enable other buttons                               */
          STOPflag = 0;
          SetInputMode (handle, p1_STOP, 0);
          SetInputMode (handle, p1_START, 1);
          SetInputMode (handle, p1_quit, 1);
          SetInputMode (handle, p1_DFT, 1);
        }
/* Plot first 16 components of DFT magnitude on last n data       */
/*   values, on DFT button                                        */
        if (ctl == p1_DFT)
        { for (k=0; k<=16; ++k)
          { real = 0;
            imag = 0;
            for (i=0; i<=n-1; ++i)
            { real = real + voltage [i] * cos (2*pi*i*k/n);
              imag = imag + voltage [i] * sin (2*pi*i*k/n);
            }
            dftmagn [k] = sqrt (real * real + imag * imag);
          }
          DeletePlots (handle, p1_dftgraf);
```

```
            PlotY (handle, p1_dftgraf, dftmagn, 16, 4, 3, 0, 1, 0);
        }
    }
}
```

The initial statements in the program for example 9 assign arbitrary integer constants to the panel and control identifiers. When the panel is created, a constant prefix menu item must be filled. We have chosen this identifier prefix as p1. Likewise each control object is given a constant name, such as sc1 for the strip chart and STOP for the stop button. A library routine such as LoadPanel refers to the entire panel and therefore references p1. The functions SetInputMode, PlotStripChart, DeletePlots, and PlotY operate on a single control referenced by the panel name p1, followed by an underscore and the control identifier. For example, p1_sc1 appears in the PlotStripChart function and p1_dftgraf in the PlotY graphing function.

The control buttons are enabled and disabled by the SetInputMode function. If the mode number following the control button identifier is zero, the button is disabled; otherwise it is enabled. The status of the button is sensed as a result of the GetUserEvent function. The variable ctl, referenced by its address in the function, returns the value corresponding to the identifier of the button clicked. If no button is clicked, ctl is set to 1.

With the background presented, the understanding of the code is relatively straightforward. The value of ctl is set to indicate no button initially and the panel of figure 11.4 is loaded and displayed. As in previous examples, the DAS-16 board's base address and A/D channel range is initialized. The outer 'while' loop, comprised of two inner 'if' blocks, keeps the program running until the QUIT button is clicked.

The first 'if' block, entered by clicking on the START button, acquires n voltage values and displays a real-time strip chart until STOP is clicked. Notice that inside this block only the STOP button is enabled. Figure 11.15 shows the strip chart resulting from a sine-wave input from the signal generator, after START and later STOP have been clicked on. In contrast to the procedure in example 8, where n data values were plotted at once, this strip chart records each value as soon as it is read. The sequence of parameters "voltage, 1, i" in the PlotStripChart function indicate the array name, number of points plotted, and starting index value respectively.

The 'for' loop inside the block continues the strip-chart recording until a multiple of n values have been processed (n being 100 in this example). Additionally, a time delay has been inserted to guarantee that each iteration takes at least 0.01 s. The statement

```
            begintim = timer();
```

at the beginning of the 'for' loop returns the number of seconds elapsed since midnight. The final statement in the loop

DATA-ACQUISITION EXAMPLES   371

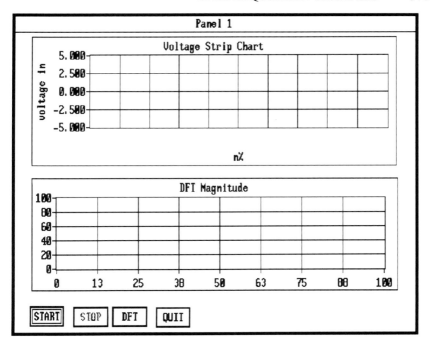

Figure 11.14 A panel created for LabWindows example 9.

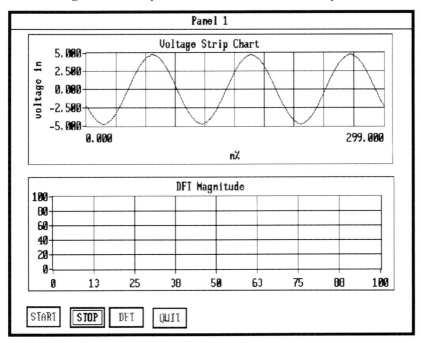

Figure 11.15 A strip-chart recording for a sine waveform.

```
            syncwait (begintim, 0.01);
```
causes the program to wait until 0.01 s have elapsed since "begintim". If the loop takes more than 0.01 s, the syncwait function has no effect. For relatively low-frequency data, the time delay is a useful way of controlling the speed of the strip chart.

After the STOP button has been clicked, the program exits the first 'if' block, whereupon the STOP button is disabled and the other three buttons enabled. A flag bit (STOPflag) remembers a single click of the STOP button. Without this flag the button would need to be clicked continuously until a multiple of n data samples had been read. By placing the GetUserFunction inside the 'for' loop we are sensing the click approximately every 0.01 s. The data continue to be read and plotted, however, until the loop index exceeds the upper limit of $n-1$.

After stopping we have the choice of restarting the strip chart, calculating and displaying a DFT, or exiting the program. In a real application we may need to acquire data continuously while performing the spectral computation. Depending on the data frequencies we may need to use DMA mode as opposed to program control.

The second 'if' block contains the same DFT calculation shown in example 7. The most recent n values acquired are used to compute the first 16 frequency harmonics. Figure 11.16 illustrates the strip chart and spectrum for the square-wave input from the signal generator. In contrast to figure 11.12, the graph is titled and a bar-graph format is used. As indicated earlier, the panel control objects give us flexibility not available for pop-up panels. For this example, the signal frequency was set to be very close to 1 Hz. Since n has been set to 100 and the delay factor is 0.01 s, one full period of the signal is being sampled. In this manner, spectral leakage is minimized.

As in example 7 the odd harmonics are again in the ratio 1, 1/3, 1/5, 1/7, ... and the even harmonics are close to zero. To illustrate the problem with spectral leakage, the program was run again with the value of n changed from 100 to 120. Figure 11.17 shows the result of the new DFT computation on 120 samples. Since the square-wave frequency was left at 1 Hz, the sampling is now done over 1.2 periods of the signal. The non-integral record length results in the distortion of the spectrum.

REFERENCES

[1] National Instruments (Austin, TX) 1991 *LabWindows User Manual*
[2] National Instruments (Austin, TX) 1991 *LabWindows Standard Libraries Reference Manual*
[3] National Instruments (Austin, TX) 1991 *LabWindows User Interface Library Reference Manual*
[4] Silverman G and Dworkin B R 1991 Instrumental behavior shaping automata *Digital Biosignal Processing* ed R Weitkunat (New York, NY: Elsevier Science) ch 12

REFERENCES 373

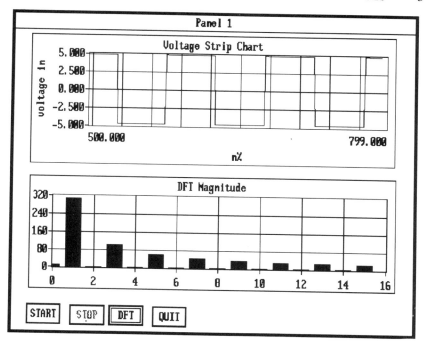

Figure 11.16 The strip-chart and DFT magnitude for a square waveform.

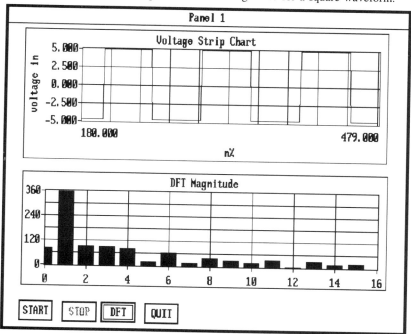

Figure 11.17 The strip chart and DFT for a non-integral number of periods.

12

The Windows Operating Environment

12.1 INTRODUCTION

Microsoft Windows is an extension for MS-DOS that presents to the user a convenient graphical interface to application programs [1, 2]. DOS's memory limitation and its lack of graphics support have contributed to the growing popularity of Windows. Even Version 5 of DOS still limits programs to 640 kbyte of RAM, whereas Windows can implement as much RAM as is available. In fact Windows can simulate virtual memory and allow applications to send parts of programs to disk when no more RAM is available. Although critics of Windows have claimed that DOS-based applications run faster, newer 386- and 486-based PC's have little problem with Windows applications. The multitasking capability of these processors and the absence of DOS's memory limitation permit several programs to run at a given time. Only one program, however, is active with control of the keyboard. As an example, data acquisition and analysis can be performed simultaneously with one program capturing the data while a spreadsheet processes the data and presents the results.

Windows uses graphical symbols called icons to represent files, programs, a cursor, and menu options. Although the Windows environment can be explored via the keyboard, use of a mouse is more convenient. Programs designed to run in this environment are called Windows applications, which can be run only after Windows is first loaded. As Windows' popularity grows, more software vendors are releasing Windows versions of their products. For example, Windows versions of both Lotus 1-2-3 and MathCAD are available. We will shortly present an example of communication between these programs to illustrate an advantage of the Windows environment.

12.2 WINDOWS APPLICATION PROGRAMS

The supervisor for using Windows, called the *Program Manager*, is generally invoked when Windows is started. However, Windows can be customized to start with one of its utility programs or with an application program. The

WINDOWS APPLICATION PROGRAMS 375

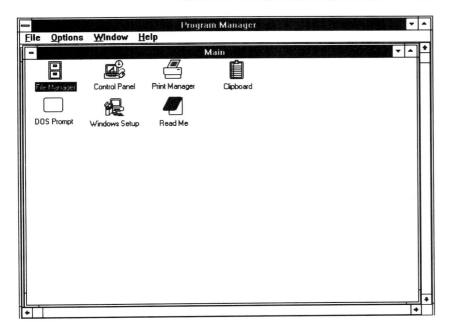

Figure 12.1 The main document window.

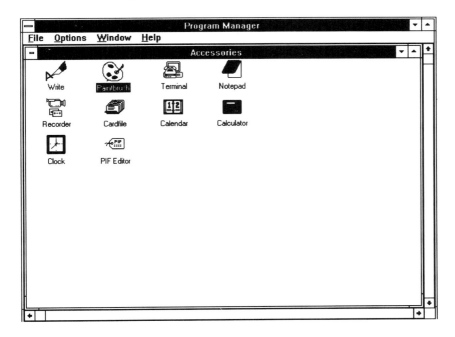

Figure 12.2 The accessories window.

Program Manager partitions programs into groups having their own document windows. Unless customized otherwise, the group opened initially by the Program Manager is the *Main* document window illustrated in figure 12.1. Although windows can be cascaded (placed side by side) or tiled (placed one behind another), we have for the moment sized our windows to cover most of the screen. In this way only one window is visible at a time.

We see that the Main group contains file and print manager programs and a temporary exit to DOS. The *Clipboard* acts as a common area for all programs, and invoking it from the Main document window will display its current content. The Clipboard, generally modified by using an *Edit* menu of Windows application programs, is one mechanism for data exchange to be discussed shortly. To start a program from the Main document window or from any other group simply requires pointing the mouse to the icon associated with that program and 'double clicking' it.

The Program Manager is initially defined with five groups including Main. The others are *Accessories*, *Games* (Reversi and Solitaire are provided), *Windows Applications*, and *Non-Windows Applications*. Applications not designed for Windows can be installed in the last group, but there may be problems with some of them when you try to run them in the Windows environment. The Accessories group shown in figure 12.2 contains several Windows applications provided with Windows itself. These include a simple word processor (*Write*), a drawing program (*Paintbrush*), a communications program (*Terminal*), along with an appointment *calendar*, *clock*, *calculator*, and other *tools*. Paintbrush incidentally was used in conjunction with the Clipboard, to capture and save screen images of the figures used in this chapter.

Figure 12.3 includes other Windows applications including B-Squared Spice (*B2SPICE* icon), Lotus 1-2-3, and MathCAD Version 3.0 (and higher).

12.2.1 B-squared Spice

B-Squared Spice, a product of Beige Bag Software, is a Windows version of PSpice. Rather than creating an input file of circuit description and command lines as required by PSpice, with B-Squared Spice the circuit is constructed graphically and the type of simulation (e.g. DC, AC, transient) is chosen from a menu. As shown in figure 12.4, a blank grid is initially displayed with B-Squared Spice and components are selected from a *Devices* menu. As with PSpice the library includes sources, as well as passive and active components.

When a component is selected a circuit symbol appears on the grid and can be placed anywhere that is desired. The various components can be connected by lines, as illustrated by the DC op amp circuit shown in figure 12.5.

An AC circuit is illustrated in figure 12.6 along with the results of an AC analysis performed. Both magnitude and phase of the output voltage are shown as functions of frequency plotted on a logarithmic scale.

WINDOWS APPLICATION PROGRAMS 377

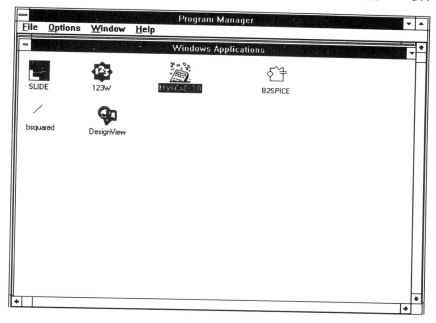

Figure 12.3 Windows applications.

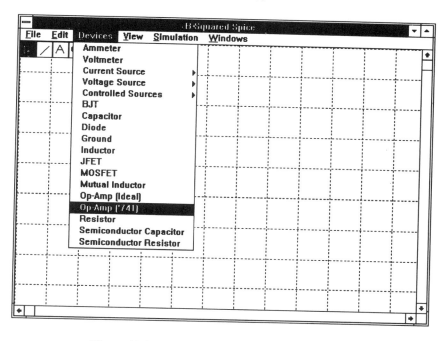

Figure 12.4 The B-Squared Spice devices menu.

378   THE WINDOWS OPERATING ENVIRONMENT

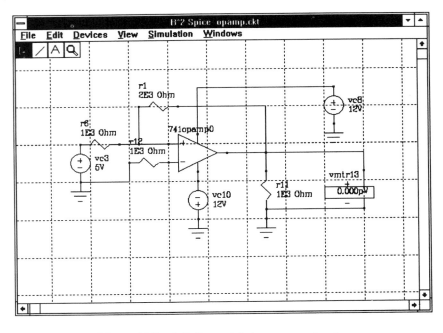

Figure 12.5   A B-Squared Spice op amp circuit.

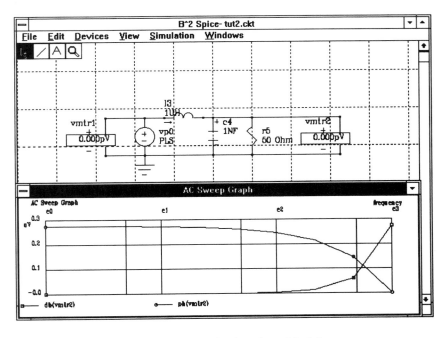

Figure 12.6   A B-Squared Spice AC circuit and graphical frequency response.

12.2.2 Lotus 1-2-3 for Windows

Figure 12.7 shows the screen for the Windows version of 1-2-3, with a worksheet entered to replicate the example in section 10.3. The various 1-2-3 commands are shown at the top, and clicking on any of them with the mouse brings down the appropriate sub-menu. The equation $y = x^2$ is tabulated in the range of x from 0 to 1.

12.3 DATA TRANSFER BETWEEN WINDOWS APPLICATIONS

Data transfer between Windows applications can be performed by using either the Clipboard, file transfer, or via dynamic data exchange (DDE). File transfer was illustrated for non-Windows applications in section 8.8.

12.3.1 Using the Clipboard

We will first illustrate the transfer of data between two applications with the aid of the Windows Clipboard. Since the Clipboard can be used to move either text or graphics, a Lotus graph for the worksheet shown will be transferred to MathCAD for Windows. Figure 12.8 shows the graph created for the simple function $y = x^2$ without title or axis headings. Clicking on the Edit option of the menu and then on *Copy* results in copying the graph to the Clipboard. Recall that the Clipboard is not the exclusive property of 1-2-3 but is shared by all of the applications.

After exiting 1-2-3, MathCAD is next selected from the Windows Applications group. Alternatively, both programs could have been running concurrently as explained previously. Figure 12.9 shows the initial MathCAD blank worksheet. Notice that the icons down the left-hand edge of the screen permit entry of various symbols such as summation and integral signs, without the need to remember the keys defined for those symbols. As with 1-2-3, an Edit menu is available in MathCAD for Windows. Selecting the *Paste* option results in the Clipboard content being transferred to the worksheet at the current cursor position. In figure 12.10 we see that the Lotus graph previously copied to the Clipboard is indeed pasted to the MathCAD worksheet.

12.3.2 File transfer

We will next illustrate file transfer, a technique that can be used with non-Windows applications also. In section 8.8 file transfer was illustrated between MathCAD and Lotus. Here we will show how a Lotus worksheet can be imported into another program for the purpose of enhancing graph presentation. The graph enhancement program used is a trial-size version of *SlideWritePlus* for Windows, Advanced Graphics Software, Inc. [3]. When SlideWritePlus is started, the blank worksheet shown in figure 12.11 is displayed. As observed,

380   THE WINDOWS OPERATING ENVIRONMENT

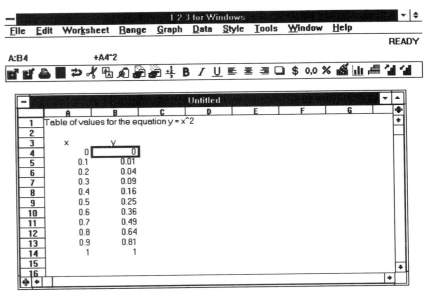

Figure 12.7   1-2-3 for Windows with a table for the equation $y = x^2$.

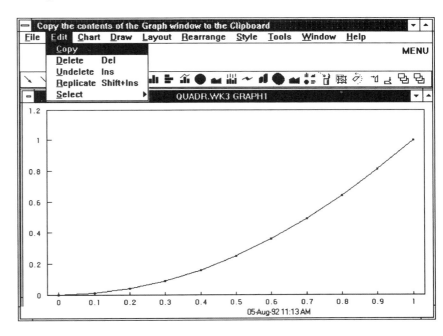

Figure 12.8   A Lotus 1-2-3 graph copied to the Clipboard.

DATA TRANSFER BETWEEN WINDOWS APPLICATIONS 381

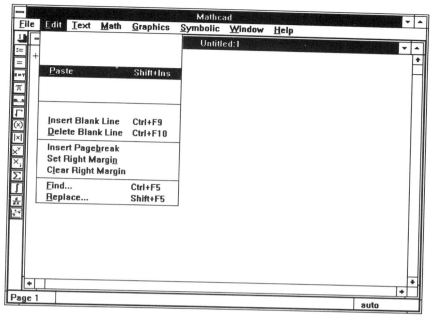

Figure 12.9 MathCAD for Windows with Paste from the Clipboard selected.

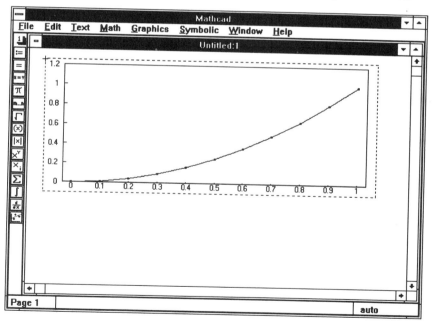

Figure 12.10 A Lotus graph transferred to MathCAD via the Clipboard.

the worksheet contains rows and columns as in a 1-2-3 spreadsheet. Data can be entered manually or via an equation.

As with other Windows applications, selection from the command line (*File*, *Edit*, *Chart*, ...) results in pull-down menus with further choices appearing. To set the stage for importing and graphing Lotus data, the new chart option from the File menu is chosen and *Graph* is selected as the chart type. Then *Import/Link Data* is selected from the File menu, resulting in the pop-up screen shown in figure 12.12 appearing. The simple worksheet for the function $y = x^2$, used in our previous example, was saved in a file quadr.wk3 in the subdirectory \123w\sample. As shown on the pop-up screen, the path and file name are properly specified and *Lotus type* is selected. When the mouse is clicked on OK a Lotus Graph-Import pop-up is displayed, from which Get Lotus Graph is selected. The worksheet then reappears with the file data entered, as shown in figure 12.13.

The graph of the function, shown in figure 12.14, is displayed when selecting the *Redraw chart* option of the *View* menu. The range and subdivisions of the x- and y-axes are set by default. They will be modified as part of the process of graph enhancement.

The Chart selection from the command line results in the Pull-down menu shown in figure 12.15 being displayed. As shown, the graph type can be changed, axes can be scaled as desired, and titles and legends can be added. Some of these features could have implemented with 1-2-3 instead.

The final enhanced graph is shown in figure 12.16. The X Axis Scaling and Y Axis Scaling pop-ups permit entering of minimum and maximum values, and selecting numbers of major and minor divisions. We have scaled both axes to the range 0 to 1 and to 10 major divisions, with grid lines displayed. Selecting *Series* and *Legends* results in a pop-up permitting colour selection and other options being displayed. A *Print Table* option was chosen, resulting in the corresponding x- and y-values being displayed below the graph. A title was inserted by choosing *Titles* from the Chart pull-down menu, and the *Draw Toolbox* shown to the left of the graph was used to insert additional text and an arrow pointer as well.

12.3.3 Dynamic data exchange

Use of the Windows Clipboard and file transfer permits transfer of blocks of data or graphics, but these techniques are limited by their inability to update data that are changing 'on the fly'. Windows incorporates a third very powerful technique called dynamic data exchange (DDE) for this purpose. Under DDE two application programs are linked together, one acting as a server and the other as a client. When the link is active, a change in the server's data results in the client's data being immediately updated.

A spreadsheet program such as the Windows version of Lotus 1-2-3 can function as a client with a link set up to a server. The DDE technique will be

DATA TRANSFER BETWEEN WINDOWS APPLICATIONS   383

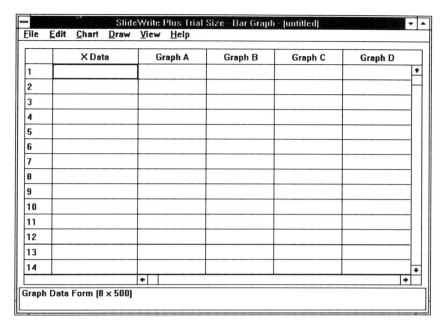

Figure 12.11 SlideWritePlus for Windows Worksheet.

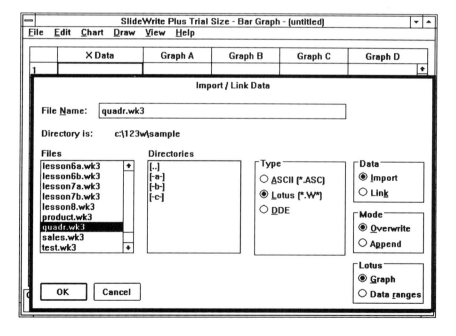

Figure 12.12 The Import/Link Data pop-up screen.

384 THE WINDOWS OPERATING ENVIRONMENT

	X Data	Graph A	Graph B	Graph C	Graph D
1	0	0			
2	0.1	0.01			
3	0.2	0.04			
4	0.3	0.09			
5	0.4	0.16			
6	0.5	0.25			
7	0.6	0.36			
8	0.7	0.49			
9	0.8	0.64			
10	0.9	0.81			
11	1	1			
12					
13					
14					

Figure 12.13 A SlideWritePlus worksheet with 1-2-3 data imported.

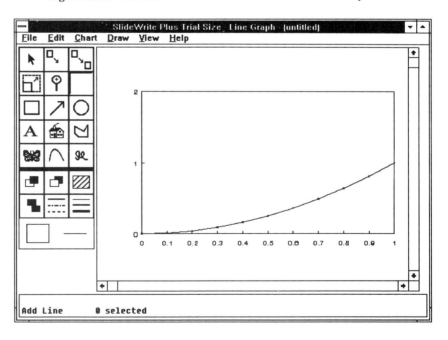

Figure 12.14 Lotus data graphed by SlideWritePlus.

DATA TRANSFER BETWEEN WINDOWS APPLICATIONS 385

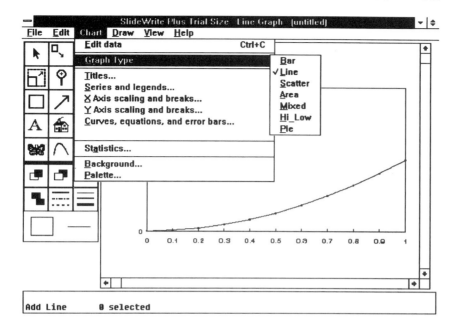

Figure 12.15 A chart pop-up menu for graph enhancement.

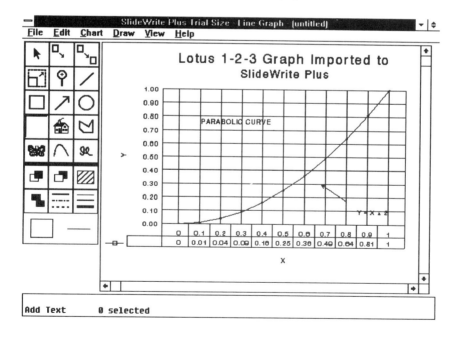

Figure 12.16 A SlideWritePlus-enhanced graph.

illustrated with the aid of a *Demonstration Server* driver program, written in a Windows version of BASIC called Visual Basic. The Demonstration Server can be linked to 1-2-3 by selecting *Link Options* from the Edit menu item, as shown in figure 12.17. The worksheet is blank except for arbitrary titles for the linked data, inserted into cells A2 and A4.

Choosing Link Options results in the pull-down menu shown in figure 12.18 appearing. Separate links can be created for different data items supplied by the server. The link information can be edited as well as temporarily deactivated or permanently deleted. For this example two links are established to the Demonstration Server. Establishing a link requires that both client and server be identified with an Application file name and a *Topic* name. For this example the Application file is an executable file named DEMO81S, and the Topic name is DemoServer1. A separate Item name identifies the data item unique to each link. For LINK2, as shown in figure 12.18, the Item named DataDisplay is linked to cell C4 in the worksheet. LINK1 was set up with the Item name Textbox and is linked to cell C2.

When the Demonstration Server is run, a graphical user interface is displayed. In both figures 12.19 and 12.20 window sizing is used to allow display of both the server and the client (Lotus). The *Data Display* item displays time of day in a numerical format (two digits each for hours, minutes, and seconds). *Clock Display*, to be utilized shortly, displays time of day in a text format. Both items are controlled by the ON and OFF buttons. The Textbox allows entry of character strings and is not subject to button control. In figure 12.19 the OFF button was last pressed and the Textbox and Data Display items are shown transferred to the worksheet. If the Textbox is changed in the Server the change immediately appears in the spreadsheet. After clicking the Server's ON button the Data Display number changes every second in both server and client.

In the final example we will use the changing data from the Demonstration Server to vary the phase of a sinusoidal waveform created with the spreadsheet. With the link active, the waveform can be graphed and continually updated as the linked data change. Figure 12.21 shows a portion of the modified worksheet along with the Demonstration Server in the clock OFF state. The two previous data links have been replaced by a single one linking the server's ClockDisplay item to cell C1. The following numbers and formulas were entered and copied to the ranges shown.

Cell	Formula	Copied to
C2	@SECOND(C1)	---
A5	0	---
A6	+A5+1	A7..A69
C5	@SIN((A5+C$2)*@PI/30)	C6..C69

The seconds value from the linked ClockDisplay data, returned to cell C2, increments by one every second and resets to zero after reaching 60. The

DATA TRANSFER BETWEEN WINDOWS APPLICATIONS 387

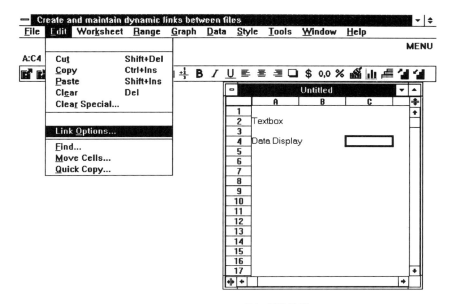

Figure 12.17 Setting up DDE links in Lotus 1-2-3.

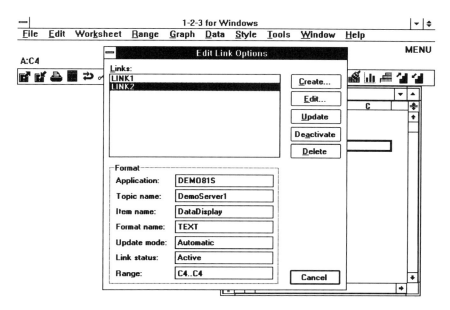

Figure 12.18 The Edit Link options menu.

388   THE WINDOWS OPERATING ENVIRONMENT

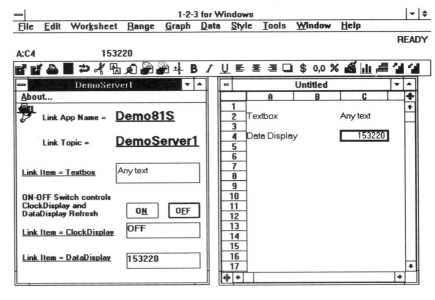

Figure 12.19   The Server linked to the Spreadsheet (`clock OFF`).

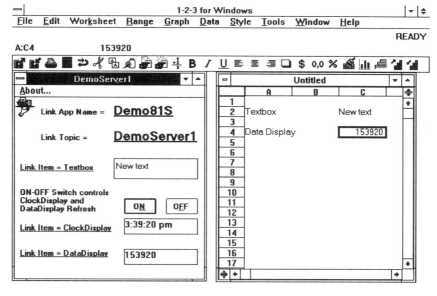

Figure 12.20   The Server linked to the Spreadsheet (`clock ON`).

DATA TRANSFER BETWEEN WINDOWS APPLICATIONS 389

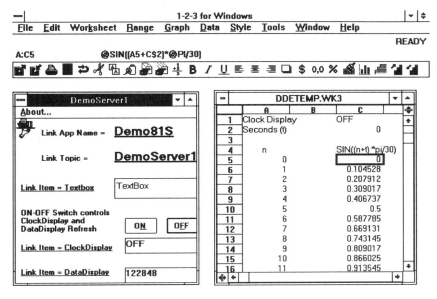

Figure 12.21 A sinusoid with the phase dependent on Linked Data (clock OFF).

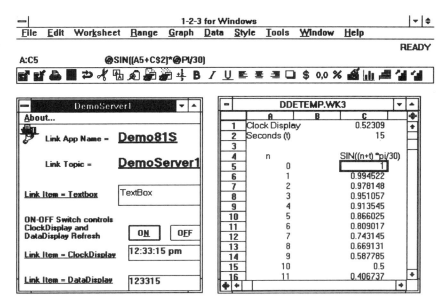

Figure 12.22 A sinusoid with the phase dependent on Linked Data (clock ON).

390 THE WINDOWS OPERATING ENVIRONMENT

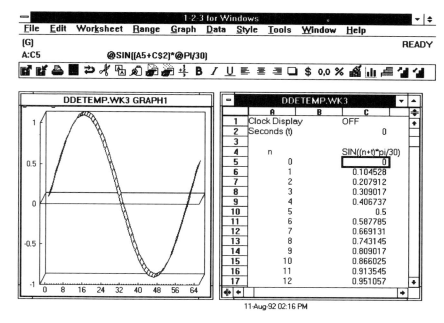

Figure 12.23 A graph of a sinusoid (clock OFF).

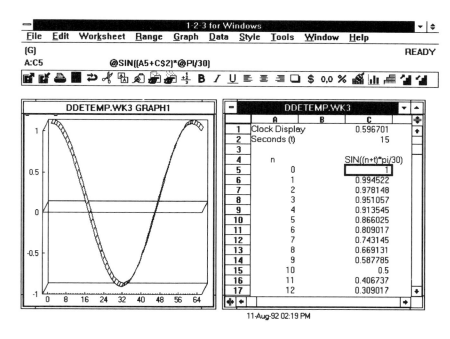

Figure 12.24 A graph of a sinusoid (clock ON).

dynamically changing value in this cell is referenced as C$2 in the sinusoid formula (recall from the discussion in section 10.3 that $ fixes the cell number when the formula is copied). Therefore, as C1 changes dynamically, so does C2 as well as C5 to C69.

Figure 12.22 is a snapshot of the same screen as shown in figure 12.21, after the Demonstration Server's ON button is clicked. The ClockDisplay item is displayed as a fraction of the day in the spreadsheet (cell C1), but the numeric value of seconds is returned as expected (cell C2). Notice that the value of 15 s results in a 90° phase shift of the sinusoid as compared to that for 0 s.

Another advantage of DDE is that a graph can be displayed and observed to change dynamically as long as the link remains active. A Lotus graph was next created for the worksheet, plotting the sinusoid as a function of n, the point number. The initial graph is shown alongside the worksheet in figure 12.23, for the ClockDisplay OFF case. After the Demonstration Server's ON button is clicked and the Lotus graph window is restored, the phase of the sinusoid is observed to change dynamically. The effect is that of a relatively slowly moving strip-chart recording. In figure 12.24 the graph is observed to have shifted by 90° for a 15 s offset, as expected.

REFERENCES

[1] Townsend C 1990 *The Best Book of Microsoft Windows 3* (New York, NY: SAMS)
[2] Everex Systems Inc. (Fremont, CA) 1990 *Microsoft Windows 3.0 User's Guide*
[3] Wittering N 1991 *A Guide to Slide Write Plus* (Carlsbad, CA: Advanced Graphics Software Inc.)

13

Fully Integrated Applications

13.1 INTRODUCTION

This chapter presents the results of two practical applications, which depict how: (1) a spreadsheet may be used to acquire data from an experiment; and (2) LabWindows was used to make a laboratory measurement. These experiments represent the level of maturity that can currently be found in laboratory environments. They make extensive use of prewritten software procedures (libraries) which can be found in many software/hardware support packages. Using such routines a user may develop user-friendly application programs that are essentially menu driven; the need for extensive HLL or assembly language programming is greatly reduced. Computer-based instrument configurations for experimental or production applications can be developed in an efficient manner.

In both examples, a PC with the following general characteristics is recommended:

- *Processor*: 386 or later (486 preferred).
- *Operating system*: DOS 3.0 or later.
- *Main memory*: 4 Mbyte or more (2 Mbyte minimum).
- *Secondary storage*: hard disk, 40 Mbytes or more; additional floppy disk recommended.
- *Extended memory*: 2 Mbyte or more.
- *Pointing device*: mouse strongly recommended.

13.2 DATA ACQUISITION USING LOTUS AND LOTUS MEASURE

The important need in laboratory and process control environments is the requirement to acquire data in a specified time interval (real time). In this example, spreadsheet software has been applied directly to the acquisition of data from an optical instrument. In particular, the software packages used were Lotus and Lotus Measure (see chapter 10).

In this example, the PC instrument was used to interface with a data-acquisition board to acquire data from an optical instrument that measures the

DATA ACQUISITION USING LOTUS AND LOTUS MEASURE

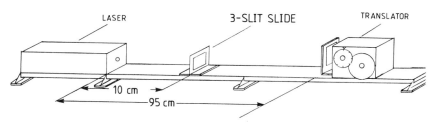

Figure 13.1 Optical apparatus for laser light measurements.

light intensity from a laser source. The optical apparatus consisted of a sliding stage with an optic fibre cable attached to the stage and it is shown in figure 13.1. Using appropriate macros, the instrument can receive and process data from the external environment.

13.2.1 Hardware configuration

A PC with facilities similar to those cited above was the basic platform. To that system a data-acquisition facility was added—a Metrabyte 16-channel high-speed analogue and digital interface (DAS-16). This self-contained board is added to the PC and functionally is connected to the system bus; it is plugged into one of the available slots in the PC. The use of such resources as the DAS-16 enables the PC to be a fast high-precision data-acquisition and signal analysis instrument. The DAS-16 facility includes an interface box and ribbon cable for convenient connection to the experimental set-up. This is shown schematically in figure 13.2.

An overview of the features of the data-acquisition board used here was presented in section 11.3.1. More detailed features and characteristics include the following:

- *Number of data channels*: provision for accepting either 8 or 16 channels of information from an experiment; if two amplifiers internal to the DAS-16 are used to provide true differential signal detection, then only eight data channels are available.
- *Range of input signals*: unipolar operation (0 to either 1, 2, 5, or 10 V); bipolar operation (± 0.5, ± 1, ± 2.5, ± 5, ± 10 V); user selected (user may change a resistor to obtain other ranges).
- *Sampling rate*: determined from software command (50 000 samples s^{-1} maximum).
- *Software configuration*: five switches on the board determine the following software-related parameters:
    base address: the address of the board within the computer's memory address space; this permits more than one data-acquisition board to be used with the same computer;
    channel configuration: either 8 or 16 channels;

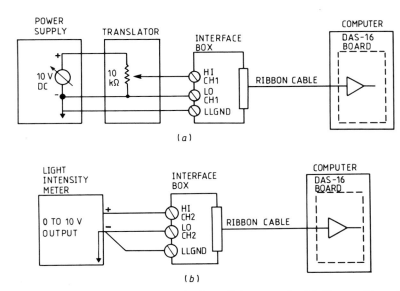

Figure 13.2 Interface hardware for the laser light example. (a) The position circuit connection. (b) The light intensity.

Table 13.1 DAS-16 hardware switch configurations.

Switch	Setting
Base address	200 (hexadecimal)
Channel configuration	8 channels
Mode	Bipolar
DMA select	1
Gain	±10 V

mode: unipolar/bipolar;

DMA level select: priority of data transfer between memory and external devices;

gain: sets the input voltage range in conjunction with the mode switch.

Table 13.1 indicates the settings for this example.

13.2.2 Software considerations

Two software application programs (packages), Lotus and Lotus Measure (Measure) were used to interface with the DAS-16 board. Measure permits the user to acquire data from the DAS-16 board and direct it into a Lotus spreadsheet without having to enter the data manually. As shown in chapter 10, Lotus is a spreadsheet with capabilities including data calculation and tabulation, and graphical presentation, for analysis of system operation.

The menu incorporates the menu structure, macro functions, and help facility that is found in Lotus. Its commands can be incorporated into user-developed macros. One of its most useful (user-friendly) features is the facility that it provides to permit the user to review or change the acquisition set-up through the device-setting screen. This menu-selected screen can direct the acquisition board to perform real-time conversion of input data directly into engineering units. The Measure software package can handle up to three preset stages wherein the user can arrange for raw data to be stored and tabulated using different parameters. Key menu items that control DAS-16 functions include:

- *ID settings*. This screen display includes several items including ID_name, type of input (type of input associated with the name), board (which DAS board to call in a multiboard environment), channel number (to be read), destination and conversion formula.
- *Stage setting*. This allows three stages of data acquisition and up to 999 loops through one or more of the stages. Each stage has an individual set of data inputs (IDs), sampling rate, number of samples, and trigger. Each stage represents a period of data acquisition. The rate divisor provides a method of applying different sampling rates to each ID within a stage. The different sampling rate for each ID is based on the sampling rate divided by the ID's divisor value.

Tables 13.2 and 13.3 below show the set-up for the current example. The *board channel number* indicates which DAS-16 board to call; the *channel number* indicates which channel is to be read. The PC can have a total of four DAS-16 boards (designated 0 to 3) installed; the channels can then be denoted as 0 to 15 (8 differential or 16 single ended). *Gain*, *range* and *formula* work together when data are being sampled and stored. The gain setting specifies the multiplication that occurs on the analogue voltage before it is converted into a digital quantity. The formula parameter defines an operation to be performed on the data after convertion to digital quantities. The raw data are considered to be integers between 0 and 4095 that must be converted to represent values in volts (or some other appropriate unit) using the assigned formula. The data being sampled are stored in a range of cells. The *range function*—see table 13.2— allocates a range of cells to store the data after they have been sampled, converted to digital quantities (properly scaled) and converted to voltage values. The *view function* specifies that the input values will be viewed while the data are being sampled.

The *trigger* parameter value determines the way in which to start the data acquisition, and there are five ways in which this can be accomplished. Once started, the underlying program is armed for the start and end stage and the number of additional loops. In this case, the trigger type selected is the keyboard and the "GO" command initiates execution of the acquisition sequence.

Table 13.2 ID setting screen parameters for the example.

Item	Position displacement component	Light intensity component
ID name	DOOR 1	DOOR 2
Type	ANALOG	ANALOG
Board number	0	0
Channel number	1	1
Gain	NONE	NONE
View	YES	YES
Range	A1..A100	B1..B100
Size	100	100
Formula	((([]*0.000487805-1)*10/2.5)-0.314	(([]*0.000487805-1)*10)

Table 13.3 Stage-setting screen parameter settings.

STAGE	ONE DIVISOR	ID/GROUP
1	1	DOOR 1
2	1	DOOR 2
SAMPLING RATE	10/SEC	
NUMBER OF SAMPLES	100	
STAGE TOTAL	100	
GRAND TOTAL	200	
TRIGGER	KEYBOARD	
ADDITION LOOP	0	
START STAGE	1	
END STAGE	1	

13.2.3 The laser experiment

Using the facilities described above, an experiment was designed and developed to sample light intensity from a laser source using a three-slit grating (refer to figures 13.1 and 13.2). Figure 13.1 shows the set-up including a He–Ne laser, linear translator and slide holder. The laser is placed at one end of the optical bench and turned on with the shutter covered allowing it to stabilize for 10 min. The fibre optic cable is attached to the translator detector holder and positioned in such a way that it (just) touches the aperture slide. The translator is located on the optical bench 95.0 cm from the laser and the photo aperture is attached to the front holder of the translator.

A three-slit grating, in its holder, is located at 10 cm from the front of the platform. After all components are mounted on the optical bench, the fibre optic cable is aligned so that it senses the light intensity coming from the laser.

DATA ACQUISITION USING LOTUS AND LOTUS MEASURE 397

The sequence for invoking the software consists of the following steps:
1. Invoke the Lotus and Lotus Measure software.
2. "/" key: the *Lotus* menu appears.
3. "F" and "R" keys: retrieves the program to be executed.
4. "Alt-F8" key: calls the *Measure* menu to load the *Stage Module menu*. (This loads the information in table 13.2 which has been programmed and stored prior to the execution of the experiment.) Selecting "stage-settings", "N" (for name), "R" (for retrieve), and typing "Door 2" will call the configuration shown in tables 13.2 and 13.3. This information is needed to specify the devices that are connected and the data to be acquired.
5. "Esc" "Esc" (Escape key depressed twice): leaves the menu and displays the macro program that was chosen, at which point the program is ready to acquire data. (Two macros were written for taking either 20 or 40 samples; they are quite similar and only one is noted, namely the one for 40 samples:

```
\I(NITIALIZ  {LET COUNT,-1}~
   \S(TART)      {APP2}
                 GQ
                 {LET COUNT,COUNT+1}~
                 {PUT C1..C40,0,COUNT,F14}~
                 {PUT D1..D40,0,COUNT,F15}~
   \G(RAPH)      /GXC1..C40~AD1..D40~V
   COUNT         40
   DISPLA        2.7634456718
   INTENS        0.039807388
   \E  (RASE)    /REA1..D200~
```

—see chapter 10 for further details regarding Lotus 1-2-3 macros.)
6. "Alt-S" (starting location of the macro): initiates acquisition.

As noted in table 13.3, 100 samples for each ID are taken at a rate of ten per second. The sampled data include the displacement of the translator stage and the laser's light intensity. (The translator stage is moved by hand as the experiment proceeds.) The sample values are converted using the formulas noted above and stored in cells A1 to A100 (indicating the displacement position) and B1 to B100 for corresponding laser intensity values. Displacement values have been converted to centimetre equivalents by the formula, and the light intensity is converted to volts. Once the data are received and stored, an average is taken for each column and stored in cell rows "C" and "D" and these averages depend on whether 20 or 40 repetitions are used. Once the averages appear, one epoch of the experiment has been completed; the counter is updated and computer awaits the next acquisition command.

The translator is moved to a new position and the complete acquisition cycle repeated by depressing "Alt-S" which initiates the macro. This cycle is repeated for either 20 or 40 epochs, whichever program has been initially selected. The experimental results can be used to establish a relationship (if any) between the laser light intensity and the position of the translator. The three-slit slide should create an interference pattern similar to the one shown in figure 13.3(a). Once the experiment is complete, the user can automatically generate a graph of the results by invoking the graphing macro ("Alt-G"). (Other macros included in the design include: "Alt-I" (initializes the program (epoch) counter to either 20 or 40); and "Alt-E", which erases the data before starting the test.)

13.3 MEASUREMENT OF OPTICAL FIBRE BANDWIDTHS USING LABWINDOWS

This example utilizes the LabWindows software package to control laboratory instrumentation so that the bandwidth of an optical fibre may be measured. As discussed in chapter 11, LabWindows is one example of an integrated instrument software package which, when used in conjunction with an IEEE 488 resource (card) within the PC, is designed to facilitate development of instrument 'drivers' and thereby simplify control of laboratory and industrial instrumentation. The user has a choice of writing the main body of the program in either Quick BASIC or C. LabWindows coding features can be used each time that the operator has to write code in order to communicate with a piece of test equipment over the IEEE 488 link. The drivers are written by selecting various options offered in the menu-driven software.

13.3.1 Description of the experiment

A square wave of known frequency is supplied to a laser source. The laser output is coupled to an optical fibre under test. A photodetector (with appropriate response characteristics) receives the optical pulses, and an A/D converter digitizes the signal. The computer performs a discrete Fourier transform on both the input and output waveforms—one full cycle of the waveforms. From this, the gain (in decibels, dB) can be computed at each harmonic frequency, and the bandwidth (measured to the -3 dB point) of the optical fibre may be determined.

The main body of the program is written in Quick BASIC and the LabWindows software package was utilized to implement all necessary code to communicate with a Tektronix 2400 series oscilloscope. The physical link between the PC and the scope was provided by the IEEE 488 card (National Instruments IEEE 488 PCII card) and associated software. A block diagram of the experiment is shown in figure 13.4 and the complete program listing is provided in figure 13.5.

Important features of the program include the following:
- Inputs:

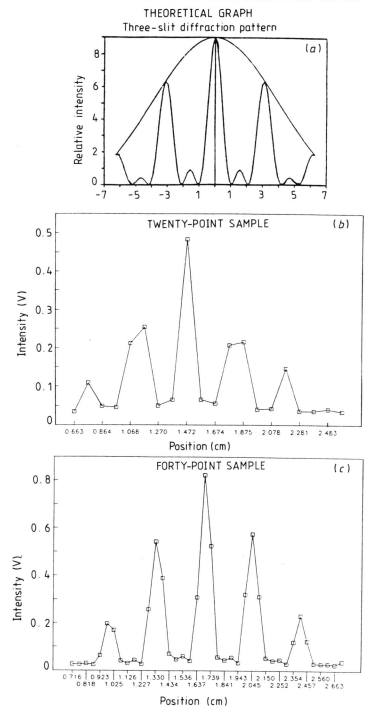

Figure 13.3 *(facing page)* The optical interference pattern produced by the example. (a) The theoretical pattern. (b) The twenty-epoch case. (c) The forty-epoch case has finer positional resolution.

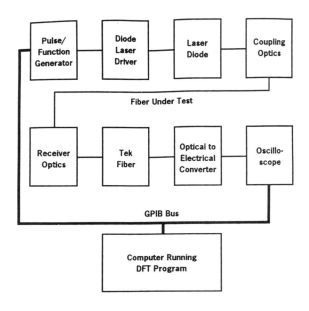

Figure 13.4 A block diagram of the bandwidth measurement example.

1. The operator is prompted to enter the fundamental frequency of the input square wave (2–4 MHz).
2. Software receives the input/output samples from the TEK 2400 Digital Oscilloscope.

- Outputs
    1. Input frequency (MHz).
    2. Number of samples in one (1) period.
    3. Gain versus frequency data.
    4. Bandwidth (dB).
    5. Snapshots of input and output waveforms.
    6. Magnitude versus frequency plots for both input and output DFTs.
    7. A frequency response plot (gain versus frequency).

To execute the program:
- Invoke LabWindows.
- Using either the mouse or ALT + first letter of command.

MEASUREMENT OF OPTICAL FIBRE BANDWIDTHS 401

```
1   REM PROGRAM BNDWDTH.BAS
2   REM LAST UPDATED DEC 8, 1991
3   REM 2:00 PM
4   REM
5   REM THIS PROGRAMS FUNCTION WILL BE TO FIND THE BANDWIDTH OF AN
6   REM OPTICAL FIBER. THE PROGRAM WILL CONTROL A DIGITAL STORAGE SCOPE
7   REM OVER AN IEEE 488 LINK.
8   REM
9   REM ***** INITIALIZATION *****
10  REM ALL VARIABLES AND SUBROUTINES ARE DECLARED
11  REM
12      DECLARE SUB WAVE(W#(),N,START,M)
13      DECLARE SUB DFT(W#(),N,START, MAG#(),FREQ#(),F,PI)
14      M=1024
15      DIM X#(1025),Y#(1025),MAGX#(600),FREQ#(600),TIM#(1025),MAGY#(600),GAIN
    #(600)
16      PI=3.1415926
17  REM
18  REM INITIALIZE SCOPE
19  REM
20  CALL tek2400.init (0)
21  REM
22  REM CONFIGURE CH 1 VERTICAL MODE
23  CALL tek2400.set.vertical.mode (2)
24  CALL tek2400.set.bw (0)
25  CALL tek2400.set.vertical (1, 2.000, 1, 0, 0)
26  REM
27  REM CONFIGURE HORIZONTAL MODE
28  CALL tek2400.set.horizontal (0, 50.0e-9, 50.0e-9, 20.0e-9, 512)
29  REM
30  REM SET CH 1 TRIGGER         5ns,   50ns,   20ns,
31  CALL tek2400.set.a.trigger (2.5, 1, 1, 0, 0, 0)
32  REM
33  REM READ CH 1 (INPUT) WAVEFORM
34  CALL tek2400.read.waveform (1, x#(), t..00002#, t..00003#)
35  REM
36  REM CONFIGURE CH2 VERTICAL MODE
37  CALL tek2400.set.vertical.mode (2)
38  CALL tek2400.set.vertical (2, 0.050, 1, 0, 1)
39  REM
40  REM READ CH 2 (OUTPUT) WAVEFORM
41  CALL tek2400.read.waveform (2, y#(), t..00004#, t..00005#)
42  REM
43  REM CLOSE SCOPE
44  CALL tek2400.close
45  REM
46  REM ENTER INPUT FREQUENCY
47      PRINT "ENTER INPUT FREQUENCY (2 TO 4 MHZ)"
48      INPUT F
49      PRINT " "
50      PRINT " "
51  REM ARRAY TIM#() IS OUR TIME REFERENCE
52      TIM#(0)=0
53      FOR K=1 TO 1024
54          TIM#(K)=TIM#(K-1)+1/(M*F)
55      NEXT K
56  REM
57  REM PERFORM DFT ON INPUT AND OUTPUT WAVEFORMS
58      CALL WAVE(X#(),NX,STARTX,M)
59      CALL DFT(X#(),NX,STARTX,MAGX#(),FREQ#(),F,PI)
```

```
60      CALL WAVE(X#(),NX,STARTX,M)
61      CALL DFT(Y#(),NY,STARTY,MAGY#(),FREQ#(),F,PI)
62  REM
63  REM OUTPUT THE INPUT FREQUENCY AND THE NUMBER OF INPUT SAMPLES
64      PRINT "INPUT FREQ (MHZ) =      ",F
65      PRINT "NUMBER OF INPUT SAMPLES = ",NX
66      PRINT " "
67      PRINT "FREQUENCY            GAIN"
68      PRINT "---------          ----------"
69      PRINT " "
70      FOR I=0 TO 24
71         GAIN#(I)=10*LOG(MAGY#(I)/MAGX#(I))
72         PRINT FREQ#(I),"      ",GAIN#(I)
73      NEXT I
74  REM CALCULATE GAIN AT -3 db POINT
75      BWMAG=GAIN#(0)-3
76  REM THIS SEGMENT CALCULATES THE BANDWIDTH
77      FOR I=0 TO 23
78         IF (GAIN#(I)>BWMAG AND GAIN#(I+1)<BWMAG) THEN
79            BW=(FREQ#(I)+FREQ#(I+1))/2
80            I=23
81         ENDIF
82         IF (GAIN#(I)=BWMAG) THEN
83            BW=FREQ#(I)
84            I=23
85         ENDIF
86         IF (GAIN#(I+1)=BWMAG) THEN
87            BW =FREQ#(I+1)
88            I=23
89         ENDIF
90      NEXT I
91  REM OUTPUT THE -3 db GAIN, AND THE BANDWIDTH
92      PRINT " "
93      PRINT "GAIN (-3db)      = ",BWMAG," db"
94      PRINT "BANDWIDTH (-3db) = ",BW," MHZ"
95      PRINT " "
96  REM
97  REM GRAPHING THE INPUT (CH 1 SCOPE)
98      CALL SetTitle ("INPUT")
99      CALL SetAxName (1, "AMPLITUDE VOLTS")
100     CALL SetAxRange (1, -5, 10, 10)
101     CALL SetTxAlign (0, 3)
102     CALL SetAxName (0, "TIME (uSEC)")
103     CALL SetAxRange (0, 0, 2, 10)
104     CALL SetTxAlign (2, 1)
105     CALL GrfCurv2D (TIM#(), X#(), 1024)
106     CALL GrfLReset (0, 0, 0, 0)
107 REM
108 REM GRAPHING THE OUTPUT (CH 2 SCOPE)
109     CALL SetTitle ("OUTPUT")
110     CALL SetAxRange (1, -1, 1, 10)
111     CALL GrfCurv2D (TIM#(), Y#(), 1024)
112     CALL GrfLReset (0, 0, 0, 0)
113 REM
114 REM GRAPHING THE DFT OF THE INPUT
115     CALL SetTitle ("INPUT DFT")
116     CALL SetAxName (0, "FREQ (MHZ)")
117     CALL SetAxRange (0, 0. 100, 10)
118     CALL SetTxAlign (2, 1)
119     CALL SetAxName (1, "MAGNITUDE VOLTS")
```

```
120     CALL SetAxRange (1, 0, 200, 10)
121     CALL SetTxAlign (0, 3)
122     CALL SetCurv2D (1)
123     CALL GrfCurv2D (FREQ#(), MAGY#(), 25)
124     CALL GrfLReset (0, 0, 0, 0)
125 REM
126 REM GRAPHING THE DFT OF THE OUTPUT
127     CALL SetTitle ("OUTPUT DFT")
128     CALL GrfCurv2D (FREQ#(), MAGY#(), 25)
129     CALL GrfLReset (0, 0, 0, 0)
130 REM
131 REM GRAPHING THE FREQUENCY RESPONSE
132     CALL SetTitle ("FREQ RESPONSE")
133     CALL SetAxName (1, "GAIN db")
134     CALL SetAxRange (1, 0, 2, 10)
135     CALL SetTxAlign (0, 3)
136     CALL GrfCurv2D (FREQ#(), GAIN#(), 25)
137     CALL GrfLReset (0, 0, 0, 0)
138 REM
139 REM THIS SUBROUTINE DEFINES ONE PERIOD OF A PERIODIC WAVEFORM
140 REM
141     SUB WAVE(W#(),N,START,M)
142     SUM=0
143     FOR I=0 TO M
144         SUM=SUM+W#(I)
145     NEXT I
146 REM
147     CNTR=SUM/M
148     Y=1
149     FOR I=1 TO 1022
150         IF ((((W#(I)<CNTR) AND (W#(I+1)=CNTR)) OR ((W#(I)=CNTR) AND (W#(
    I+1)>CNTR))) AND (Y=1)) THEN
151             START=I+1
152             Y=2
153         ELSE
154             IF ((W#(I)<CNTR AND W#(I+1)=CNTR) OR (W#(I)=CNTR AND W#(I+1)
    >CNTR) AND Y=2) THEN
155                 Y=3
156             ELSE
157                 IF ((W#(I)<CNTR AND W#(I+1)<=CNTR) OR (W#(I)<=CNTR AND W#(
    I+1)<CNTR) AND Y=3) THEN
158                     N=I-START
159                     I=M
160                 ENDIF
161             ENDIF
162         ENDIF
163     NEXT I
164     END SUB
165 REM
166 REM ***** DISCRETE FOURIER TRANSFORM *****
167 REM THIS ROUTINE CALCULATES THE DFT ON ONE PERIOD OF THE WAVEFORM
168 REM
169     SUB DFT(W#(),N,START,MAG#(),FREQ#(),F,PI)
170     FOR I=1 TO 49 STEP 2
171         XR=0
172         XI=0
173         L=N+START-1
174         K=-1
175         FOR Y=START TO L
176             K=K+1
```

404   FULLY INTEGRATED APPLICATIONS

```
177            A:=2*PI*I*K/N
178            XR=W#(Y)*COS(A)+XR
179            XI=W#(Y)*SIN(A)+XI
180         NEXT Y
181            J=(I-1)/2
182            MAG#(J)=(SQR((XR^2)+(XI^2)))/N
183            FREQ#(J)=I*F
184         NEXT I
185      END SUB
186      END
187
188
189
```

Figure 13.5   A listing of Quick BASIC source code for the fibre optic example.

Description	String of commands
Load BNDWTH.BAS	FILE,LOAD,BNDWDTH.BAS,LOAD
Load Tek2400 Instrumentation library	INSTRUMENTS,LOAD,INSTR,LOAD, TEK2400.FP,LOAD
Set pulse generator to desired frequency	Set manually to square wave of desired frequency
Run program	PROGRAM,RUN

Experimental results are shown in figures 13.6 and 13.7; figure 13.7 is a corresponding graph of the response indicating that the bandwidth of the fibre under test is approximately 60 MHz and figure 13.6 is a tabulation of the response.

```
 1  ENTER INPUT FREQUENCY (2 TO 4 MHZ)
 2  ? 2
 3
 4
 5  INPUT FREQ (MHZ) =        2.0
 6  NUMBER OF INPUT SAMPLES = 503.0
 7
 8  FREQUENCY              GAIN
 9  ---------              ----------
10
11   2.0                  -43.240627
12   6.0                  -43.106917
13  10.0                  -42.610394
14  14.0                  -42.300812
15  18.0                  -41.464834
16  22.0                  -40.890387
17  26.0                  -40.557034
18  30.0                  -39.545154
19  34.0                  -39.07912
20  38.0                  -37.470707
21  42.0                  -37.620958
22  46.0                  -38.369166
23  50.0                  -40.071497
24  54.0                  -40.159662
25  58.0                  -43.745589
26  62.0                  -45.866443
27  66.0                  -48.96151
28  70.0                  -53.155409
29  74.0                  -52.972111
30  78.0                  -56.130002
31  82.0                  -54.459066
32  86.0                  -56.309317
33  90.0                  -60.309111
34  94.0                  -68.595194
35  98.0                  -59.699657
36
37  GAIN (-3db)    =  -46.240627    db
38  BANDWIDTH (-3db) =   64.0    MHZ
39
40
```

Figure 13.6 Tabulated results of the fibre optic cable bandwidth experiment.

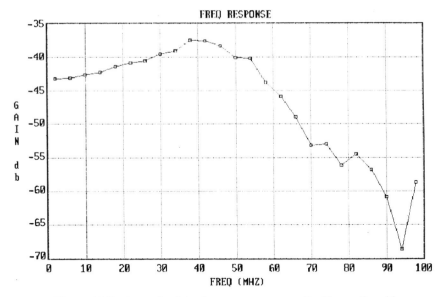

Figure 13.7 A graph of the frequency response of a fibre optic cable.

14

New Tools for Laboratory Environments

14.1 INTRODUCTION

Can machines think? This fascinating question has challenged humans for many decades, even centuries. In 1938 the British mathematician Alan Turing showed that a simple computational model, the *Turing machine*, was capable of universal computation. (This became one of the bases for the stored-program computer.)

Could stored-program computers be capable of simulating or emulating arbitrary human actions? With information about the state of every neuron and the mechanics of neuron firing, it would seem possible to compute the next state of someone's neurons and thus be able foretell that person's actions. In principle it seems possible to program machines to act and think like people. The name generally applied to the study of such phenomena is *artificial intelligence* (AI).

In 1950 Turing proposed a test—the *Turing test*—which could be used to determine how close a machine might come to acting human; an outline of the test is sketched in figure 14.1. In this test, the interrogator is asked to determine which is the computer and which is the human responder through a dialogue (presumably including a series of queries). Turing asserted that in fifty years (2000) it would be possible to make computers play the imitation game so well that 70% of interrogators would not make the right identification after five minutes of questioning.

In 1967 Marvin Minsky speculated that a computer with a properly organized set of one million facts about everyday life should be able to exhibit very great intelligence. Today's computers can easily store many times the amount of information suggested by Minsky and yet no one has succeeded in constructing an 'intelligence' seriously equivalent to that of an average human.

In one sense it can be argued that human intelligence will always exceed that exhibited by an automaton. From the work of the 20th-century thinkers Gödel and Turing, there are problems that defy logical solution and consequently the possibility of mechanical simulation. Kurt Gödel showed that, given a powerful, consistent system of logic, one can construct a true proposition that cannot be proved within that system. In addition, Turing's design of a *universal computer* included well defined problems that could not be solved, even in principle, by

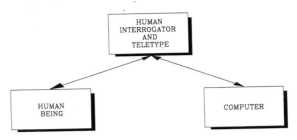

Figure 14.1 A sketch depicting the Turing test.

a program. For example, he proved that no program can determine whether another program will enter an endless loop when it is executed.

Why is it so difficult to elicit some form of intelligent behaviour from machines? Is it poor engineering, overly ambitious goals, or something fundamental that puts goals beyond our reach? One limitation follows from the view that a computer processes symbols without regard to their meaning; a person processes everything within a framework of interpretations. Stored-program computers may have the wrong structure for being able to 'compute' like people. As soon as someone finds a way of mechanizing a class of human actions, people tend to stop regarding those actions as 'intelligent'. They often want machines to take over the 'tedious routine'. Experimentation within the laboratory is often 'routine' in nature; it is therefore a good candidate for automation *including the interpretation of results*. In this sense, the computer may exhibit intelligence that is close to that of a human observer and may even exceed it—the computer is capable of virtually unlimited vigilance while human observers are not good at sustained error-free scrutiny of data.

Let's not ask how to make computers behave more like people; instead let's explore how to design computers to help people to function more effectively. Some of the things computers are exceptionally good at are executing algorithms, retrieving information, processing and filtering signals, assisting with communications, and monitoring processes. It is in this context that we can describe new 'intelligent' applications within a laboratory setting.

Turing discussed two approaches to building machines that might emulate intelligent thought. The first is to pursue very abstract activities, such as logical deduction or playing chess. The second is to provide the machine with sophisticated sensory organs and a mechanism for storing complex codes denoting moments of experience; by repeated exposure to sensory patterns, these machines could come to recognize familiar patterns and perform associated actions. Most AI research has followed the former path, perhaps because many of the early workers were mathematicians, skilled in abstract reasoning. Turing's second approach suggests another alternative. Rather than start at the top of the hierarchy of biological functions, start at the bottom, reproducing the control functions and behavioural patterns of insects, birds, mammals, and primates.

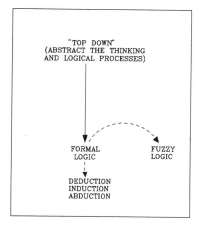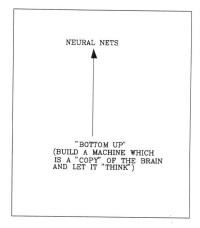

Figure 14.2 Two approaches to building an automaton that imitates human behaviour.

Use these as building blocks for the higher-level functions of the human being, ultimately coming to understand the mechanisms that give rise to intelligence and abstract thought in the human brain. Such an architecture would be capable of reproducing many of the behaviours found in the nervous system.

Within the laboratory, automata of this type can learn, understand, interpret and arrive at conclusions in a manner that would be considered intelligent if a person were doing it. The two formulations are shown in figure 14.2. In one form, the machine is organized to implement formal logic including deduction, induction, and abduction. Formal predicate logic determines the truth or falsity of a statement. However, some statements may be 'partially' true or false. To provide for these possibilities, a relatively new implementation of AI has evolved; it is known as *fuzzy logic*.

The second form of AI is known variously as adaptive, associative, connectionist, or neural network, among others. These are 'highly parallel' computational elements, which carry out information processing by means of its overall state response to an input. (This is sometimes thought of as feedback amplifiers connected in parallel.)

Expert systems and neural nets will be described below.

14.2 ARTIFICIAL INTELLIGENCE AND EXPERT SYSTEMS

AI encompasses a broad range of technologies including image/pattern recognition, speech recognition and generation, and robotics. One of domains with potential for laboratory environments relates to those computer systems that are designed primarily for symbol manipulation. In particular, they are configured to simulate problem solving in narrowly defined disciplines; because of this they are often referred to as *expert systems* [1]. There are many examples

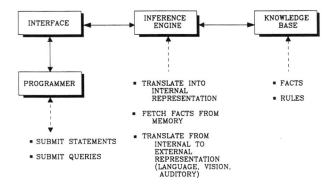

Figure 14.3 A block diagram of an expert system.

of such systems: an antibiotic prescription system—historically, one of the first practical examples of such designs; chemical compound classification; maintenance of electronic instruments; maintenance of telephone systems; marketing (computer) aids; real-time scheduling of (telescope) experiments.

Conventional software programs follow a precise step-by-step set of instructions. In contrast to this, an expert system solves problems in a 'procedural' way. Such programs use a combination of knowledge (facts, axioms, rules, informal heuristics (so-called rules of thumb, short cuts, experience)), formal logic, and some form of human interface by which the user can enter facts, and/or query the system. A system of this type is shown in figure 14.3.

The *inference engine* and the *knowledge base* form the heart of expert systems while the human interface provides the means by which the user may gain access to the system. (Often, the interface includes graphical aids for user queries or data-base modifications—additions, deletions, alterations of facts and rules.) The inference engine translates the user entries into a form that is compatible with the internal representation of the information. It also translates conclusions into some form of natural representation (language, speech, or vision) for the user. 'Fetching facts' from the data-base requires the computer to carry out a series of logical 'search' procedures. Figure 14.4 is a more detailed functional representation of an expert system. In general, at any one instant the data-base of such machines contains the 'state of the world' and/or the goals to be achieved by the processes (e.g., the differential diagnosis defined by the symptoms). These facts may be stored in one or more data structures such as arrays, lists, tables, predicate (logical) expressions, property lists, or semantic nets. The data are grouped according to ultimate purpose. For example:

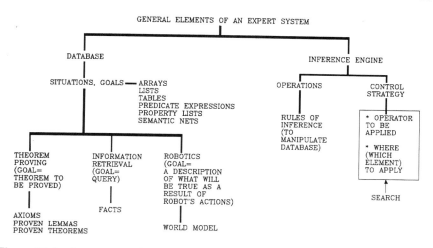

Figure 14.4 A summary of components of the data-base and inference engine of an expert system.

Purpose	Information
Theorem proving	Axioms, proven lemmas, proven theorems
Information retrieval	Facts
Robotics	World model

The inference engine includes two kinds of entities: the operations or *rules of inference*; and the *control strategy* defined as the procedure to be followed for choosing which operation to apply and on which element of the data-base to apply it. A simplified *flow diagram* which describes the way in which expert systems arrive at a goal is shown in figure 14.5. (This is the graphic equivalent of a DO-UNTIL pseudo-code structure.) Not shown in the figure is the way in which a goal is determined in the first instance. Moreover, this same architecture can be used to satisfy subgoals, which in turn can be used to resolve larger questions.

Representation of information within the data-base consists of stylized (abstracted) versions of the world and must satisfy several constraints. (Even with such structure, information can be misrepresented because it has neither context nor redundancy, two elements that characterize human expression—an example of this will be provided below.) Figure 14.6 shows the necessary properties for internal representation of information.

- *No referential ambiguity.* It must be clear which 'object' is being referenced without confusion. For example, we cannot input a 'fact' such as: 'The patient has a pain in his lower back'. (Who is the 'he'?) One might use 'Mary' but this too is ambiguous; there may be many people with that name. Changing it to 'Mary Johnson' may still not clarify matters (for the computer). One solution is to create a unique name, like Mary_1. Such representations are

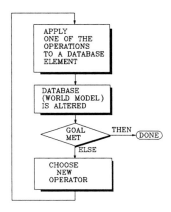

Figure 14.5 How an expert system functions.

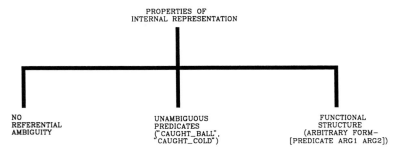

Figure 14.6 How information is represented within the knowledge base.

called *instances* or *tokens*.
- *Unambiguous predicates*. A *predicate* is an assertion of a fact about one or more entities. As an example, consider the following imaginative dialogue between an expert system and a user:
    User: Mary_1 caught ball_10.
    User: John_3 caught cold_5.
    User: Who is ill?
    Expert: Mary_1 caught ball_10 and John_3 caught cold_5. Information within the data-base must distinguish between the 'caught' predicates. Once again, this may be dealt with by introducing separate instances of 'caught': catch_illness; catch_object. The information cited above would then become: 'John_3 catch_illness cold_5'. (Normally, some delimiters are used to distinguish different pieces of information. One such is the use of parentheses—(Mary_1 catch_object ball_10).)
- *Functional structure*. Natural language may not always be consistent in word order (e.g. the direct object is sometimes put at the front—'A ball was caught by Mary'). For historical reasons, the predicate in internal representation is

written first (e.g. (catch_object Mary_1 ball_10)).

In spite of such specificity, translation from natural language (facts) to internal representation may not be easy. Consider two 'facts' within the data-base:

1. All dogs have tails.
2. Every dog has a tail.

Two interpretations are possible:

A. 1 and 2 represent the same fact.
B. 1 means that every dog has at least one tail or each dog has several tails.

Expert systems should be able to draw inferences—that is, to come to believe a new fact on the basis of other information. There are several types of inference:

- *Deduction.* This is a logically correct inference. A deduction from a true premise is guaranteed to result in a true conclusion. (This might be called 'perfect logic' as we cannot draw a false conclusion; other logic is fallible.)
- *Induction.* Logical inferences are drawn from particular facts or individual cases. (A classic instance of this comes from the following sequence: Robins have wings; Wrens have wings; ... *Birds have wings*.)
- *Abduction.* This form of logic is closest to the way in which humans draw inferences. This can be thought of as 'plausible inference'. (Such an inference is imperfect and may produce incorrect conclusions.) Abductive logic conforms to the following pattern:

  Given: (if a b)
            b is true
  Infer: a is true.

(The first statement above is translated as follows: 'If a is true then it follows that b is true'.) This type of reasoning leads to the 'best guess' result. For example, one way to generate a diagnosis would be to: determine the symptoms; find *all* diseases that have such symptoms; choose one of the 'more likely' ones (perhaps the 'most likely'). To correlate this with the formalism shown above make the following (rough) assignments: 'a' = diagnosis (disease); 'b'= the symptoms. (Expert systems of this type rely on *Bayesian statistics*.)

Within expert systems of the type being considered, the inference engine includes facilities to draw inferences; this is generally accomplished by means of the *predicate calculus*—a way of calculating the truth of propositions. It includes:

- A language for expressing propositions.
- Rules for inferring new facts from those that are given.

These elements are summarized in figure 14.7 and include: *formulas, connectives, variables* and *quantifiers*.

Formulas contain

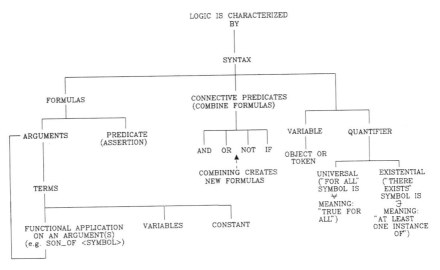

Figure 14.7 The structure of the predicate language.

- predicates;
    - predicates require one or more arguments;
    - each argument must be filled by a term;
        - terms include:
            - constant symbols
            - variables
            - functional applications (a function with zero or more arguments like son_of Mary where the function 'son_of' has one argument, namely, Mary in this case—arguments of functions are themselves terms).
- connectives;
    - we can build more complicated formulas from atomic formulas by combining them with *connectives* (and, or, not, if).
    - Example. If *p* and *q* are formulas then
        - (and *p q*), (or *p q*), (not *p*), (not *q*), (if *p q*)
        - are formulas.
    - The relations between a connective predicate and its arguments
        - are expressed in terms of a *truth table*.
- variables and quantifiers;
    - A *variable* is an *object* or *token*.
    - *Universal quantifiers* certify that something is true for all possible values of a variable.
    - An *existential quantifier* certifies that at least one instance of the argument exists.

p	q	(and p q)	(or p q)	(not p)	(if p q)
t	t	t	t	f	t
t	f	f	t	f	f
f	t	f	t	t	t
f	f	f	f	t	t

Figure 14.8 Truth tables for predicate connectives.

The truth tables for the *and, or, not* and *if* connective predicates are reproduced in figure 14.8; predicate values are either true (t) or false (f).

Example 14.1. Translating a natural language fact into internal representation.

Consider the statement: 'All elephants are grey'. Within the *knowledge base* this would be represented as:

(∀(x) (if (inst x elephant) (colour x grey)))

In this internal representation 'inst' stands for 'instance of' and 'colour' is a predicate. The internal representation therefore translates (into 'natural language') as follows: 'For all x, if x is an instance of an elephant, the colour of x is grey'. (The *if* connective joins the predicates (inst x elephant), which plays the role of p, and (colour x grey), which assumes the q-position in the general formulation.) The variable 'x' would be replaced by an appropriate token or object when using this fact for logical inference. Some computer languages (not discussed here) are well suited for such programming. In particular, LISP ('*lis*t *p*rocessing') is an example of such an *o*bject-*o*riented *p*rogramming ('OOP') language; there are many others.

Using the predicate calculus. One general procedure for drawing logical inferences is shown in figure 14.9. This is a 'goal-driven' formulation of the logical inference process and implements *abductive* reasoning (as described above). The data-base is first searched to see whether or not the conclusion already exists; if it does, the query has been resolved and the inference is immediately established. (Searching procedures are discussed below.) If the conclusion is not in the data-base it is decomposed into a series of subgoals, which can subsequently be proved (using the same inference procedure as shown in the figure—the algorithm is inherently *recursive*). When all subgoals are satisfied the original query is proved. Within abductive reasoning, a subgoal is established as follows:

Recall abductive logic:
    Given: (if $p\ q$) is true
                 q is true
    Conclude: p is true.

(Notice from the truth table shown in figure 14.8 that (if $p\ q$) and q may

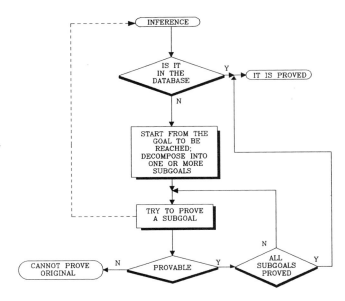

Figure 14.9 A flow diagram for logical inference (goal oriented, backward search).

be true but corresponding states of p may be either true or false; this can give rise to a logical fallacy. As an example, consider the following (rather simplistic) argument:

(if (patient has broken leg) (patient has pain in leg))
(patient has pain in leg)
Conclusion (abductive): (patient has broken leg)
Obviously, a pain in the leg may be due to many other causes.)

However, when using abductive (backward) reasoning, we first ascertain whether or not p already exists in the data-base. If it does not, we search the data-base for an *if* connective in which the unknown p matches up to the q of the connective. If it does, then the p corresponding to that q—call it p'—becomes a subgoal. If the subgoal can be established then the original p can be confirmed as being true.

A 'classic' example of this is shown in figure 14.10. The data-base predicates are numbered 1 to 9. The query or statement to be resolved is 'Was Marcus loyal to Caesar?'. Notice that not every predicate (statement) in the data-base is needed to resolve the query.

Other knowledge representative techniques. Other forms of knowledge representation, in addition to the predicate structure, have been developed. Their characteristics are (briefly) surveyed here:

- **Procedural representation.** Knowledge about the world is encoded as *procedures*—small programs that 'know how to do specific things'. (For example, in a parser for a system to understand natural languages,

The Knowledge Base
1. (inst Marcus man)
2. (inst Marcus Pompeian)
3. (∀(x) (if Pompeian (x)) (Roman (x)))
4. (inst Caesar ruler)
5. (∀(x) (if (Roman (x)) (or (loyal_to (x,Caesar)) (hate (x, Caesar)))))
(Romans were either loyal to Caesar or hated him.)
6. (∀(x) (∃(y) (loyal_to (x,y))))
("Everyone has someone who is loyal to him/her.")
7. ((∀(x,y) (if (and (person (x) ruler (y) try_kill (x,y))) (not loyal_to (x,y))))
8. (try_kill (Marcus, Caesar))
9. (∀(x) (if man(x)) (person(x))))
(All men are people.)

To establish: Was Marcus loyal to Caesar?
A. There is no predicate of this type in the database (DB). Either Marcus was loyal or he was not loyal. Start, by attempting to prove:

      (not (loyal_to (Marcus, Caesar)))

B. This is not in the DB. Notice the q of statement 7. This matches (after generalizing the tokens Marcus and Caesar to x, and y), so substitute the variables Marcus and Caesar for x and y:

(∀(Marcus, Caesar)
    (if (and (person (Marcus) ruler (Caesar) try_kill (Marcus, Caesar)))
      (not loyal_to(Marcus, Caesar))))

C. The following subgoals ("qs") can be established directly from the DB.

    ruler (Caesar) is true from statement 4.
    try_kill (Marcus, Caesar) is true from statement 8.

D. One subgoal remains to be resolved:

    person (Marcus)

E. Use statement 9 by matching the q with person (Marcus), and fixing x to Marcus.

    (∀(Marcus) (if man (Marcus) (person (Marcus))))
    man (Marcus) is true by statement 1 in the DB.
    Therefore it follows that person (Marcus) is true.

F. All subgoals in B have been resolved (true), therefore the Expert System would conclude that
    (not loyal_to(Marcus, Caesar)) is true.

Figure 14.10 An example of goal-driven reasoning that can be implemented with an expert system.

'knowledge' that a noun phrase may contain articles, adjectives, and nouns is represented in the program by (sub)routines that know how to process articles, nouns, and adjectives. The permissible grammar for a noun phrase is not stated explicitly.)
- **Semantic nets.** An explicit psychological model of human associative memory. A net includes:
  - nodes representing objects, concepts, events;
  - links representing nodal interrelations.

Important associations can be made explicitly and succinctly; relevant facts about objects or concepts can be inferred from the nodes to which they are directly linked without a search through a large data-base. Inferences drawn

by manipulation of the net are not assuredly valid as in the logic-based representation.
- **Production systems.** Based on models of human cognition, the data-base consists of rules, called *productions*. The rules have a form of condition–action pairs. Current research emphasizes the ability to develop self-modifying (learning) systems.
- **Natural representation.** A robot camera encodes a visual scene as an array representing average brightness from the visual field (*direct* representation). This is useful for finding boundaries of objects in the scene.
- **Frames.** A data structure that includes declarative and procedural information in predefined internal relations. *Slots* are maintained for facts typically known about the object or concept, and an attached procedure for finding additional knowledge. A sample frame might appear as:

```
DOG frame
self: an ANIMAL; a PET
breed:
owner: a PERSON
(if needed: find a PERSON with pet=myself)
Name: a PROPER NAME (DEFAULT = Rover)
```

A likely frame is selected from the data-base to aid in the process of understanding or reasoning. The expert system attempts to match a frame to the 'query' frame.
- **Scripts.** This is 'frame-like'; a structure that describes a stereotyped sequence of events in a particular context. They are useful in 'real-world' situations where there are patterns to the occurrence of events. One could make inferences because scripts would include answers to the user's queries.
- **Search methods.** As shown in figure 14.4, one element of the inference engine is needed to decide the next step in any solution process. This part of the expert system is the *control strategy*; it is normally referred to as the *search* process. Searching embodies deciding which operator is to be applied and where it is to be applied—which element of the data-base. Sometimes such control is 'highly centralized' within the program (system) in a separate control executive that decides how problem-solving resources should be expended. Sometimes control is diffuse, being spread among the operators themselves.

In general the object of expert systems is to achieve some goal(s) by applying an appropriate sequence of operators to an initial 'task domain' situation ('state of the world'). Each application of an operator modifies the situation in some way. If several different operator sequences are worth considering, the representation often maintains data structures showing the effects on the task situation of each alternative sequence.
- *Forward reasoning* (also called *data-driven*, or *bottom-up*). The application

of operators to the data structures in which the problem state is brought forward from its initial configuration to one satisfying a goal condition.
- *Backward reasoning* (also called *goal-directed*, or *top-down*). Decompose the problem to be solved into one or more subgoals that are easier to solve than the original problem. The solution of these subgoals is sufficient to solve the original problem.
- *Means–ends analysis*. This is a combination of forward and backward searching. The current goal is compared to the 'current state of the world' (task domain) and a difference between them is extracted. The difference is used as a forward operator to reduce the difference. Subgoals are set up to change the problem state in such a way that the relevant operator can be applied. After the subgoals are solved, the modified situation becomes a new starting point from which to solve for the original goal.

Tree structures are commonly used to implement the control strategy for the search. The tree is used to represent the set of problem states produced by operator applications. The tree consists of:

- *The root node*: the initial problem situation or state.
- *Successor node(s)*: the new states, which can be produced by the application of one operator to the root node. Subsequent operator applications produce successors to these nodes, and so on.

When the order in which potential solution paths are explored is arbitrary, using no domain-specific information to judge where the solution is likely to be, the search is called a *blind search*. Searches must be contained or they will lead to *combinatorial explosion* of which chess-playing is a good example. They may be contained in several ways: recast the problem to be solved; use heuristic ('rule-of-thumb') knowledge; stop after a finite number of iterations or a finite amount of computation time; 'prune' (eliminate) spurious nodes. Briefly stated, blind searches fall into one of two categories:

- *Breadth-first* search. Expand the nodes in the order of their proximity to the start node, measured by the number of paths ('arcs') between them. This considers every possible operator sequence of length n before any sequence of length $n + 1$. This method is time consuming but is guaranteed to find the shortest possible solution—a sequence of applied operators that transforms the starting configuration to the goal configuration. A flow diagram describing how to carry out a breadth-first search is shown in figure 14.11. The underlying data structure for the solution consists of two lists, a list of expanded nodes and a list of unexpanded nodes.
- *Depth-first* search. The most recently generated, or deepest node is expanded first. Example: automation within the laboratory will require 'intelligent' robots capable of manipulating objects. Consider a problem in which a robot must order objects as shown in figure 14.12(a). Only one operator is available to the robot, namely, *move(x, y)* which translates as 'move object x onto another object, y'. Some conditions must exist to carry out the *move*

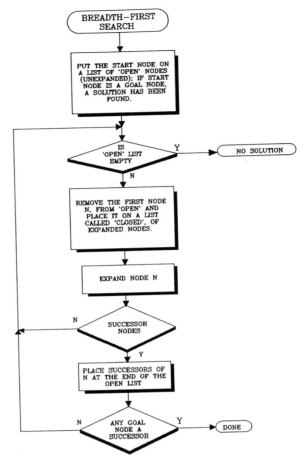

Figure 14.11 A flow diagram (algorithm) for breadth-first searching.

successfully:
- x, the object to be moved, has nothing on top of it.
- If y is a block, there must be nothing on top of it.
- The operator is not to be used to generate the same state more than once; otherwise, the search might end up in an endless loop.

A breadth-first search (see the flow diagram in figure 14.11) for such a problem produces the tree shown in figure 14.12(b). The nodes are states S0 to S10; node S1 is the successor of state S0 and is reached by following the command 'move object 1 to the base'. The nodes are generated and expanded in the order given by their numbering (S0, S1, S2, ..., S10). When the solution (S10) is found, the list of expanded nodes contains S0 to S5 and the open list still contains S6 to S10.

ARTIFICIAL INTELLIGENCE AND EXPERT SYSTEMS

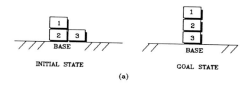

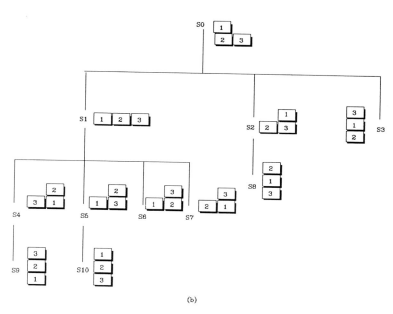

Figure 14.12 Robotic reordering of objects. (a) A drawing of the task. (b) A breadth-first search tree.

14.2.1 Examples of expert systems in the laboratory

Instrumentation. Molecules may be characterized by the way in which they interact with light; this provides a way to determine the composition of substances. The technique of spectrophotometry is the means by which the absorption of light by a substance of interest leads to the analysis of its contents. A block diagram of a typical spectrometer system is shown in figure 14.13(a) [2]. A light source provides incident radiation to an interferometer (which includes the movable mirror shown in the figure). The interferometer may be a Michelson-type one, which is shown schematically in figure 14.13(b). Consider for a moment that the light is monochromatic. Part of the incident beam passes through the beam splitter and is reflected back by the fixed mirror. Another part of the incident energy is reflected by the beam splitter and subsequently by the movable mirror. Reflected light from both mirrors returns to the beam

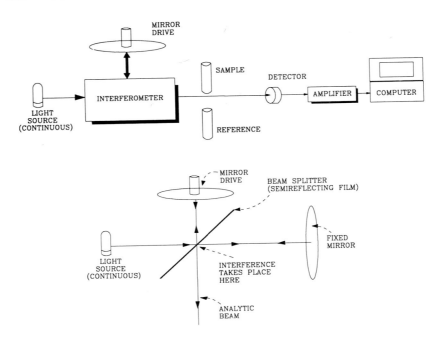

Figure 14.13 (a) A block diagram of a spectrometer. (b) A schematic diagram of a Michelson interferometer.

splitter; if the difference in path lengths (between the fixed and movable mirrors) is an integer number of wavelengths, $m\lambda$—where m is an integer and λ is the wavelength, the beam will be reinforced (*constructive interference*). When the path length is $(m + \frac{1}{2})\lambda$, destructive interference occurs and the beam is obliterated. Interference of the light varies sinusoidally as the movable mirror is displaced at constant velocity yielding alternating light and dark bands called *interference fringes*. If the light source is polychromatic, a plot of beam intensity versus the position of the movable mirror is called an *interferogram*. Equations that describe operation of a spectrometer include the following. Let

δ = path difference between fixed and movable mirrors
σ = wave number (= $1/\lambda$, λ is wavelength of light)
R = beamsplitter reflectance at σ
T = beamsplitter transmittance at σ
I_A = light intensity of analytic beam
I_T = light intensity incident on the beamsplitter.

Then, when $R = T$,
$$I_A = 0.5 I_T (1 + \cos 2\pi\sigma\delta).$$

Normally, $R \neq T$, in which case

$$I_A = 2R_\sigma T_\sigma I_T(1 + \cos 2\pi\sigma\delta).$$

The 'AC' component of this term

$$2R_\sigma T_\sigma I_T \cos(2\pi\sigma\delta)$$

plotted as a function of δ constitutes an interferogram. A 'continuous' light source maintains a constant light intensity over a wide range of wavelength (e.g. 'white light'). The response of the sample (in time) as the movable mirror is changed produces the spectromatic signature of the unknown as a function of wavelength (wavenumber). Ideally, the absorption (spectrum) follows Beer's law (for each analyte); it adds a multiplicative term of the form 10^{-A} to the above equation, where A is the absorbance at each wavenumber. The total analytic signal is:

$$x_{\sigma,\delta} = 2G(\sigma) R_\sigma T_\sigma I_T 10^{-A} \cos(2\pi\sigma\delta).$$

If the source is continuous, all of its wavelengths will be present simultaneously and each will independently undergo interference and absorption by the sample at each retardation (d). (Beer's law states that each infinitesimally thin section of the sample containing 'dn' absorbers will absorb a fraction of the incident power, dP/P, and therefore

$$-\frac{dP}{P} = -kn \quad \text{or} \quad P = P_0 \times 10^{\varepsilon bc}$$

where ε = molar absorptivity; b = cell thickness; c = concentration of absorbers (mol cm^{-3}).

Computer-based analytic instruments constructed on the principles cited above are commercially available. They include such facilities as:

- Menu-driven input for setting system parameters.
- Interactive graphics for displaying data.
- A command language with the ability to create macros (programs); these invoke data-processing 'methods' that perform quantitative spectrum analysis, spectral search analysis, data manipulation (smoothing, interpolation, etc) on the spectrum.
- On-line help.

While these are 'user-friendly' facilities, expert system technology can be used to enhance their capabilities further. The following limitations have been identified in existing commercial spectrometers:

- A large number of parameters must be set for data collection and for the different (spectral) analysis methods. The on-line help may only explain the meaning of a parameter. It does assist an operator in deciding how to choose a value in a specific situation. An expert system could advise the operator on

how to accomplish a measurement goal using the available resources of the instrument and would also suggest specific parameters based on the type of experiment to be carried out.
- After a while, the number of available macros will be so large that it will be impossible to maintain or even keep track of them.
- The methods used in spectral search analysis have 'rigid' matching algorithms; a symbolic spectrum interpretation technique based on the qualitative (rather than quantitative) description of the spectra would aid in modelling substances. Descriptions would be based on the type, probability, and concentration of different components in a sample.

An expert system would prompt the operator with questions relevant to the particular group of parameters or operational mode of the instrument. (The operator could answer the questions using (graphical) input panels that include a menu-based interface.) The system would then start a reasoning process using its local rule-base. This results in a recommendation regarding parameter values or processing methods. The operator may accept or reject these suggestions; if accepted, they would be transferred to the instrument. An example of a rule in the rule-base might appear as [3]:

```
(RULE aromatic
    (sample type is ?type))
(PREDICATES-DO
    (eq ?type 'aromatic))
(ACTIONS-DO
    (set_parameter  'lowcollectfreq 600)
    (set_parameter  'highcollectfreq 5200))
(ACTIONS-UNDO
    (set_parameter  'lowcollectfreq 400)
    (set_parameter  'highcollectfreq 4800))
```

This rule is represented in terms of an object-oriented program statement and is interpreted as follows: if the sample type is aromatic—having the presence of at least one benzene ring—then frequency parameter limits (low and high collection frequencies) are proposed to be set to 600 and 5200 respectively. If the substance under test in not aromatic then the (default) settings are restored (400 and 4800). (The "?type" object in the program would be replaced at the time at which this rule is tested in the reasoning process.)

The goal of the spectrometer is to identify the contents of a sample. An 'expert chemist' can recognize the 'signature' (spectrum) of a particular material in the sample by identifying peaks with various attributes at different frequencies. A common technique for automating this is accomplished by comparing the observed spectra with a library of references in a sequential manner. Recently, symbolic interpretation methods have been suggested in order to provide higher flexibility for the identification process [4]. The spectrum is characterized by location, intensity, and widths of the peaks and is processed by using a decision tree. The exhaustive search in the tree reveals the presence of different components. The tree is described in the form of a large set of formal rules. A typical decision point is:

```
(IF (PEAK (3050 2990))
    THEN (SET 0.3)
    ELSE (IF (PEAK (3150 2990))
         THEN (SET 0.1)
         ELSE (IF (SITUATION oil)
              THEN (SET 0.05))))
```

with the following meaning: If there is a peak in the sample in the range 3050–2990 cm^{-1}, set the probability of the aromatic class to 0.2, or else if there is a peak in the sample in the range 3150–2990 cm^{-1}, then set the probability of the aromatic class to 0.1, or else if the sample contains oil, then set the probability of the aromatic class to 0.05.

Consultation on laboratory data. While there are many examples of AI in support of medical diagnosis [5], information derived directly from laboratory tests suggests a completely integrated system for diagnosis as well as assessment of the evolution of patient state and choice of therapy [3]. Analytic results need to be transformed into useful information in a clinical environment. One area in which enhanced interpretation of laboratory data has been demonstrated is concerned with disorders of fluid–electrolyte balance in relation to management of the critically ill patient. Laboratory data on the patient derived from blood and urine samples in addition to bedside observations are integrated into a mathematical model of the disease process in order to arrive at suitable treatments. Mathematical models built on differential equations describe the underlying physiological processes. They can be used to predict the evolving patient state in response to both pathology and therapeutic inputs. Using this information the system can provide a diagnosis and suggested treatment regimens (for patients with fluid and electrolyte disorders) using a knowledge-based approach with subsequent verification by the mathematical model. The expert system includes the following software modules: the user interface; a patient-specific data-base; a diagnostic module; a treatment module; and a dynamic mathematics model [6].

- *The user interface.* A menu-driven system allows the user to enter new (patient) data, display existing data, perform a diagnosis, receive therapy recommendations based on the existing data, update a patient's file, and access the mathematics model (or exit).
- *The patient-specific database.* This includes both bedside and laboratory data. Examples from laboratory data include plasma concentrations of: sodium; potassium; albumin; creatinine; and urea. Urine information embraces the concentrations of sodium, albumin and potassium. Additional laboratory

data provide information on plasma and urine osmolality, and haemoglobin. Bedside information includes: blood pressure; central venous pressure; and temperature.

- *The diagnostic module*. This accepts patient-specific data and produces a differential diagnosis in the form of a disease state. It is built upon a rule-based system that defines the conditions required for a disease state to be present. Information regarding the patient's data is characterized by such descriptors as normal, low, moderately high. This module includes rules for determining the disease state. One such rule is:

   ```
   Hypoalbuminaemia (Nephrotic Syndrome)
   plasma sodium low, plasma osmolality normal, plasma
   albumin low, urine albumin high.
   ```

- *The treatment module*. Treatments are suggested according to symptoms; the user responds to machine-generated queries such as the state of the cardiovascular and renal systems, dietary status, and currently prescribed drugs as well as the diagnosed disease state. One example of the suggested treatment (for a cardiovascular state) is:

   ```
   If blood pressure normal and central venous pressure
   low, then prescribe colloid to raise the central venous
   pressure.
   If blood pressure low and central venous pressure low
   then prescribe colloid to raise central venous pressure
   and inotrope (dobutamine or dopamine) to raise blood
   pressure.
   ```

- *Dynamic mathematical model module*. The output is in the form of a graph of blood pressure and pulse rate plotted against time. The model performs as many simulated hours or days as directed.

The system was capable of diagnosing well defined disorders; however, it was unable to identify multiple disorders—something that might be difficult for a human expert as well as the automata.

14.2.2 Fuzzy logic systems

Machine-driven formal logic of the type outlined previously can automate rules such as: $p \to q$ (if p is true, then q is true). The key requirement here is the truth of the predicate; there must be no uncertainty about definitions or ambiguity about classes of objects. *Fuzzy logic* may be applied to resolve those instances where such constraints are not satisfied [7]. Shown in figure 14.14 is a graph that specifies the extent to which a value on the abscissa is a member of the class on the ordinate. (Also shown is a second curve, which corresponds to

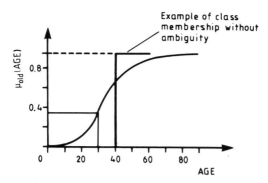

Figure 14.14 A graph of class membership.

the case where there is no ambiguity about 'being old' and which characterizes the case for 'fuzzyless' logic.) The degree of membership is designated as μ. For example, in the curve shown, a person aged 30 is a member of the (fuzzy) set *old* with a degree of membership of 0.35. (For consistency with formal logic systems, membership values range from 0 to 1 with 0 signifying false and 1 signifying true.)

Important contributions to the theory of fuzzy logic were made by Lofti Zadeh who defined the way in which fuzzy sets could be combined. The following (fuzzy) set operations are noted:

A .AND. B = $\min(\mu_A, \mu_B)$
A .OR. B = $\max(\mu_A, \mu_B)$
.NOT. A = $1 - \mu_A$

Example 14.2. Class Membership. A person belongs to the set *heavy* with a membership of 0.91 if he/she weighs 200 pounds. This is represented by $\mu_{HEAVY}(200) = 0.91$. A person belongs to the set *fat* with a membership of 0.75 if he/she has a waist measurement of 38 inches. This is represented as $\mu_{FAT}(38) = 0.75$. The degree of membership of the class

(Waist of 38 is .not. fat) .and. (weight of 200 is heavy)

is expressed by

(1 - m_FAT) .and. (m_HEAVY)
= min ((1 - .75),.91)
= 0.25

There is only 25% truth to the statement that the person with a 38-inch waist who weighs 200 pounds belongs to a group that is not fat but (and) heavy.

We can devise fuzzy rules that are based on the *if–then* rules of the formal logic but with an important difference. An *if–then* rule has the following form:

if {condition} then {action}

This requires that the *action(s)* are executed if the *condition* portion of the rule is true. In fuzzy logic there is a critical difference in that the *action(s)* are carried out: *the action is executed with the degree of membership of the rule's condition*. If $\mu_{condition} = 0.1$ then the action is executed with the same μ, namely $\mu_{action} = \mu_{condition} = 0.1$.

Example 14.3. Balancing a weight (pendulum) at the end of a motor shaft [8].

Figure 14.15 includes all of the information that is needed for this example. An automaton is to be built to maintain a weight (pendulum) balanced at the end of a rod. (Consider an analogous problem of maintaining a stick balanced at the end of one's nose.) Figure 14.15(a) is a sketch of the arrangement. If the weight (W) falls, current is applied to the motor (M) to drive it in a direction opposite to the motion, to restore balance. The problem inherently contains some complications well suited to (expert) solution with fuzzy logic. For example, we must resolve the question 'How much should we compensate for changes in W?'. Three factors are considered important: *theta*, the angle between W and the vertical axis; *deltheta*, the speed at which W is falling; and *IM*, the amount of corrective current to M (how much to drive M). The rule-base includes a number of heuristic rules including the following:

- If (W is a little off centre) then (small IM)
- If (W off centre by a large degree .and. deltheta increasing rapidly) then (large IM)

It is the fuzzy 'adjectives' (little, large) that inclines the solution to fuzzy logic. The rules and membership class functions are the repositories of an expert system. For this example we consider five membership classes: *NM*, negative medium; *NS*, negative small; *Z*, zero (includes some PS and NS values); *PS*, positive small; and *PM*, positive medium. The input variables are *theta* and *deltheta* and the output variable is IM. Typical membership functions for *theta* are shown in figure 14.15(b). (They are representative of typical membership functions, which often have a 'triangular' or 'trapezoidal' shape.) For example, if the angle between the vertical and W is 20° it belongs to membership classes Z, and PS with a 50% belief that it belongs to both Z and PS. The heuristic rules may be reinterpreted; for example, a sample rule appears as:

    if (theta PS .and. deltheta Z) then (IM NS)

A number of rules may be invoked ('fired') according to the input conditions. A sample rule matrix is shown in figure 14.15(c). The rule matrix operates as follows: The 'and' conditions are classes of *theta* and *deltheta*; enter the table at the appropriate combination; entries in the table are membership classes of IM that are 'fired'. If, for example, *theta* belongs to both Z and PS, and *deltheta* belongs to both NS and Z, then all the membership classes (of IM) enclosed by the heavy line will contribute to the current to be delivered to the motor. The

ARTIFICIAL INTELLIGENCE AND EXPERT SYSTEMS

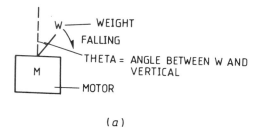

(a)

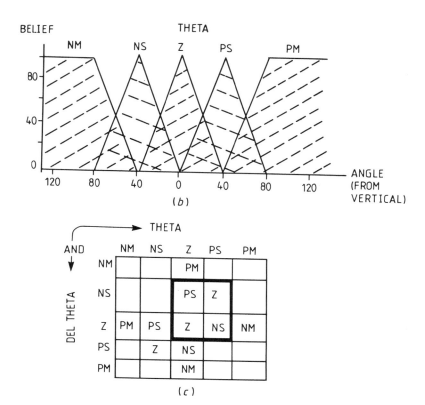

(b)

(c)

Figure 14.15 A fuzzy logic example. (a) A sketch of the system to be controlled. (b) Membership functions. (c) The rule matrix.

motor will only respond to a single value of current. Thus, fuzzy logic requires that the actions that are triggered (by the associated conditions) are combined so as to produce a single result. The single result is referred to as being *crisp* in contradistinction to the fuzzy classifications.

Obtaining crisp results. Three methods for resolution of fuzzy characteristics

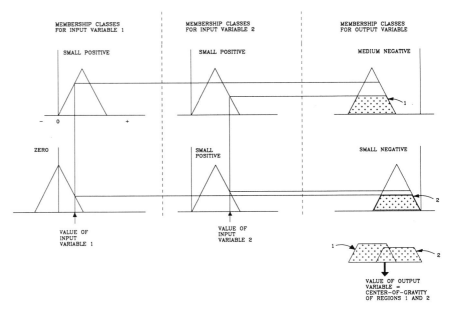

Figure 14.16 The centre-of-gravity method for obtaining crisp answers to fuzzy problems.

are common: finding the centre of gravity; finding the maximum height—select from the range of output values the one that receives the greatest degree of belief from fuzzy processing; and finding the average output values. Consider, here, the centre-of-gravity method and refer to figure 14.16 [9]. This generic problem includes two input variables—var_1 and var_2—and one output variable—out—that is to be controlled. The inputs have various membership categories with two shown: small positive (SP); and zero (Z). The output variables can fall into two membership categories, small negative (SN); and medium negative (MN). As a result of the input conditions, two rules are 'fired':

If ((var_1 is SP) .and. (var_2 is SP)) then (out is MN)
If ((var_1 is Z) .and. (var_2 is SP)) then (out is SN)

The values of var_1 and var_2 are superimposed on the membership classifications. The top row represents the results of combining (var_1 is SP) .and. (var_2 is SP) using the fuzzy rule that governs the operation ($\min(\mu_A, \mu_B)$). The shaded region marked '1' includes all class members that satisfy the combined belief. Similarly, region 2 represents the fuzzy combination of ((var_1 is Z) .and. (var_2 is SP)). Regions 1 and 2 are extracted and that value of output that is the centre of gravity (weighted average) becomes the crisp answer.

Stenosis (narrowing) of the main arteries supplying the myocardium is the most common underlying cause of heart attacks. Using nuclear cardiology it is possible to detect abnormalities in the distribution of blood flow to the myocardium. Regions with less than normal blood flow have a perfusion defect. The relation between the percentage of perfusion defect and severity or location of stenosis is not well defined. Fuzzy sets can be used to represent perfusion defects and to generate expert rules to help in diagnosis [10]. The locations and extents of perfusion defects are anatomically related in a 'fuzzy' way to the site of stenosis within a specific coronary artery.

14.3 NEURAL NETS

One of the new computer architectures that has become increasingly useful in the processing of information is the neural net. Such arrangements consist of a multiplicity of cooperating computing elements. Each unit is a relatively simple, and inexpensive computational element, which may be implemented in hardware or simulated on a general-purpose computer. The main characteristic of the net is the way in which it operates; procedures and calculations are carried out simultaneously ('in parallel') in contrast to what happens in a general-purpose machine, which completes tasks in a serial manner. (Suggestions for using an approach like this in order to implement intelligent automata came from Turing as noted in figure 14.2.)

The historical roots of this technology can be traced to a number of sources—as far back as 1958—principal among these being attempts to model the central nervous system of humans and the brain in particular [11]. Early modelling attempts raised a number of problems and criticisms that had to be resolved; recently neural nets have been employed to great advantage in pattern recognition applications. They have powerful potential for addressing such problems in the laboratory where analysis of data might be greatly speeded.

The neuron. The individual computing elements of the neural net are based (very) roughly on the physiological model of the neuron, the fundamental element of the human nervous system; a schematic representation of this is shown in figure 14.17 [12]. The neuron is electrochemical in nature and the interior of a cell typically rests at a potential of −65 mV relative to the surrounding fluid. If an appropriate voltage or stimulus is applied to the cell, one or more 'pulses' are generated within the cell. Such signals are called *action potentials*; these propagate along the cell's *axon* to the *dendrites* of other cells. Dendrites receive information—normally from a multiplicity of axons—and deliver these pulses to the *soma* or cell body. The action potentials must be transmitted across a physical barrier known as a *synapse*—this is accomplished as each pulse arrives at the synapse (along the axon); a chemical transmitter substance is released into the synaptic channel and *diffuses* across the very narrow channel to the dendritic receptor where it is converted once again into an electrical signal. (The process is electrochemical in nature and a considerable amount of research is being undertaken to obtain complete understanding of the

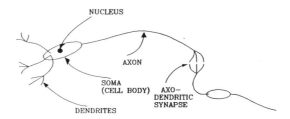

Figure 14.17 A schematic diagram of a nerve cell (neuron).

mechanism.)

The amplitude of the action potential remains constant even as the size of the stimulus that initiated the response is increased above a suitable threshold. However, the number of pulses per second increases with stimulus size. Within the nervous system information is transmitted as 'frequency modulation'; this is summarized in figure 14.18. As shown in figure 14.18(c), nerve impulse frequency is linearly related to the strength of the stimulus over some portion of the stimulus intensity.

During the process of transmission across the synapse, the pulse train is 'demodulated' or reconverted into the stimulus pattern that originally gave rise to the action potentials in the first place. Within the cell, stimulus information from all dendrites is summed (integrated) to provide a summary stimulus to the cell receiving the information; this is converted once again into a (new) pulse train as discussed above and transmitted out on the cell's axon to other cells. This process exists and continues throughout the animal's nervous system.

Not all stimuli received by the cell body necessarily excite the cell into action; some axo-dendritic connections may tend to reduce excitation—such signals are *inhibitory*. Within the nervous system, combinations of excitatory and inhibitory signals can be combined to perform calculations that help to ensure survival of the species—provide for obtaining food, etc. For example, figure 14.19 shows an 'idealized' schematic diagram of the visual system (the *compound eye* consists of an array of lens/receptors) of some animals. Specialized neurons called photoreceptor cells receive light from the environment (figures 14.19(a), (b)) and convert this information into a pulse train whose frequency varies according to the intensity of the light stimulus impinging on their visual fields (figure 14.19(c)). If the geometry of the eye was 'perfect', as shown, then the optical pattern consisting of bright/dark bands—the 'step'—would produce pulse trains from the receptors that mirror the optical stimulus in a similar manner.

Figure 14.20 depicts a more realistic representation of the eye; the visual fields of the photoreceptors overlap and their pulse train output is proportional to the average light falling on their visual field. This produces the optical distortion shown in figure 14.20(c); a bright band is converted into a 'smear'. However, within such animal systems the photoreceptors have additional neuronal connections from neighbouring units. This is shown as axonal branches

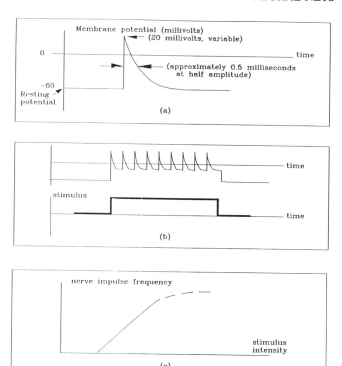

Figure 14.18 Information transmission in the nervous system. (a) A single-action potential (pulse); voltage regions above the resting potential are depolarized, regions below are hyperpolarized. (b) A typical pattern of response to a stimulus. (c) The variation of the nerve impulse frequency with stimulus intensity.

in figure 14.21(b); these connections are inhibitory with the magnitude of the inhibition being proportional to the distance to the neighbours. (All photoreceptors are connected in this way even though the figure shows only one set of such interconnections.) When these inhibitory inputs are factored into the net value of stimulus 'seen' by each photoreceptor, the pulse train frequencies will appear as they do in figure 14.21(c). The edge is again distorted but in this case it appears to be enhanced or exaggerated. Such arrangements help these animals to locate or identify light/dark changes, which is fundamental for sighting food or predators, among other things. (For example, the visual system of the frog includes a compound eye and frogs use this to 'see' flying insects on which they subsist. While humans do not have a compound eye a similar phenomenon can be noted; this is 'Mach bands' in which a light/dark edge includes thin regions of brighter and darker intensity than are present in the stimulus pattern, just adjacent to the 'edge'.) Neuronal nets performing an enormous variety of computation are found throughout the nervous systems of

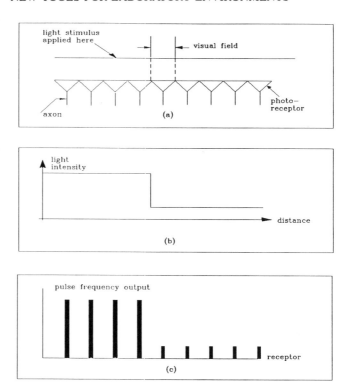

Figure 14.19 An idealized visual system of animals with compound eyes. (a) A schematic representation. (b) An example of a light/dark stimulus pattern. (c) The response of the eye to the simulus of part (b).

animals.

A functional model of a neuron is shown in figure 14.22. In this model, inputs and outputs are voltages whose amplitudes are considered to be proportional to the frequency of the pulse trains found in animal systems as described above. Each input is first multiplied by its corresponding weight; inhibitory connections may be modelled by using a negative weighting factor. The weighted signals are summed and subsequently passed to an element that performs a non-linear transformation (amplification). A variety of transformation devices can be used; one of the more widely used elements transforms the sum according to the following formula:

$$\frac{1}{1+e^{-Gx}}$$

where G is a parameter that may vary according to the application and x is the input to the threshold device (the output of the summer). This is a 'first-order' representation of a neuron; it includes integration (summation), weighting, and a non-linear element, which also form parts of the neuron. However, it is to

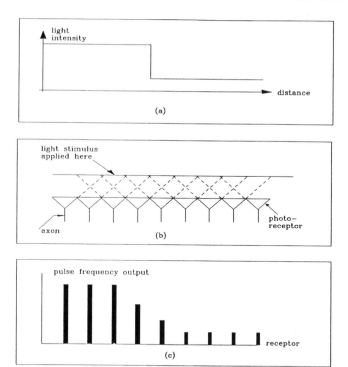

Figure 14.20 Effects of overlapping visual fields. (a) A light/dark stimulus pattern. (b) The geometry of overlapping visual fields. (c) The photoreceptor response showing distortion of the pattern.

be noted that the operation of this model departs markedly from the behaviour of a biological neuron. Nevertheless, its application to problems in parallel-distributed processing has proven to be so useful that this model may be studied in its own right, not necessarily as the equivalent of a biological neuron.

14.3.1 Neural net architectures

A computer program consists of a series of steps that determine how the input (data set) is converted into a resulting output. Thus the program can be viewed as a 'mapping' between input and output. Some programs do not explicitly define the steps; rather they specify a procedure—as described above for expert systems—for invoking a set of rules. (When using expert systems, some problems become difficult when the computer cannot find a sequence of rules to define the problem-solving procedure—'the search for the answer fails'.)

Neural nets do not involve either of the algorithmic mechanisms cited above (programs that specify either steps to be carried out or procedures for finding answers). The 'algorithm' used in neural net technology consists of a 'learning'

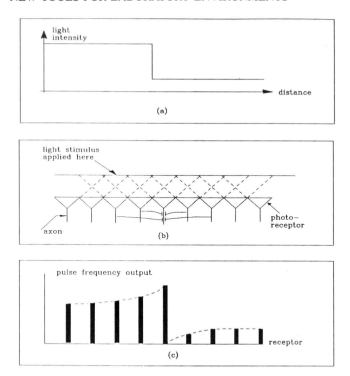

Figure 14.21 The effect of lateral inhibition. (a) A light/dark stimulus pattern. (b) Neuronal interconnection showing (partial) inhibitory connections from neighbouring photoreceptors. (c) The photoreceptor response showing resulting 'enhancement' of the light/dark boundary.

procedure by which the net itself (ultimately) establishes the mapping between the inputs and the outputs. (Instead of the machine including a sequence of steps supplied by the programmer that carries out operations on data supplied by the user, the neural net can be thought of as supplying the program for associating inputs (supplied by the user) and outputs.) A neural net consists of a system of processing elements of the type shown in figure 14.22. One example of a neural net is depicted in figure 14.23 where each circle represents one of the computational elements shown in figure 14.22. Each column of processing elements is referred to as a 'layer' and a neural net may include one, two, or more layers. (Three-layer configurations are the most frequently encountered with the first layer called the *input*, the second called the *hidden* and the third referred to as the *output* layer.) The processing elements are called *nodes* and each receives inputs from either an external source, from another node, or some combination of these. The inputs are processed in accordance with the model shown in figure 14.22; thus the output of each node is given by

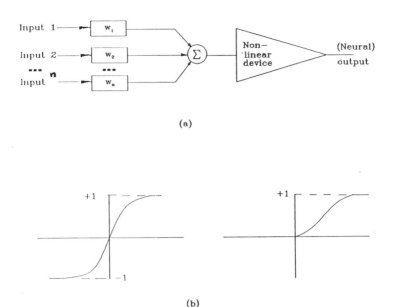

Figure 14.22 A simple model of a neuron. (a) A functional representation. (b) Samples of typical (non-linear) theshold device characteristics.

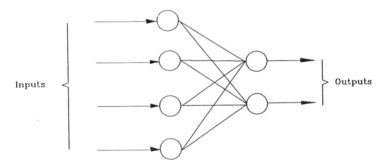

Figure 14.23 A simple neural network. Each circle represents a single neuron as shown in figure 14.22.

$$e_o = f\left(\sum_{i=1}^{n} w_i e_i\right)$$

where each e_i is the input to the node and w_i is the corresponding weighting factor; f is the non-linear ('threshold') function applied to the sum, typically like the one cited above.

A number of more complex neural net arrangements are possible in addition to the simple two-layer network shown in figure 14.23; some of these are

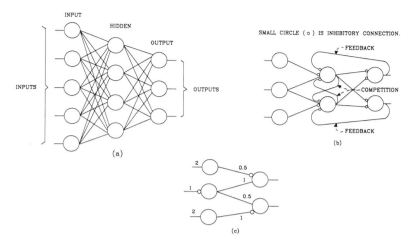

Figure 14.24 Examples of neural net interconnections. (a) A three-layer neural net. (b) Competition and feedback. (c) The system with weights assigned.

Table 14.1 Neural network classification.

Feed forward	Non-linear	Supervised	Backprop Neocognitron
		Unsupervised	Neocognitron Kohonen Counterpropagation
	Linear	Adaline Perceptron Linear associator	
Feedback	Trained	Adaptive resonance	
	Constructed	Hopfield BAM	

represented in figure 14.24. A generic form of the three-layer network is shown in figure 14.24(a); no weights appear on the interconnections and none of them are inhibitory (negative weighting factors). Figure 14.24(b) shows some additional neural net organizational features, namely *feedback* and *competition*. Connections within a layer provide competition while connections from a layer to its predecessor define a feedback connection. Weights have been added to a simple network in figure 14.24(c).

14.3.2 Neural net operation

Neural networks may be classified according to table 14.1.

If a neural net's output is not dependent on previous values then it is considered to be *feed forward*. When a network's are dependent to some degree on its previous output values it is said to be a *feedback network*. Linear models are characterized by a linear transfer function acting on the weighted sum of signals; non-linear models include a transfer function, which conforms to the equation previously cited. In *constructed* models the weights are calculated using a vector operation known as the outer product (of every input pattern with itself or with an associated input); after 'construction', a partial or inaccurate input pattern may be presented to the network, which produces a 'correct' interpretation. Hopfield networks and bidirectional associative memories (BAM) are examples of this type of network. (They will not be described any further [13].)

Most neural network applications use the non-linear, feed-forward model. (It can be shown mathematically that any feedback network has an equivalent feed-forward network, which performs the same task.) While some research progresses on *unsupervised learning*, the most useful applications have evolved from *supervised learning* (or training). (Unsupervised learning does not have a teacher; the network is presented with a number of inputs and it 'organizes itself' in a way that produces classifications of the input.) In supervised learning, the teacher is aware of the correct response for a given input. The computer examines the error between the correct output and the network's response. An error signal is fed back through the network, altering the weighting factors in such a way that the same error does not reoccur when the given input is presented again. The most famous form of this technique is called *back propagation*.

Proper design of feed-forward neural nets that operate on back-propagation learning require the following steps:

- Define the problem to be solved.
- Choose information; gather data.
- Train the network.
- Test the network.
- Run the network.

14.3.3 Training the network

The network is trained on a series of input/output pairs, which are presented to the neural net. Outputs in the (output) layer are compared against their 'target' values. If there is an error between the observed and the desired value then each weight on the input to that neuron is adjusted. One algorithm for accomplishing this is the back-propagation adjustment, which is summarized in figure 14.25. An arbitrary sequence of three cells is shown; in the first view prior to exposure to one of the training patterns the network is characterized as having three output values, O_y, O_x, and O_z, and two weights between cells, W_{yx} and W_{xz}. One of the training patterns is applied and an error between the desired and actual outputs at z is detected. The weight W_{xz} is adjusted according to the following:

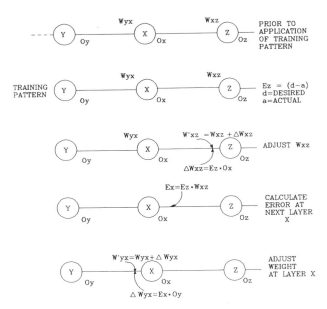

Figure 14.25 A back-propagation supervised learning scheme for error adjustment.

$$\Delta W_{xz} = E_z O_x$$
$$W'_{xz} = W_{xz} + \Delta W_{xz}$$

where E_z is the observed error (desired output − actual output). This adjustment is repeated for the neuronal cell x using the formulas shown below—the algorithm derives its name from the fact that the error (and weight adjustment) is 'propagated' back from the output layer to intermediate layers:

$$E_x = E_z W_{xz}$$
$$\Delta W_{yx} = E_x O_y$$
$$W'_{yx} = W_{xy} + \Delta W_{yx}.$$

This error calculation and weight adjustment is carried back for all cells in the chain. (Some methods include a smoothing or *learning rate* factor, which improves the way in which the network approaches its final configuration; the adjustment formula in this case is

$$\Delta W = \eta \delta O$$

where η is the learning rate parameter, δ is the appropriately reflected error signal and O is the output of the proper (prior) cell.)

The set of training patterns is repeatedly presented to the network and the weight adjustments described above are duplicated. The training set is normally a 'small' subset of all the patterns that the neural net is expected to recognize. Once the training set is identified without errors, the network may be tested on an expanded group to verify that it is working properly. The weights are analogous to the stored program of the traditional computer program. The network as a whole operates to arrive at the 'essential identification rule' for input patterns; noise and other errors are minimized during identification.

14.3.4 Applications

Visual pattern recognition has been extensively studied for a variety of uses [14]. For example, a chemical drawing consisting of characters and graphics is representative of an underlying chemical compound. A scanner is used to read the image into the personal computer. The image is processed mathematically (using a technique known as finding the two-dimensional Fourier transformation) and the result is fed into a neural network for recognition and subsequent translation into bonds and atomic symbols. The characters and graphics have Fourier signatures, which can be recognized by the neural net [15].

In a similar way sound waves derived from non-destructive testing of concrete can be analysed using a neural net to determine whether or not any flaws exist. The concrete is struck sharply with a hammer—and a sound *impulse response* is obtained. The Fourier transform provides the frequency characteristics of the beam. The value of the response at each measured frequency is presented to a neural net that has been previously trained to recognize flaws of various kinds. The same technique can be applied to electrocardiograms, electroencephalograms, and many other waveforms.

Hypertension is an extremely difficult condition to detect and manage. (In fact, there is considerable variability in the medical community regarding what actually defines hypertension.) A neural network expert system for diagnosing and treating hypertension has been developed [16]. Hypertension is a widespread pathological condition occurring in about 15–20% of the world's population. It usually involves a 'sustained' elevation of the diastolic blood pressure above 90 mm Hg. A single (indirect) measurement of the blood pressure can lead to both false positive and false negative diagnoses. For a more definitive diagnosis, the dynamics of blood pressure over the course of the day must be considered. A neural net expert system known as a *hypernet* has been developed to diagnose and recommend treatment for hypertension. A simplified block diagram of the network is shown in figure 14.26. The network is made up of three modules:

- A module (network) that produces a 24-hour blood pressure reference record (one reading each hour) for normal subjects of the same age and sex. (Blood pressure variations are functions of age and sex as well as daily changes.)
- A module whose inputs are the patient's clinical report. The output is the degree of compatibility with each of the four most common chemical treatments.

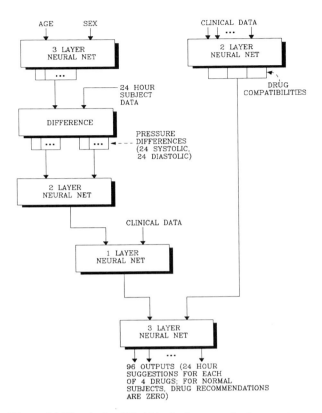

Figure 14.26 A simplified block diagram of a hypernet.

- The third module, which includes the following inputs: differences between the reference series generated by the first module and the patient's 24-hour readings; clinical (anamnestic) data; and outputs from the drug compatibility module.

The resultant network generates the following outputs: a suggested dosage for each drug for each of the 24 hours (24 × 4 or 96 neurons in the output layer). The entire system was trained and tested on 300 clinically healthy subjects and 85 'suspected' hypertensive subjects. A test set was tried once the neural net had learned to recognize appropriate patterns. This included 35 subjects: 10 had no sign of blood pressure excess (nor signs of organ damage, a consequence of high blood pressure); and 25 were suspected of hypertension. A diagnosis was considered correct if the system and a specialist agreed on treating or not treating. Hypernet correctly diagnosed 33 of the 35 (94%). In addition to the high performance in diagnosis, the system exhibited a number of other characteristics often found in neural nets; it rejected 'noise' and other corruptions generated from measuring and/or typing errors; once it has all the data, it can reach its conclusions in less than one second, far more quickly than a classical expert system. Additional learning might also be possible to improve the accuracy and reliability further.

The ability of a neural net to recognize patterns provides the most immediate application to laboratory environments. Examples similar to those cited above can be expected to appear with more frequency in the near future.

14.4 FUTURE DEVELOPMENTS

Operating speed, reliability, and information integrity remain long-term goals of computer-based instrumentation. The ability to share information among users can also be added to this list. The techniques described in this chapter will continue to find new applications in the laboratory. Several emerging hardware architectures and software methods can be expected to play important roles in implementing these goals. Some of the most prominent are now listed.

Reduced-instruction-set computers (RISC). Computers of this kind are characterized by a fixed instruction size (in terms of the number of bytes per instruction) as well as a limited repertoire of instructions. Each instruction is decomposed into a sequence of suboperations, which are then carried out by separate units within the processor. This type of architecture resembles an assembly line in which each station carries out a part of the task. As one station carries out its responsibility for a given instruction, leading stations may carry out their own activities on the next instruction. Throughput—the number of instructions completed per unit of time—is thereby increased. A number of PCs with RISC architectures have begun to appear in the laboratory, and prices continue to fall as competing systems are introduced into the marketplace. (Economical RISC-based systems can now be purchased for around $3000.) Clock speeds for such systems now exceed 100 MHz.

While the speed of such processors has increased dramatically over the past few years, the operating speed of the associated software (code) has lagged. This is primarily because software engineers have not taken full advantage of RISC architectures. (To date, programs have simply been 'ported' from older (CISC) machines to the new processors.) Thus, we are likely to see further developments in programming and computational algorithms such as the use of distributed and multiprocessing organizations for problem solutions.

Object-oriented programming and design (OOP and OOD). With the availability of high-speed circuitry, the software development cycle has become an obstacle to rapid laboratory system implementation. The software development cycle includes the steps of:

- requirement specification—a high-level description of the problem to be solved;
- design specification—a detailed specification of the algorithm;

- implementation—which deals with coding;
- testing—which deals with the quality assurance and testing of software;
- maintenance—which maintains the integrity of the program.

To speed this cycle up the underlying conceptual model of the laboratory has become *object oriented*. An object is characterized by its attributes and by its associated operations. (A common example might include integers (whose attributes include those of ordinary numbers over a given range), and the well known operations of addition, subtraction, multiplication and division. Integers can thus be interpreted as objects.) There are several potential advantages associated with OOP:

- Better conceptualization and modelling of the real world.
- The resulting program is more 'robust'. Languages supporting OOP include various forms of built-in checking to avoid run-time errors.
- Performance is enhanced. Knowledge of the types of object facilitates compile-time optimization.
- It separates implementation (code) from specifications. This makes modifications easier as only a portion of the code is affected.
- It allows extensibility of the system wherein software is reusable.

A number of HLLs currently include, or will include in the near future, support for OOP. These include C++ and Ada9X.

Data sharing. The organization of data-bases using OOD will enhance the ability of users to maintain information as well as to share such data with other interested parties. Data-bases are essential components of intelligent machines (when coupled to an appropriate inference engine) and the efficiency with which such structures can be built and/or expanded is important for rapid response to laboratory problems.

Data have come to constitute an important 'asset' of any enterprise and will receive considerable attention (in terms of its management) over the next few years. Of particular interest are the emerging client/server architectures. In such architectures, users may be widely dispersed and communication systems will thus receive much attention. Issues to be resolved include: data integrity; security; and system reliability.

REFERENCES

[1] Rich E and Knight K 1992 *Artificial Intelligence* 2nd edn (New York, NY: McGraw-Hill)
[2] Strobel H A and Heineman W R 1989 *Chemical Instrumentation: a Systematic Approach* 3rd edn (New York, NY: Wiley)
[3] Shamsolmaali A, Carson E R, Collinson P O and Cramp D G 1988 A knowledge-based system coupled to a mathematical model for interpretation of laboratory data *IEEE Eng. Med. Biol. Mag.* June

[4] Woodruff H B and Smith G M 1981 Generating rule for PAIRS—a computerized infrared spectral interpreter *Anal. Chem. Acta* **133** 545–53
[5] Kulikowski C A 1988 Artificial intelligence in medical consultation systems: a review *IEEE Eng. Med. Biol. Mag.* June
[6] Dickinson C J, Ingram D and Ahmed K 1987 *MACPEE Graphical Version: a Simulation of Heart, Peripheral Circulation, Kidneys, Body Fluids, Electrolytes and Hormones* (Oxford: IRL)
[7] Kosko B 1992 *Neural Networks and Fuzzy Systems* (Englewood Cliffs, NJ: Prentice-Hall)
[8] Williams T 1991 Fuzzy logic simplifies complex control problems *Comput. Design* March 90–102
[9] Schwartz D G and Klir G J 1992 Fuzzy logic flowers in Japan *IEEE Spectrum* July
[10] Cios K J, Shin I and Goodenday L S 1991 Using fuzzy sets to diagnose coronary artery stenosis *Comput. Mag.* March
[11] Rosenblatt F 1958 The perceptron: a probabilistic model for information storage and organization in the brain *Psych. Rev.* **65** 368–408
[12] Stevens C F 1966 *Neurophysiology: a Primer* (New York, NY: Wiley)
[13] Lawrence J 1992 *Introduction to Neural Networks and Expert Systems* (Berkeley, CA: California Scientific Software)
[14] Fukushima K 1988 A neural network for visual pattern recognition *Comput. Mag.* March
[15] Grunlend G H 1972 Fourier process for hand printed character recognition *IEEE Trans. Comput.* **C21-2** 195–201
[16] Poli R, Cagnoni S, Livi R, Coppini G and Valli G 1991 A neural network expert system for diagnosing and treating hypertension *Comput. Mag.* March
[17] Sztipanovits J and Karsai G 1988 Knowledge-based techniques in instrumentation *IEEE Eng. Med. Biol. Mag.* June

Index

AC analysis, SPICE, 226–227
AC-to-DC converter, 313–315
Accumulation, signal detection in noise
    DSPlay, 316–317
    LabWindows, 347–349
    MathCAD, 276–278
    Spreadsheets, 335–343
Accuracy, numeric, 39, 40
A/D—see Data conversion
Address, memory, 51, 54, 56
    Single-address, 51
    Three-address, 51
ALU, 52
AM—see Modulation, amplitude
AM waveform, DSPlay, 306–308
Amplifier, operational—see Op amp
Analogue filter simulation
    MathCAD, 263–266
    SPICE, 224–227, 254–256
Application programs, Windows, 375, 377–392
Architecture, hardware, 14–17, 49–105
    Automated environments, 88–97
    Functional components, 49, 50
    Multiple processor, 86–88
Artificial intelligence, 407–444
(See also Fuzzy logic, Neural nets)
    Abduction, 413
    Backward reasoning, 419
    Control strategy, 411
    Deduction, 413
    Forward reasoning, 418–419
    Frames, 418
    Induction, 413
    Inference engine, 410
    Knowledge base, 410
    Means–ends analysis, 419
    Predicate, 412
    Predicate calculus, 413–416
    Procedural representation, 416–417
    Production systems, 418
    Scripts, 418
    Search, breadth-first, 419
    Search, depth-first, 419–420
    Semantic nets, 417–418
    Tree structures, 419
Average—see Mean value
Averaging—see Accumulation

BASIC programs, LabWindows, 360–363
B-squared Spice, Windows, 377–379
Binary number systems, 25–29
    Number conversion, 27
    Negative integers, 25–27
Bus, computer, 50, 85–86

C language
    LabWindows examples, 344–358, 363–374
    Sorting routine, 279
Capacity, information, 44, 45
Causality, 156

INDEX

Central processing unit (CPU), 49, 50–55
CISC, 55, 56
Clipboard, Windows, 377, 380–382
Codes, 29–36
  alpha-numeric, 34
  efficiency, 34–36
  error checking, 33, 34
  Gray, 30–33
  weighted, 30
  unit-distance, 30–33
Communications, 88–97
  GPIB, 95–97
  IEEE 488, 91–95
  RS-232, 88–91
Companding, 43, 44
Compile—see Program, translation
Computer, choosing a platform, 100–104
Computer, single-chip, 97–100
Computer-based instrumentation, 3–17
  Architecture, 14–17
    Dedicated, 15
    Multiple-instrument, 15
    Multiprocessor, 15, 16, 86–88
    Remotely controlled, 15
    Tightly coupled, 15
  Examples, 6–14
    Chemical analysis, 11–13
    Mass spectrometry, 9–11
    Neurophysiology, 6–9
  Role, 3, 4
    Computer interface, 5, 6
    Data handling, 4, 5
    Documentation, 6
    Process control, 5
Conditioning, signal, 196–199
Control panel, LabWindows, 368–374
Conversion, code, 27, 29, 32
Convolution, digital
  DSPlay, 312–314
  MathCAD, 265–267, 274–275
  MATLAB, 291–292

CPU—see Central processing unit
Curve fitting
  MathCAD, 281–286
  MATLAB, 296–297
  Spreadsheets, 332–334

D/A—see Data conversion
DAS-16, LabWindows
  Hardware, 359–360
  Programming, 360–366
Data acquisition, 393–399
Data acquisition, LabWindows, 358–374
Data analysis
  LabWindows, 348, 350–352
  MathCAD, 277–286
  MATLAB, 293–296
  Spreadsheets, 325, 330–334
Data conversion, 202–212
  Analogue-to-digital, 203–211
    Binary counting, 203
    Dual slope, 206–207
    Flash converter, 205–206
    Practical converters, 207–211
    Successive approximation, 203–205
  Digital-to-analogue, 211–212
  DSPlay, 310–312
  SPICE, 239–243, 248–251
Data structures, 148–155
  Array, 152
  Heap, 152
  Linked list, 151–152
  Module, 155
  Pointer, 149
  Queue, 154–155
  Record, 153–154
  Stack, 150–151
DC circuit simulation, SPICE, 221–224
DDE—see Dynamic data exchange
DFT—see Discrete Fourier transform
Dhrystones, 55, 56
Diagnosis, medical, 425–426

INDEX 453

Programming, assembly language, 133–138
  Addressing modes, 135–137
    Absolute, 136, 137
    Immediate, 136, 137
    Register, 135, 137
    Register indirect, 136, 137
  Directives, 134
  Fields, 134
  Operations, 136
Programming, languages, 4, 126–131
  Identifiers, 127
  Pascal, 126–131
Programming, packages, 138–144
  Database, relational, 138–141
    Attribute, 139–140
    DML, 139
    Primary key, 140
    Table, 139
    Tuple, 139
  Also see SPICE, MathCAD, MATLAB, DSPlay, Lotus, 1-2-3, LabWindows
PSpice—see SPICE

Rectifier circuit, SPICE, 244–246
Register, flags, 53
Register, local, 53
Resistor model, SPICE, 248–249
Resolution, 39, 40
RISC, 55, 56, 147
Robot, 419–421
Root finding, MathCAD, 267–268
RS-232, 88–91
  Baud rate, 89
  CTS, 91
  Current loop, 89
  DCE, 91
  DSR, 91
  DTE, 91
  DTR, 91
  Full-duplex, 90
  Half-duplex, 90
  RTS, 91

  Simplex, 90
  Voltage level, 89

Sampling, information, 44, 45
Sensitivity analysis, SPICE, 240–243
Signal conditioning—see Conditioning, signal
Signal detection in noise
  DSPlay, 315–320
  LabWindows, 347–349
  MathCAD, 274–278
  MATLAB, 292–294
  Spreadsheets, 335–343
SlideWritePlus, Windows, 378, 380, 383–386
Sorting data
  LabWindows, 350–352
  MathCAD, 278–280
  Spreadsheets, 330–332
Spectral leakage
  DSPlay, 303–305
  LabWindows, 372–374
Spectrometer, 421–425
SPICE, 220–260
  Analogue filters, 224–227, 254–256
    AC analysis, 226–227
    Transient analysis, 224–226
  DC circuits, 221–224
  Fourier series, 251–254, 256–262
    FM waveform, 258–260
    Square-wave, 252–254, 256–258
  Modelling of devices, 243–251
    Logarithmic amplifier, 246–248
    Rectifier circuit, 244–246
    Resistor model, 248–249
  Operational amplifier (op amp) circuits, 228–243, 246–251
    D/A converter, 239–243, 248–251
    Differential amplifier, 233–238
    Idealized subcircuit, 228–229
    Integrating circuit, 232–233
    Inverter circuit, 229–231
    Monte Carlo analysis, 248–251
    Sensitivity analysis, 240–243

454 INDEX

UA741 subcircuit, 230–231
PROBE utility, 254–260
Spreadsheets, 322–343
   Data analysis, 325, 330–334
      Curve fitting, 332–334
      Sorting, 330–332
      Statistical analysis, 325, 330–332
   Fourier series, square-wave, 328–330
   Graphing, 327–328
   Macros, 324, 338–343
   Math operators, 325
   Signal detection in noise, accumulation, 335–343
Square wave, Fourier series
   DSPlay, 309–311
   MathCAD, 263–264, 269–272
   MATLAB, 289–290
   SPICE, 252–254, 256–258
   Spreadsheets, 328–330
Square wave, LabWindows DFT programs, 364–366, 370–374
Standard deviation
   LabWindows, 348, 350–351
   MathCAD, 281–282
   MATLAB, 295
   Spreadsheets, 325, 330–332
Statistical analysis of data
   LabWindows, 348, 350–351
   MathCAD, 280–282
   MATLAB, 293–296
   Spreadsheets, 325, 330–332
Strip-chart recording, LabWindows, 365, 367–368
Summing circuit, D/A converter, 239–241, 249–251

Telemetry, 200–202
   Carrier, 200
   Medium, 200
   Modulation, 200
Time domain analysis, 156
Time invariance, 156
Transducers, 173–196
   Electrochemical, 190–192
   Force/pressure, 183–190
      Capacitive, 188–190, 191
      Mechanical, 188
      Piezoelectric, 186–188
      Resistive strain, 185–186
      Strain gauges, 185–186
   Optical, 179–183
      Fibre, 183
      Photodetector, 181–183
      Phototubes, 180–181
   Shaft encoder, 193–194
   Temperature, 174–179
      Platinum, 177–178
      Semiconductor, 178–179
      Thermistor, 175–176
      Thermocouple, 176–177
   Transformer, variable, 194–196
Transient analysis, analogue filter
   MathCAD, 263–265
   SPICE, 224–226

UA741 op amp subcircuit, Spice, 230–231

Variance
   LabWindows, 348
   MathCAD, 281–282
   Spreadsheets, 325, 330–332
VGA adaptor, 79–82
Visual BASIC, 383

Whetstones, 55, 56
Windows, 119–120, 375–392
   Application programs, 375, 377–392
      B-squared SPICE, 377–379
      Lotus, 1-2-3 for Windows, 375, 377–378, 380–392
      MathCAD, 375, 377–378, 380–382
      SlideWritePlus, 378, 380, 383–386
   Clipboard, 377, 380–382
   Control menu box, 120

Desktop, 119
Dynamic data exchange (DDE), 383, 387–392
File transfer, 380, 383–386
Icons, group window, 120
Icons, program, 120
Menu bar, 120

Program manager, 375–377
Size button, 120
Title bar, 119
Window frame, 119
Work area, 119
Word size, 50